CAMBRIDGE LIBRARY COLLECTION

Books of enduring scholarly value

Life Sciences

Until the nineteenth century, the various subjects now known as the life sciences were regarded either as arcane studies which had little impact on ordinary daily life, or as a genteel hobby for the leisured classes. The increasing academic rigour and systematisation brought to the study of botany, zoology and other disciplines, and their adoption in university curricula, are reflected in the books reissued in this series.

Life and Letters of Sir Joseph Dalton Hooker

Sir Joseph Dalton Hooker (1817–1911) was one of the most eminent botanists of the later nineteenth century. Educated at Glasgow, he developed his studies of plant life by examining specimens all over the world. After several successful scientific expeditions, first to the Antarctic and later to India, he was appointed to succeed his father as Director of the Botanical Gardens at Kew. Hooker was the first to hear of and support Charles Darwin's theory of natural selection, and over their long friendship the two scientists exchanged many letters. Another close friend was the scientist T. H. Huxley, and it was the latter's son, Leonard (1860–1933), who published this standard biography in 1918. The first volume describes Hooker's early life and his career up to 1860. It includes many letters to Darwin as the two men discussed the new theories and the publication of *On the Origin of Species*.

Cambridge University Press has long been a pioneer in the reissuing of out-of-print titles from its own backlist, producing digital reprints of books that are still sought after by scholars and students but could not be reprinted economically using traditional technology. The Cambridge Library Collection extends this activity to a wider range of books which are still of importance to researchers and professionals, either for the source material they contain, or as landmarks in the history of their academic discipline.

Drawing from the world-renowned collections in the Cambridge University Library, and guided by the advice of experts in each subject area, Cambridge University Press is using state-of-the-art scanning machines in its own Printing House to capture the content of each book selected for inclusion. The files are processed to give a consistently clear, crisp image, and the books finished to the high quality standard for which the Press is recognised around the world. The latest print-on-demand technology ensures that the books will remain available indefinitely, and that orders for single or multiple copies can quickly be supplied.

The Cambridge Library Collection will bring back to life books of enduring scholarly value (including out-of-copyright works originally issued by other publishers) across a wide range of disciplines in the humanities and social sciences and in science and technology.

Life and Letters of Sir Joseph Dalton Hooker

VOLUME 1

EDITED BY LEONARD HUXLEY

CAMBRIDGE UNIVERSITY PRESS

Cambridge, New York, Melbourne, Madrid, Cape Town,
Singapore, São Paolo, Delhi, Tokyo, Mexico City

Published in the United States of America by Cambridge University Press, New York

www.cambridge.org
Information on this title: www.cambridge.org/9781108031004

This edition first published 1918
This digitally printed version 2011

ISBN 978-1-108-03100-4 Paperback

LIFE AND LETTERS OF
SIR JOSEPH DALTON HOOKER, O.M., G.C.S.I.

VOLUME I.

J. D. Hooker.

From the Portrait by George Richmond (1855).

Vol. i. *Frontispiece*]

LIFE AND LETTERS

OF

SIR JOSEPH DALTON HOOKER

O.M., G.C.S.I.

BASED ON MATERIALS COLLECTED AND
ARRANGED BY LADY HOOKER

PORTRAITS AND ILLUSTRATIONS

BY LEONARD HUXLEY

AUTHOR OF 'LIFE AND LETTERS OF T. H. HUXLEY,' ETC.

VOLUME I

LONDON
JOHN MURRAY, ALBEMARLE STREET, W.
1918

DEDICATED

TO

THE MEMORY OF MANY FRIENDSHIPS

PREFACE

THERE seems to be, if I may be allowed to say so, a touch of personal appropriateness in the fact that the writing of Sir Joseph Hooker's Life has fallen to the son of his close friend. The work has thus been doubly a labour of love and remembrance, and by good fortune it traverses a biographical field some part of which has already been worked over by me. If, however, I cannot claim to be a professed student of botany, something of my defect has been remedied by the kindness of others. The proofs have been most carefully read through by Sir David Prain, the present Director of Kew, and Miss Matilda Smith, Kew's botanical artist, who moreover has verified many references at Kew and supplied material for biographical notes not easily accessible elsewhere.

Sir Joseph Hooker, for all that he accused himself on occasion of being a bad correspondent, was in reality an indefatigable letter writer. Indeed, he declares somewhere that the busier he was, the longer and fuller his letters were likely to be—and he was always busy. Apart from a vast official correspondence and regular weekly letters to various members of his family, there are extant over 700 sheets copied from his letters to Charles Darwin, whose own share of the correspondence, typed out, fills more than 800 pages. No other single correspondence compares with this; but it is easily balanced by the total of letters to the next half dozen or so among his multitudinous correspondents, to name only Bentham and Harvey, Anderson and La Touche, Mr. Duthie and my father. Add to this his journals of travel, his various books, his scientific essays—the first written at nineteen, the last at ninety-four—the material to draw upon has been superabundant. Nor must the 'Life and Letters of Charles Darwin' (briefly cited as C.D.) and the 'More Letters of Charles Darwin' (M.L.) be forgotten. They are a mine of information about the scientific interests of the period and the personal relations between the two friends, and my grateful acknowledgments to Sir Francis Darwin are repeated here.

One more name must be mentioned in this place, a name which also appears on the title-page. In gathering materials, in collating letters, in furnishing personal information, the task undertaken with such thoroughness by Lady Hooker has been no light one. But if her careful 'spade-work' has meant much for the book, to the writer her active sympathy has meant even more.

L. H.

October 1917.

CONTENTS

OF

THE FIRST VOLUME

ILLUSTRATIONS

TO

THE FIRST VOLUME

LIFE OF SIR JOSEPH DALTON HOOKER

CHAPTER I

EARLY DAYS

A LIFE whose span is almost a century may well be witness of great changes: the ninety-four years of Sir Joseph Dalton Hooker's life are the more intensely interesting because he himself was one of the chief workers in bringing about such changes. Indeed, the century almost covered by his life saw a greater revolution than any of our era except, perhaps, that of the Renaissance. Once more the civilised world was born anew: it was the century of the New Renaissance. The revolution in thought was paralleled by a revolution in the means of civilised life. The two influences united in effecting the most profound readjustments alike in social values and in the outlook of the human mind. Power over nature transformed the way of life: the insight into nature which secured that power, equally freed inquiring minds from the barriers imposed by the established guides of thought, who only permitted nature to be interpreted through the perspective of creed.

Against those barriers the flood of natural knowledge had been slowly piling itself up, only awaiting the hand that should open a channel and a fresh impulse and a common direction to these chained-up currents. Mechanical aids, such as the magnifying lens, had opened the way to new investigations of life since the seventeenth century. From the needs of medicine sprang the organised knowledge both of botany and of animal life: first the herbal and the history book of animals, full of strange lore; then the gradual searching out of living framework and vital processes, which finally took rank and order as the anatomy and physiology of animals and plants. That these researches awakened doubts of the conventional creeds as applied to nature is evidenced by the familiar sneer at the dangerous folk who recognised the constancy of natural law in the workings of the human frame—*ubi tres medici, ibi duo athei.* Chemistry began to emerge experimentally from the mists of alchemy some half-century before Hooker's birth. Geology took operative shape yet later: with Lyell's 'Principles' in 1839 the first step was built of the stairway that actually led to the theory of evolution. The succession of differing forms of similar creatures in a fossil state provoked challenge of the doctrine of immutability of species; indeed, as has been well said, if the theory of evolution had not existed, Geology would have had to invent it. By the fifties, also, botany, in its search for a natural system of classification, was ripe for the acceptance of an evolutionary explanation.

If the interest awakened by scientific men is proportioned to the degree in which their researches and discoveries come home to 'men's business and bosoms,' giving new colour or shape to the eternal questions of the making of the heavens and the earth, the nature of matter, the play of subtle forces, the laws of life and disease, man's place in the universe, his origin and his destiny, then in every province of physics and astronomy, in medicine and its fellow sciences, the nineteenth century saw great and memorable figures stand out: but most memorable the central group, who, touching most nearly upon life and its place in the universe, awoke the loudest opposition and achieved the greatest triumph.

Charles Lyell pointed the way to Darwin: after the appearance of the 'Origin of Species,' Thomas Henry Huxley was chief champion in the support and spread of evolution on the one hand, and, on the other, of freedom of scientific thought and speech. It was Hooker's privilege to be Darwin's sole confidant for near fifteen years, his generous friend, his unstinting helper, his keen critic, and ultimate convert in the light of his own work and the material he could so abundantly furnish.

The story of Joseph Hooker's life-work is, in one aspect, the history of the share taken by botany in establishing the theory of evolution and the effect produced upon it by acceptance of that theory. He began with unrivalled opportunities, and made unrivalled use of them. As a botanist, he was born in the purple, for in the realm of botany his father, Sir William Hooker, was one of the chief princes, and he had at hand his father's splendid herbarium and the botanic garden which he had made one of the scientific glories of Glasgow University.

Joseph Hooker's earliest recollections are preserved in an autobiographical fragment, set down late in his life. Noteworthy among the events that emerge from childish forgetfulness, like hill-tops above a sea of mist, is the early love of nature and especially of plants, inborn in him and indeed inherited from both lines of his parentage. His father and his mother's father were both botanists, and singularly enough they both began their studies as such with the mosses, quite independently of one another; so that, being confessedly 'a born Muscologist,' he playfully dubs himself 'the puppet of Natural Selection.'[1]

> I was born [he writes] June 30, 1817, at Halesworth, Suffolk, being the second child and son of William Jackson Hooker and Maria, *née* Turner, of Great Yarmouth. My brother was older than myself and my parents had subsequently three daughters. I was named Joseph after my Grandfather Hooker, and Dalton after my godfather, the Rev. James Dalton, M.A., F.L.S., Rector of Croft, Yorkshire, a student of carices and mosses and discoverer of Scheuchzeria in England.
>
> My memory reverts to a very early age—when only three years old to my father's house at Halesworth, and incidents connected therewith, amongst others the gardener, in mowing a damp meadow behind the house, slicing the frogs with his scythe, and my brother running along the top of the garden wall to my mother's alarm. He died in 1840. Curiously enough I have no recollection of a magnificent dog,

[1] Anniversary dinner of the Royal Society, Nov. 30, 1887.

a Newfoundland I believe, that my father kept, and which was notorious for its thefts from the butchers' shops of the town.

My Grandfather Hooker's house in Magdalen Street, Norwich, I remember even better, where my grandmother used to show me the glazed drawers of his insect cabinet. On leaving Halesworth for Glasgow, my father sold his insects to Mr. Sparshall of that city, a well-known collector. The collection is now in the Norwich Museum. Also I well remember his little garden and greenhouse of succulent plants, and on seeing a Coccinella on a post, repeating to it the stave:

Bishop Bishop Barnabee
When will your marriage be?
If it be to-morrow's day,
Take your wings and fly away.

Of my Grandfather Turner's house in Yarmouth, I remember being carried there in my nurse's arms early in 1821, on the eve of my mother taking myself, brother and sisters to Glasgow, where my father, who had taken up his Professorship in the previous summer, was awaiting us. My grandfather occupied the house of Gurney's Bank, of which he was a resident Director. I remember distinctly the railings before the Bank, its drawing-room, and my aunts' seizing me from my nurse, dancing with me round the room, and striking the harp to amuse me. Also I remember the walls of the room being covered with pictures of which my grandfather had a small but very choice collection. This collection was sold after my grandfather's death in 1858. Some of the pictures, notably the Titian, a Hobbema and, I think, a Greuze and one or more Cotmans are in the Wallace Collection.

Of the journey from Yarmouth to Glasgow by post horses I have a distinct recollection, during which my mother caught ague in crossing the Fens, with which she was troubled for many years. Of incidents I can only remember my brother running to eat a cake of white soap, mistaking it for an apple. I also distinctly remember the picturesque place, Inn of Beattock Bridge, in Dumfriesshire, but why I cannot tell.

My next memory is the arrival in Glasgow by night, and going into lodgings (No. 1, Bath Street) which my father had taken pending his obtaining possession of a new house which he had purchased in West Bath Street (No. 17), in which lodgings I found my Grandfather and Grandmother Hooker, who had accompanied or followed my father to Glasgow with a mass of furniture from the Halesworth and Norwich houses, on some bedding from which I slept, for the first night, on the floor.

Of the following years I have little of note to record beyond having an excellent governess, a Miss Turnbull, of whom I was very fond, and a mild attack of scarlet fever when I was six. No doubt I had other illnesses of childhood, as I had the credit of being the leader in contracting them.

At the age of five or six, my early leaning towards botany was shown by a love of mosses, and my mother used to tell an anecdote of me, that, when I was still in petticoats, I was found grubbing in a wall in the dirty suburbs of the dirty city of Glasgow, and that, when she asked me what I was about, I cried out that I had found *Bryum argenteum* (which it was not), a very pretty little moss I had seen in my father's collection, and to which I had taken a great fancy.[1]

At a later period, when still in my early teens, I took up the study of these beautiful objects, and formed a good collection of the Scottish species in the Highlands and elsewhere; and my first effort as an author was the description of three new mosses from the Himalaya.[2]

Of this early love of botany and kindred eagerness for travel, he continues in the Royal Society speech already quoted:

A little older, and when still a child, my father used to take me excursions in the Highlands, where I fished a good deal, but also botanised; and well I remember on one occasion, that, after returning home, I built up by a heap of stones a representation of one of the mountains I had ascended, and stuck upon it specimens of the mosses I had collected on it, at heights relative to those at which I had gathered them. This was the dawn of my love for geographical botany.

Another little circumstance connected with a moss had also its influence on my future career. You may remember

[1] This is the better version of the tale, as given in the Royal Society speech above mentioned.

[2] See p. 22.

a passage in Mungo Park's 'Travels' in search of the source of the Niger, when he describes himself so faint with hunger and fatigue, that he laid himself down to die; but being attracted by the brilliant green of a little moss on the bank hard by, said to himself: If God cares for the life of that little moss, He surely will not let me perish in the desert. Park put a piece of it in his pocket-book, and, fortified by the thought, went on his way. He soon arrived at a hut occupied by poor black women, who fed him, and sang him to sleep with impromptu words, pitying the poor white man far away from his home and friends.[1] A scrap of that moss was given to my father by Mungo Park, or a friend of his, and was shown to me. It excited in me a desire to read African travels, and I indulged in the childish dream of entering Africa by Morocco, crossing the greater Atlas (that had never been ascended) and so penetrating to Timbuctoo. That childish dream I never lost; I nursed it till, half a century afterwards, when, as your President has told you to-day, I did (with my friend Mr. Ball, who is here by me, and another friend, G. Maw, F.L.S.), ascend to the summit of the previously unconquered Atlas.

When still a child, I was very fond of Voyages and Travels; and my great delight was to sit on my grandfather's knee and look at the pictures in Cook's 'Voyages.' The one that took my fancy most was the plate of Christmas Harbour, Kerguelen Land, with the arched rock standing out to sea, and the sailors killing penguins; and I thought I should be the happiest boy alive if ever I would see that wonderful arched rock, and knock penguins on the head. By a singular coincidence, Christmas Harbour, Kerguelen Land, was one of the very first places of interest visited by me, in the Antarctic Expedition under Sir James Ross.

'The spirit of a youth that means to be of note, begins betimes,' and heredity and early training are strong among the directing factors for such a spirit. As has been said, Hooker's father, William Jackson Hooker, was one of the first botanists of his age; his grandfather, Joseph Hooker, spent much of his leisure in the cultivation of rare plants; his

[1] The incident of the moss occurs in chapter xix of Park's *Travels*, after he had been robbed by a party of Foulahs; the negro women's compassion is an earlier incident of chapter xv.

maternal grandfather, Dawson Turner, of Yarmouth, banker, botanist, and antiquarian, was especially interested in the cryptogams, made collections, and published sumptuous volumes.

The Hookers, who claimed lineal descent from John Hooker, *alias* Vowell, the historian, and uncle of Richard, the 'Judicious,' author of the 'Ecclesiastical Polity,' were a Devonshire family settled in Exeter, who dropped their original name of Vowell in the sixteenth century.

There is a very old parchment genealogical tree taken from the Heralds' College in 1597, continued since and completed from other sources, which traces the Hooker ancestry for five centuries. The first name of the series, Seraph Vowell, hailing from Pembroke, suggests a Welsh origin in Ap-Howell. The second in descent, Jago Vowell, marries Alice Hooker, daughter and heiress of Richard Hooker of Hurst Castle, Hampshire, whose family name is adopted with his own. Hence the constant repetition in the genealogy of 'Vowell *alias* Hooker.'

Though offshoots of the Hookers, especially after the Civil War, are found as successful traders at Crediton or as far afield as London, where one became Lord Mayor, the Hooker family is most closely associated with Exeter, where it is still represented. Thus a John Hooker was M.P. for Exeter in 1470; Robert Hooker, youngest born and sole survivor of twenty brothers and sisters, in 1529, and his son, another John, in 1571. This latter John was the first Chamberlain of Exeter, and wrote a book on the antiquities of Exeter, still preserved in the city archives. He exemplified the active business capacity of many of his name by founding the first 'Guild of Merchant Adventurers' under a charter from Queen Mary. It was not long before the Devon Merchant Adventurers were typified by his kinsman, John Oxenham, Drake's comrade, and the first Englishman to sail on the Pacific. Adventure also took John Hooker with Sir Peter Carew to Ireland, where he became a member of the Irish Parliament in 1568.

But the world owes him a greater debt. He supplied the means for educating his nephew Richard, the 'Judicious' Hooker. Next after the Chamberlain comes the Vicar of

Caerhayes in Cornwall, from whose son Valentine the modern Hooker family traces its descent. Post-Reformation Hookers tended to Puritanism. In the Laudian persecutions the Rev. Thomas Hooker escaped to America, and there founded a family which has won its own meed of distinction in Church and State. 'Fighting Joe Hooker,' for instance, gained his by-name in the War of North and South.

Another Hooker is recorded as fighting under Fairfax and Essex in our own Civil War, afterwards settling down at Crediton.

Among the 2000 clergy who were driven from their livings after the Act of Uniformity were several Hookers. One is mentioned as minister of the Presbyterian chapel at Crediton, another at Chumleigh. The chapel registers show that many of the name became Nonconformists. Zeal for the Protestant cause led some to join in Monmouth's ill-starred rebellion; those who escaped the scaffold at Exeter ended their lives as slaves in Barbados.[1]

The Joseph Hooker already mentioned, seventh in descent from John, migrated from Exeter and set up in business at Norwich, where his son William Jackson was born in 1785. Lydia Vincent, Joseph Hooker's wife, claims special notice for her artistic heritage. George Vincent,[2] her cousin, studied under 'Old Crome' with Cotman[3] and J. B. Crome, and during his short career, was one of the lights of the Norwich School. Lydia's sister had married William Jackson of Canterbury—indeed Jacksons and Vincents intermarried for several generations—and their only son was godfather to his cousin William Jackson Hooker, to whom he afterwards left the Jackson property.

[1] Based on *Devon Worthies*, by the late Robert H. Hooker of Weston-super-Mare, who erected the beautiful statue of the Judicious Hooker in the Cathedral Close at Exeter.

[2] George Vincent (1796–1836 ?), the landscape painter, was born and educated in Norwich. A pupil of John Crome, he exhibited, chiefly Norfolk views, at Norwich between 1811 and 1831, and in London 1814–31, where he lived from 1818. His etchings date between 1821 and 1827.

[3] John Sell Cotman (1782–1842) was a landscape and portrait painter, chiefly in water-colours. He studied in London in 1798 and exhibited there 1800–6 and again 1825–39. He was Drawing Master in Norwich 1807–34, and in King's Coll., London, 1834–42; etched plates of Norfolk buildings and antiquities 1811–39, and published etchings of 'Architectural Antiquities of Normandy' made in 1817–20 (see vol. ii. p. 197).

The Vincent strain is responsible for Joseph Hooker's great feeling for art. The power of draughtsmanship came also from the Cotmans through his mother, Maria Turner, for her grandmother (Dawson Turner's mother) was Elizabeth Cotman, but the faculty thus transmitted was that of the copyist rather than the art-lover.

William Jackson Hooker, inheriting love of the garden and books from his father, of art from his mother, was one of those who came into the world with the true spirit of the naturalist, a characteristic he transmitted in full measure to his son. Like all such, his love for the outdoor world took him into field and wood and intimacy with the life of nature; in his school-days he collected insects and flowers and read books on natural history, and early got to know the flowers and mosses, the liverworts and lichens and freshwater algæ round his home in the heart of that county which possesses two-thirds of the species of British plants. No sordid cares, such as often overshadow a young man's future, prevented him from indulging his bent; for at the age of four he inherited a competency from his cousin-godfather, William Jackson of Canterbury, and as he grew up, he resolved to devote himself to travel and natural history. A keen sportsman, he made a fine collection of the birds of Norfolk; close relations with Kirby and Spence [1] and Alexander Macleay [2] spurred his pursuit of entomology.

His science and his scientific drawing both won early notice. When he was twenty he discovered, near Norwich, a species of moss (*Buxbaumia aphylla*) previously unknown in Britain; and three years later Sir J. E. Smith, in dedicating to him the genus Hookeria, made special mention of his illustrations of Dawson Turner's Fuci and of the difficult genus Jungermannia. The latter genus, be it noted, was an especial favourite of his. He published a monograph on the British Jungermanniæ

[1] William Kirby (1759–1850), entomologist, nephew of J. J. Kirby: educated at Caius Coll., Cambridge, was an original Fellow of the Linnean Society 1788. He published a famous Introduction to Entomology (1815–26) with William Spence (1783–1860), F.R.S. 1818, Hon. President of the Entomological Society, to which he bequeathed his collection of insects.

[2] Alexander Macleay (1767-1848), F.R.S. 1809, entomologist and Colonial Statesman; was Colonial Secretary for New South Wales 1825–37.

in 1816, and, as will be seen hereafter, his son, finding any on his travels, never fails to mention the fact in his letters home.

In his earlier days, William Hooker travelled afield botanising in Scotland and the Isles, no slight undertaking in 1807 and 1808; and in 1809 made his celebrated voyage to Iceland, where he witnessed a bloodless revolution (see p. 108), and on his homeward way lost his collections and all but lost his life by the burning of his ship. But he was unable to carry out his wider plans of visiting Ceylon and Java, S. Africa and Brazil, though he visited France, where he made acquaintance with the great botanists in Paris and Switzerland, a centre of botanical and geological interest.

In 1815 he married Maria, the eldest daughter of his friend Dawson Turner, and at his father-in-law's advice, embarked his remaining fortune in a brewery, in which the Turners and Pagets were interested. This promised to recoup the loss of large sums which he had sunk in the bottomless depths of the Spanish Funds. It was an enterprise, however, for which his aptitudes were little suited, and the business went steadily down. But this loss of fortune was the beginning of his greater career. Had the friendly alliance of Hooker, Turner, and Paget prospered, he would have remained an amateur—if a distinguished amateur—in science, and would never have achieved the special eminence which was to shape his son's career and be continued in it. A growing family and diminishing revenue made him look out for some botanical post that should both give scope to his special powers and bring in an income. Through the influence of his friend Sir Joseph Banks,[1] botanist, explorer,

[1] Sir Joseph Banks (1743–1820), President of the Royal Society, became a botanist in a burst of schoolboy enthusiasm. His ample inheritance enabled him to travel and to become a munificent patron of science. His most famous expedition was that with Captain Cook in the *Endeavour*, when he took with him, at his own expense, Dr. Solander, the pupil of Linnæus, two draughtsmen, and two attendants. In 1778 he was elected P.R.S., and held the office till his death, exercising a generous but rather autocratic sway over the scientific world, for whom his great collections and library were always open, and his house in Soho Square a gathering point. He left his library and herbarium to Robert Brown, his librarian, for life, with reversion to the British Museum, not only leaving him £200 a year, but providing for the famous draughtsman, Francis Bauer, during his life, that he might continue his drawings from new plants at Kew. As scientific adviser to George III, he also arranged for collectors to gather plants for Kew from abroad.

and chief power in the official world of English science, he was appointed by the Crown in 1820 to the newly founded Chair of Botany in Glasgow, in succession to Dr. Graham,[1] who, after occupying it a couple of years from its foundation, had been appointed to Edinburgh.

Here Sir William met with immediate and striking success. He established a flourishing school of botany ; raised the infant botanical garden to the front rank, supplying it and his herbarium with the products of every country with which the trading community of Glasgow was in touch. The experience gathered in Glasgow prepared his signal success in after years at Kew. Here, therefore, his sons grew up in an atmosphere of natural science, whether class-work or field-work, of long-drawn and unceasing industry, of contact with distinguished workers in natural history in general and botany in particular.

> The Professor [writes Prof. F. O. Bower in his Commemorative Oration] had established himself in Woodside Crescent, conveniently near to the garden, and doubtless his little son was familiar with it and its contents from childhood. He grew up in an atmosphere surcharged with the very science he was to do so much to advance. His father's home was the scene of manifold activities. It housed a rapidly growing private herbarium and museum. It was there that the drawings were made to illustrate the amazing stream of descriptive works which Sir William was then producing. New species must have been almost daily under examination, often as living specimens. Between the garden and the house the boy must have witnessed constantly, during the most receptive years of childhood, the working of an establishment that was at that time without its equal in this country, or probably in any other. The eye and memory will have been trained almost unconsciously. A knowledge of plants would be

[1] Robert Graham (1786–1845), M.D. He practised some years in Glasgow, and in 1818, when a separate chair of botany was established at the University, was appointed the first professor. In 1820 he became regius professor at Edinburgh, being succeeded at Glasgow by Sir William Hooker, with whom he had a scientific and personal friendship. Joseph Hooker, in turn, was within a little of succeeding him at Edinburgh, for he remained a close friend of the Hookers, often joining in Sir William's botanical excursions, and when he fell ill in 1845, he secured Joseph Hooker as his substitute and prospective successor.

acquired as a natural consequence of the surroundings, and without effort entailed by study in later years. Sir Joseph once said to me: 'You young men do not know your plants.' Certainly we did not in the way that he knew them. Few have ever known, few ever will know them in that way. Such knowledge comes only from growing up with them from earliest childhood, as he did.

The influence of Sir William's teaching, with its personal stimulus, its wealth of illustration by specimens and diagrams, its fostering of accurate observation and its botanising excursions, is well described in his son's own words taken from the address delivered at the opening of the Botanical Laboratory in Glasgow 1901. We see the boy sharing in these excursions long before he was a regular student at his father's lectures.

It was a bold venture for my Father to undertake so responsible an office, for he had never lectured, or even attended a course of lectures. But he had resources that enabled him to overcome all obstacles—familiarity with his subject, devotion to its study, energy, eloquence, a commanding presence with urbanity of manners, and, above all, the art of making the student love the science he taught. But his energies were not confined to lecturing. Feeling the want of a manual on the Scottish Flora to put into the students' hands, he published, in time for use in his second course, the 'Flora Scotica,' in two volumes, the outcome mainly of his earlier Scottish expeditions; and in readiness for his third course he produced, at his own cost, and from drawings made by himself, an oblong folio of twenty-one lithographed plates, with descriptions of the organs, etc., of upwards of three hundred plants. A copy of this work was placed before every two students in the class during that portion of the day's lecture which was devoted to the analysis of plants, obtained from the garden and placed in the students' hands for this purpose. I should mention that every student was expected to provide himself with a pocket lens, knife, and pair of forceps, aided with which he followed the demonstrations of the professor. I think it may fairly be said that these early lectures heralded the dawn of scientific botanical teaching in Glasgow University.

Another claim upon the professor's energies was due

to the fact that the botanical class was in a great measure ancillary to that of Materia Medica, a practical knowledge of which latter subject was at that time required of candidates for a medical degree, diploma, or licence by, I believe, all the examining bodies in the United Kingdom.

Now the Glasgow students of botany were, with a few exceptions, preparing themselves for the medical profession, and a considerable proportion of them at that time looked forward to service in the army, navy, India, and the colonies, where they would be thrown on their own resources for ascertaining the quality of their drugs, which had either undergone a long voyage from England or had to be replaced by such substitutes as the practitioner's knowledge of botany might enable him to discover. The professor hence devoted much time to teaching the botanical characters of the principal medical and economic plants. To this end he made large coloured drawings of them in flower, fruit, etc., which were hung in the class-room when the natural orders to which they belonged were being demonstrated, and he passed round dried specimens of them taken from his herbarium, or living ones from the garden when they were to be had, together with samples of the drugs or other products which they yielded.

It remains to allude to the class excursions, which have always been, and still happily are, a prominent feature of the botanical teaching in the Scottish Universities. Of these there were three: two, on Saturdays, were habitually to Campsie Glen and Bowling Bay respectively. The third, which was eagerly looked forward to by the most ardent of the students, took place at the end of June. It was to some good botanising ground in the Western Highlands. As many as thirty students have taken part in these larger excursions, each provided with as small a kit as possible, a vasculum, and apparatus for drying plants. They were often accompanied by students from Edinburgh, and sometimes by eminent botanists, British and foreign. In those days there were few inns in the Western Highlands, and fewer coaches, and the roads were bad. On one of my father's first excursions he provided a marquee to hold the party, which was transported in a Dutch wagon drawn by a Highland pony; and for supplies the party depended on the flocks and fowls of the cottagers. On the first excursion on which I was taken, when a boy, to Loch Lomond, there was no inn at Tarbet, and we all slept there in our clothes, on heather spread on the floor of a cottage; on another occasion when I was allowed to join the party (more for fishing than for botanising) on an excursion to Killin, we walked the whole way from the head of Loch Lomond along the old military road made in the previous century by General Wade, eulogised in the well-known distich:

If you'd seen these roads before they were made,
You'd have lift up your hands and blessed General Wade.

If I were asked what I regarded as of most importance to the student in the manner of my father's teaching as sketched above, I would answer that it taught the art of exact observation and reasoning therefrom, a schooling of inestimable value for the medical man, and one that is given in no other profession, but which ought to come, in this country, as it does in Germany, early in the education of every child. I have met many of my father's pupils abroad, in India, and the colonies, who have told me that these botanical lectures gave them the first ideas they had ever entertained of there being a natural classification of the members of the vegetable kingdom. Then with regard to the results, in a botanical point of view, the magnetism of the lecturer and the interest of the subject imbued many of his pupils with a love of science that proved permanent and fruitful. They made observations and collections for their quondam professor in the temperate or tropical climates of both hemispheres, some of them throughout their lives, which have very largely contributed to a knowledge of the flora and vegetable resources of the globe.

Not only was Sir William Hooker a great teacher and administrator, but a most prolific writer. His writings were unequalled in the number and accuracy of the plates with which they were illustrated. The number of these his son estimated at 8000, of which 1800 were from his own drawings. His systematic work covered a wide range, and, apart from its intrinsic value, has a peculiar interest here in its relation to the systematic work of his son. His publications on the plants of Parry's and Sabine's[1] Arctic voyages and on the botany of Beechey's voyage to Behring Strait, the Pacific, and China, compare with his son's Antarctic and Australian work. His 'Flora Boreali-Americana,' his 'British Flora,' his 'Niger Flora' are paralleled by work in the same fields. His ten books on ferns—for he was the leading pteridologist of his time—prelude Joseph Hooker's interest in the cryptogams, while the great series of the 'Icones Plantarum,' begun in 1837 to illustrate new and rare plants selected from the author's herbarium, which later became the nucleus of the great Kew Herbarium, was continued under his son and successor at Kew, thanks to the bequest left for this purpose by Bentham.

For the most part this work of his was a labour of love, often involving financial responsibility as well. Generous to others, and enthusiastic for his work, he thought little of his own interests in comparison with the scientific privileges offered by the position at Kew. He drew upon his private means, not only for his books, but for the ceaseless succession of botanical magazines of which he undertook the editorship, in order to secure a channel for recording the immense variety of new facts that came before him as director of large and expanding botanical gardens, facts needing to be set on record, though too scattered and disconnected for publication in anything but a 'miscellany.'

Joseph Hooker's mother, Maria Turner, brought another strongly marked strain of character and capacity into his

[1] Sir Edward Sabine, K.C.B. (1788–1883), saw active service in the American war of 1812, but after 1816 devoted nearly all his life to science, especially astronomy and terrestrial magnetism. For his researches on these subjects when in the Arctic with Ross and Parry he received the Copley medal in 1821, and subsequently extended his researches half across the world. He, assisted by Ross and others, made the first systematic magnetic survey of the British Isles, and, paying a visit to Berlin, prompted Humboldt to urge the establishment of magnetic observatories throughout the British Empire in connection with those already established elsewhere by other Governments, a proposal which led to Ross's Antarctic expedition. Sabine was President of the Royal Society from 1861–71; he had been general secretary of the British Association 1839–59, except in 1852, when he was President. His *magnum opus*, which included a complete statement of the magnetic survey of the globe, extended over thirty-six years from 1840, in his series of 'Contributions to Terrestrial Magnetism' in the *Philosophical Transactions of the Royal Society*.

inheritance. She was an accomplished woman, who not only shared her husband's tastes, but by her well-cultivated gifts was able to enter into his pursuits. Their outlook on life was similar, for both had been bred in the evangelical tradition, which she perhaps preserved the more rigidly. Like him, she had a love for music and art, and a keen interest in the sciences affected by her father, especially botany. She was widely read, and wrote with a facile pen steeped in all the copious rotundity of the Johnsonian school. From the Turner side, no doubt, she transmitted something of the business faculty that was to stand her son in good stead when he came to deal with men and affairs.

Similarity of tastes and interests had first drawn together Dawson Turner and W. J. Hooker. The younger man was speedily impressed by the great vigour and strong character of the elder, admiring his practicality the more for being himself careless of selfish interests in the enthusiasm of his pursuits. For the rest of his life Dawson Turner became his scientific friend, his intimate correspondent, his business mentor. Dawson Turner, indeed, won well-deserved success alike as banker, author, botanist, and archæologist. His mother, Elizabeth Cotman, brought him an artistic heritage. On his father's side, business and scholarship had been grafted upon a solid yeoman stock of Norfolk. For nearly two and a half centuries since the first Turner bought his modest acres at Kennington in 1570, these passed from father to son.

At the end of the seventeenth century, a younger son, Francis (1681–1719), was bred to the law, and settled in Yarmouth, where he married the daughter of the Town Clerk, Thomas Godfrey, and with obvious propriety succeeded to his office in 1710.

His only son was another Francis, who took Orders, married Sarah Dawson, and had four sons: (1) Francis, an eminent surgeon; (2) Joseph, who was Senior Wrangler in 1768, then Master of Pembroke College, Cambridge, and ultimately Dean of Norwich; (3) Richard, who, through the influence of his brother the Dean, became incumbent of Great Yarmouth; and (4) James, who became the resident partner in the firm of

Gurney & Co. when they opened a branch of their Norwich bank at Great Yarmouth.

This James Turner married, as has been mentioned, Elizabeth Cotman, and gave his mother's family name to his son Dawson (*b.* 1775).

Dawson Turner, as might be expected, went to Pembroke College, where his uncle was Master; but in his second year his father died, and he had to leave the University and take his place at the bank. But business did not exclude letters. As banker and author he was a forerunner of Grote and Bagehot and Lubbock. His library, his collection of autographs, his small but choice gallery of pictures, were all notable.

As early as 1797 he became a Fellow of the Linnean Society, and later, of the Society of Antiquaries and the Royal Society.[1]

Through the Turner connexion the Hookers gained several interesting cousinships—notably with the Palgrave family. Dawson Turner married Mary Palgrave (1774–1850), second daughter of William Palgrave, of Coltishall, and Elizabeth Thirkettle. Her younger sister, Anne Palgrave (1777–1872), married Edward Rigby, M.D., of Coltishall. Three of the Rigby daughters were married in Esthonia: Anne (1804–69) to George de Wahl, Maria Justina (1808–89) to Baron Robert de Rosen, Gertrude (1813–59) to Theophile de Rosen; Gertrude's daughter, again, in 1860 married General Manderstjerna, and the rest of her children married and remained in Russia. These second cousins of his welcomed Joseph Hooker on his visit to St. Petersburg in 1869.

Another Rigby daughter, Elizabeth (1809–93), married Sir Charles Eastlake, P.R.A. She was a close and lifelong friend of her cousin Joseph. Matilda, the eighth child and youngest of the Rigbys, married James Smith. Their

[1] Dawson Turner published important illustrated works on the British Fuci, the Mosses of Ireland, and especially the *Natural History of Fuci*, 1808–19, and, with L. W. Dillwyn, *The Botanist's Guide through England and Wales*, 1805. Later he devoted himself especially to art and antiquities. He wrote largely on the archæology of Norfolk and Suffolk, *inter alia* 'Grangerising' Blomefield's *History of Norfolk* with 2000 drawings. His chief archæological work was his *Account of a Tour in Normandy*, with fifty etchings by his wife and daughters and John Cotman.

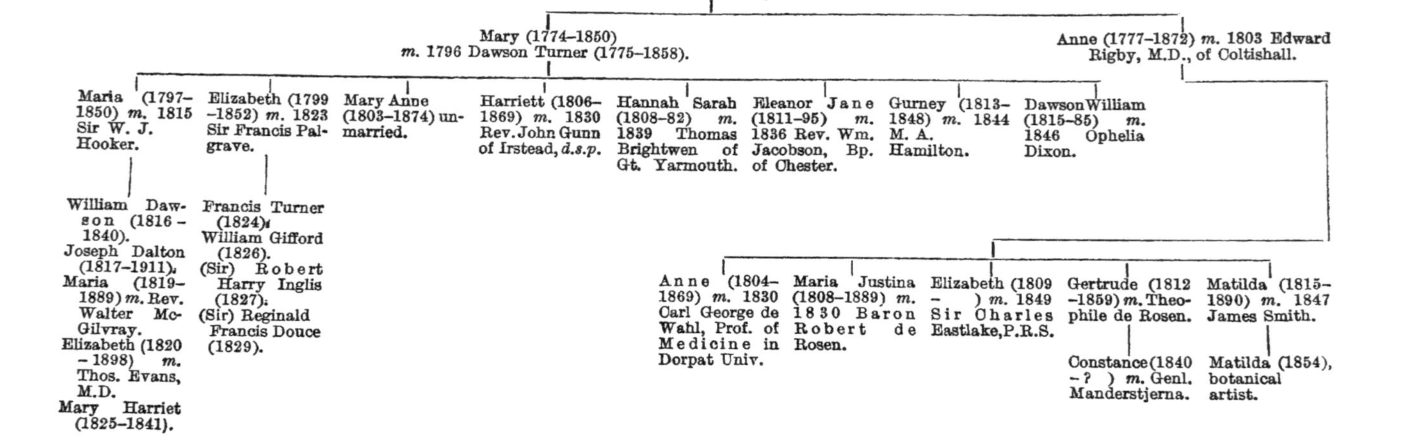
WILLIAM PALGRAVE = ELIZABETH THIRKETTLE.
Mary (1774–1850) m. 1796 Dawson Turner (1775–1858).
Anne (1777–1872) m. 1803 Edward Rigby, M.D., of Coltishall.
Maria (1797–1850) m. 1815 Sir W. J. Hooker.
Elizabeth (1799–1852) m. 1823 Sir Francis Palgrave.
Mary Anne (1803–1874) unmarried.
Harriett (1806–1869) m. 1830 Rev. John Gunn of Irstead, d.s.p.
Hannah Sarah (1808–82) m. 1839 Thomas Brightwen of Gt. Yarmouth.
Eleanor Jane (1811–95) m. 1836 Rev. Wm. Jacobson, Bp. of Chester.
Gurney (1813–1848) m. 1844 M. A. Hamilton.
Dawson William (1815–85) m. 1846 Ophelia Dixon.
William Dawson (1816–1840).
Joseph Dalton (1817–1911).
Maria (1819–1889) m. Rev. Walter McGilvray.
Elizabeth (1820–1898) m. Thos. Evans, M.D.
Mary Harriet (1825–1841).
Francis Turner (1824).
William Gifford (1826).
(Sir) Robert Harry Inglis (1827).
(Sir) Reginald Francis Douce (1829).
Anne (1804–1869) m. 1830 Carl George de Wahl, Prof. of Medicine in Dorpat Univ.
Maria Justina (1808–1889) m. 1830 Baron Robert de Rosen.
Elizabeth (1809–) m. 1849 Sir Charles Eastlake, P.R.S.
Gertrude (1812–1859) m. Theophile de Rosen.
Matilda (1815–1890) m. 1847 James Smith.
Constance (1840–?) m. Genl. Manderstjerna.
Matilda (1854), botanical artist.

daughter, Matilda, is the skilful botanical artist who succeeded Walter Fitch as illustrator at Kew.

Dawson Turner's eldest daughter, as we have seen, married W. J. Hooker. His second daughter, Elizabeth, married that Francis Cohen who on his marriage assumed the name of Palgrave with the consent of her uncles, the two surviving sons of William Palgrave, and last male representatives of the family.

Of Elizabeth's children Sir Francis Palgrave became Keeper of the Records. One of his sons, Francis Turner Palgrave, is in perpetual memory as editor of the *Golden Treasury*; another was William Gifford Palgrave, the famous traveller in the East; another, Sir R. H. Inglis Palgrave, banker and writer on financial subjects; and the fourth, Sir Reginald Palgrave, Clerk to the House of Commons. To all these first cousins Joseph Hooker was warmly attached, and with Inglis Palgrave especially, who constantly advised him on business matters, he kept up a lifelong correspondence, albeit a correspondence which seldom lends itself to quotation for general purposes.

Of the rest of the Turner family Harriett (1806–69) was the author of 'Letters from Holland.' She married, 1830, Rev. John Gunn, President of the Geological Society, Norwich.

Hannah Sarah (1808–82) made sixty portraits from drawings on stone, and fifty-one drawings for the 'Outlines in Lithography' for private circulation. She married, 1839, Thomas Brightwen of Great Yarmouth.

Eleanor Jane (1811–95) was an accomplished classical scholar. She married, 1836, Rev. Wm. Jacobson, D.D., Bishop of Chester.

Gurney (1813–48) married, 1844, Mary Anne Hamilton.

Dawson William (1815–85) Headmaster of the Royal Institution, Liverpool, married Ophelia Dixon.

The atmosphere in which the young Hookers grew up was one not only of strenuous work, but also of a certain austerity in moral and religious training, recalling the Puritan trend of their forbears. In daily example they saw that their father rose early, worked late, and seldom went out to entertainments. Like his wife, he was, as has been said, a strong Evangelical, seeing the hand of an overruling Providence in every turn of events, and accepting bereavements, or the prospect of them, with a pious resignation coloured by the warm conviction of future reunion. In the letters of both husband and wife, hopes for the future are regularly expressed with the pious qualification 'if God wills,' and present sorrows borne as 'the will of God.' Speculative thought beyond evangelical limits had no part in this household; they and theirs should uphold their own observances boldly before 'the scoffer and the sceptic.' The children were brought up simply, strictly, without indulgence—it seems, indeed, with some measure of rigidity—to be God-fearing, honourable, hardworking members of their society. If the outlook was in some respects narrow, compensation lay in the intellectual activities that found scope in varied scientific pursuits, in drawing and some music, and intercourse with men distinguished in science and travel. There is an obvious danger of young folk becoming priggish and didactic under such conditions, which tend to isolate them from the ordinary boys and girls of their world and to make them despise the thoughtless amusements and unfruitful occupations of their fellows; the saving salt for the young Hookers lay in their real enthusiasm for living pursuits and the freshness of their interests.

The family was five in number: William Dawson, Joseph's senior by fifteen months (*b.* April 4, 1816, *d.* January 1, 1840); Joseph Dalton, *b.* June 30, 1817; Maria, *b.* May 8, 1819; Elizabeth, *b.* November 15, 1820; and Mary Harriette, *b.* October 2, 1825, who died of consumption on June 19, 1841.

References to these early days are scattered and fragmentary. It is very clear that a strain of delicacy ran through the family, which showed itself in susceptibility to consumption. Joseph as an infant was 'croaky Joe,' with a tendency to cough and croupy hoarseness; William, shortly after his early marriage, was threatened with the disease, and was therefore sent to make a home and a medical career in Jamaica, where he was carried off by yellow fever, January 1, 1840. Then came nearly two years' painful anxiety over the two youngest sisters, who were at school in London under Mrs. Teed, at Little Campden House. A few weeks after Joseph set sail in the *Erebus*, in the autumn of 1839, Elizabeth fell ill, and had to winter at Hastings under the care of a great-aunt, Mrs. Walford Taylor, and to undergo a course of treatment in the next summer under Dr. Jephson at Leamington, where she was joined by Mary Harriette at the beginning of the holidays in July. Worse followed. On reaching Glasgow, Elizabeth fell back; Mary was found to be very ill. With some difficulty they were taken to Jersey at the end of September. Lady Hooker nursed them with the help of her capable and devoted eldest daughter; after much suffering, Elizabeth recovered, Mary Harriette slowly faded away.

Brothers and sisters were warmly attached to one another. Joseph's affections were not spread afield; they were the more intense for being concentrated upon his family circle—'the seven persons I really love'—and a few other friends. Writing home from the Antarctic after receiving the news of his brother's and sister's death, he accuses himself of the fault of selfishness. More justly, perhaps, he would have used the word self-centred; he always has the full sympathy of his correspondents, and his own letters show abundant care for those dear to him.

The home regime was sufficiently firm. Sir William, courtly, handsome, attractive, perhaps laid weight mainly on the duty of pure motive and honourable conduct; Lady Hooker was also a strict disciplinarian and a stickler for the forms of reverence which the manners of her young days demanded of children for their parents. When Joseph, for instance, came in from school after a long and tiring walk home, he must present himself to his mother, but was not allowed to sit down in her presence without permission, and was kept standing until it was clear that discipline had conquered inclination.

In their boyish days, William, the firstborn, was clearly the mother's favourite. He was the more clever, lively, and forthcoming. In Lady Hooker's letters to her father, Dawson Turner, Joseph as a rule appears rather as the plodder without his brother's brilliancy. William, however, with all his quickness and cleverness, had a vein of instability. The contrast between the brothers in the matter of perseverance shows itself from the first, and Joseph's determination to master whatever he undertook calls forth his mother's just praise. Later, William made a large collection of birds, while Joseph collected insects and plants. William won his literary spurs at one-and-twenty by printing for private circulation his 'Notes on Norway,' the account of a trip to Scandinavia; while Joseph, in the same year, first appeared in print with the description of three new mosses from the Himalaya in the 'Icones Plantarum' (ii. 194).

The boys went to Glasgow High School, where they received the old-fashioned, liberal, Scottish education—an education that culminates in the Arts' degree for proficiency in Latin and Greek, mathematics, logic and English literature, and moral philosophy. In after life Hooker thought the moral philosophy course had been of little value to him; his classical studies, however, were not lost even from an utilitarian point of view, and he remained always able to write Latin easily.

Sir William and Lady Hooker's letters to Dawson Turner afford a few glimpses into the boys' school-days. Thus Lady Hooker writes on June 9, 1824, after a description of Willy's lessons—to our great astonishment that little boys of seven and eight should attend a college lecture on botany:

> He and Joseph accompany their father, with Frank and Robert,[1] to the lecture every morning. It is fine exercise for them, and they return to breakfast at half-past nine o'clock, as hungry almost as my sisters and brothers used to be. I think that Joseph would be the child to please you in his learning. He is extremely industrious, though not very clever. Willy *can* learn the faster if he chooses, but while his elder brother sets his very heart against his lessons, Joseph bends all his soul and spirit to the task before him.

[1] Frank Garden and Robert Monteith lived with the Hookers for some four years, studying at Glasgow before proceeding to Cambridge (in 1827). 'Our two eldest boys,' Sir William calls them: they were eight or nine years older than his own boys.

And on December 21, 1824, she hopes that the grandfather's note accompanying a present will have a stimulating effect on the grandson who so little inherits his disposition; for:

> Willy is sadly negligent with regard to his lessons, especially his Latin ones. If we could but inspire him with a little emulation he would make great progress, for when any sufficient inducement occurs, he learns remarkably quickly and far outstrips his brother; but generally he is content to let Joseph get before him; and though we caress the latter and slight Willy [the modern mother, we hope, does not adopt this method of arousing emulation], yet William is not in the least jealous, but loves his brother as dearly as if he were not his superior.

Education, indeed, wore a stern face in those days. Poor Willy!

> I wish that I could tell you that your eldest grandson had inherited from his grandsire a little taste for learning languages. But ever since we returned from Yarmouth, the lessons, especially those in Latin, have been a perpetual source of sorrow both to the teacher and to the *teachee* (I wish I could say to the learner). Writing and arithmetic are the only departments of his education in which Willy has made any progress. But during the last ten days, a new light has seemed to dawn upon the child's mind. [He has made many good resolutions, couched in picturesque scriptural phrases.] We shall see how long they will last, but you may be sure that we bestow all manner of caresses and encouragement upon him. Indeed, we are ourselves happy in an opportunity to show a little kindness towards the poor child, who has lately received from us nothing but reproof and punishment.

Again, in 1828, Joseph being just eleven, his father writes:

> I wish you could bring the dear boy Gurney with you, and let him go to Killin in June with me and see Launden Cameron and climb the Breadalbane mountains. . . . Last year I took Willy the same route, and this year I think

of taking both him and Joseph. [Gurney and Dawson, by the way, Dawson Turner's sons, were almost of an age with William Hooker, being but three years and one year older respectively, and so more like cousins than uncles to the boys.]

In 1829: They make very fair progress with their tutor (who coached them in Latin) and are much more inclined to like lessons than they used to be.

1829: The boys beg to thank you for your kind present of 'The Boys' Own Book'; it is seldom out of their hands during playtime.

In after life Sir Joseph often talked of how he loved this book, and read it and consulted it.

In 1831 comes the first mention of their repeated stay at Helensburgh so that the children may have country air and liberty. Burnside was a delightful memory; but even more beloved was Invereck, and it became their country home in 1837. Indeed, when it came into the market in the late seventies, Hooker would have bought it had it not been so far from Kew.

As at thirteen, 'Joseph is becoming a zealous botanist,' so at fifteen, 'Joseph is contented and happy at home, and studying Orchideæ most zealously.'

In 1832, when the boys were sixteen and fifteen respectively, they entered Glasgow University, with four sets of lectures each, all in Latin and Greek for Joseph.

Joseph has paid a good deal of attention to collecting and drawing insects, though he has not nearly so much natural ability for sketching as his brother has. Mrs. Lyell sent Joseph a very nice specimen box, stored with four or five dozen of the rarer insects found near Kinnordy.

The Lyells of Kinnordy were to play a large part in Hooker's life. Charles Lyell, the elder,[1] was a botanist of distinction and

[1] Charles Lyell (1767–1849), eldest son of Charles Lyell of Kinnordy, was distinguished both as a Dante scholar and a botanist. Living at Bartley in the New Forest from 1798 to 1825, he devoted himself especially to the study of the mosses, several species of which bear his name, as well as the genus *Lyellia* of Robert Brown.

an old friend of Sir William's; and his son was that greatest of geologists who was to be the early inspirer of Darwin and his lifelong friend together with Hooker.

Later in the same year, 1832:

> Joseph is in the senior Latin and senior Greek, and next year will take logic and mathematics along with his brother. William continues ardently devoted to ornithology, and Joseph to botany and entomology. The latter is already a fair British botanist and has a tolerable herbarium, very much of his own collecting. But the orchideæ are his great favourites, and he has an eye for them, and a memory too for their names, which often surprises me. Had he time for it he would already be more useful to me than Mr. Klotzsch [his assistant].[1]

The removal to a new house in Glasgow, at Woodside Crescent, 'spirited up' the family to an access of tidying, and 'Joseph has taken in hand to arrange all his father's duplicate plants, selecting among them for his own collection, and he has been pursuing this occupation with much diligence for some weeks.'

Next year, Joseph being sixteen, his father declines an invitation for him to go to the Dawson Turners' at Yarmouth, saying, 'the expense is very considerable for a lad who is scarcely old enough to derive permanent advantage from such a journey; and both he and his brother have now entered upon studies which can scarcely with propriety be interrupted.' The permanent advantage of studying his grandfather's collections would doubtless come later, when he should be further advanced in his regular botanical work.

A little later Sir William sends his father-in-law a parcel, in which is enclosed a small box of insects which Joseph is 'very desirous of transmitting to Mr. Paget.'[2]

The same entomological enthusiasm inspires two early

[1] S. J. Klotzsch spent some years as Sir William's curator at Glasgow, and was the founder of the mycological portion of the herbarium. Subsequently he became keeper of the Royal Herbarium at Berlin. Hooker gives an amusing description of his oddities in the *Memoirs* of his father, p. xxxiii.

[2] No doubt Charles, brother of (Sir) James Paget, the famous surgeon (1814–99), and one of the seventeen children of Samuel Paget, brewer and ship-

letters to Dr. Harvey,[1] who had sent him the first part of Stephens' 'Entomology' with some specimens. As his own collection is not yet very well supplied—Scotland not being a country where insects abound—he sends, in default of a better return, some German plants given him by Mr. Klotzsch (December 3, 1833).

And again on December 11, 1835, when Dr. Harvey had promised to collect insects for him at the Cape, he sends instructions as to a new method of preserving specimens in hot climates, and continues:

> Your account of the country fills me with an ardent desire to go there; however, I suppose I must be content to live on that unnourishing diet *hope* for some years to come. I should give a great deal to be present at the opening of the boxes of insects the travellers from the interior bring down, they must bring some splendid things; pray, what becomes of them?
>
> William is particularly obliged for your anxiety about procuring birds, and, believe me, I am more eaten up with entomological zeal than ever; who knows but I may, ere I die, publish an Entomologia Capensis? That poor unfortunate Stephens is determined to go on to the end with his invaluable work; he cannot now, I hear, afford to keep his

owner, who was Mayor of Great Yarmouth in 1817. The Pagets, the Dawson Turners, and the Hookers were closely allied in a friendship of long standing.

Between 1830 and 1834 James was apprenticed to Dr. Costerton, and, with his brother Charles, wrote a book on the natural history of Great Yarmouth.

[1] William Henry Harvey (1811–66), of Irish Quaker stock, began his lifelong friendship with Sir W. J. Hooker through his discovery at Killarney of the moss *Hookeria lætevirens* (1831). After holding various posts at Dublin he went in 1835 to South Africa with his brother, on whose death he succeeded in the post of Colonial Treasurer. In 1842 he broke down in body and mind from overwork. Returning home, he became Keeper of the University Herbarium at Dublin, and in 1848 Professor of Botany under the Royal Dublin Society. He visited America in 1849–70; the Indian Ocean and Australasia in 1853–6, and on his return succeeded to the botanical chair at Trinity College, Dublin.

His work included a Flora Capensis, but he is best known as an authority on Algae, publishing a *Manual of British Algae* (Laylor, 1841), the *Phycologia Britannica, Nereis Australis, The Seaside Book* (1849), *Nereis Boreali-Americana, Phycologia Australica,* as well as on the *Antarctic Algae of Beechey's Voyage,* and to him J. D. H. refers his collection of Southern Algae. His work lay in 'discrimination, description, and illustration'; he had no share in the Darwinian movement, though ready to admit natural selection as a *vera causa* of much change, he would not go so far as to admit it as a *vera causa* of species.

wife, a salutary lesson to all not to marry, who want to devote their time to Nat. Hist.

Of Joseph's University work Sir William writes:

> William and Joseph have entered upon their College duties of the present session apparently with much satisfaction. They both take Mathematics and Moral Philosophy. Joseph in addition attends the private Greek class, and William, Surgery.

The following letter from Lady Hooker may be quoted at length for the light it throws on the family's work and successes. Lady Hooker, it will be observed, cultivated a sub-Johnsonian style; or perhaps more truly reflected that of the Swan of Lichfield, itself a reflection from the authentic Johnson.

'Your son' is an affectionate trope for son-in-law, and his 'honors' mean that he is now created Sir William, Knight of Hanover, an order which became extinct with the separation of Hanover from the British Crown.

> Saturday, May 7, 1836.
>
> Many thanks for your affectionate congratulations on your son's honors and your grandsons' prizes, on the industry of the latter, I should rather say. The hope of pleasing their relations and gaining their good opinion, goes so far with both William and Joseph (especially the former), and they value so highly (as they ought to do) your favor and commendation, that I feel particularly gratified at your having taken the trouble of writing to them upon this occasion. Your present to them is quite too munificent, as perhaps they felt,—for Joseph immediately remarked he *hoped* his grandfather was very rich, or he should not like to take so much money from him. They would, I am sure, gladly add a few lines to thank you, in their own hand-writing, but their father and I have just left them at Helensburgh, where they will spend the Sunday with their grandpapa and sisters, returning home early on Monday morning. A fortnight ago, Joseph walked 24 miles—from Helensburgh to Glasgow—rather than wait for the steamer next morning, by which delay he would have missed a lecture. Willie has gone to-day to fish in Loch Lomond,—he started at 3 o'clock this morning: Joseph has been equally earnestly employed in turning over

stones and hunting in the rejectament of the sea for beetles. His collection of insects is becoming considerable, he devotes every spare minute to it, and has opened a correspondence with several entomologists, both British and foreign. We sent you a Glasgow newspaper last Tuesday, which mentioned the prizes: in the Natural Philosophy Class, where Joseph gained one prize and worked for three, he was the youngest student of all, and much younger than the majority of those who attend the Anatomical Lectures, where he carried off the single prize which alone is given, among a class often consisting of more than a hundred individuals. These circumstances, which cannot be publicly known, ought yet to be thankfully taken into account by us, when calculating the amount of his labour and of the success which has crowned that labour. I could not help hoping that the dear boy had caught a shred of his grandfather's mantle (far be it from me, by this awkward and tattered simile, however, to imply that the garment is either worn out or cast aside by the honored wearer) when I saw him, earnestly and unprompted during his papa's absence, undertake the task of cataloguing every book in the house. All the names were written down and arranged alphabetically, and part of the fair index was made before his father returned.

Of his tastes and education, Joseph himself wrote later, towards the end of the Antarctic voyage, to his aunt, Mary Turner. The letter, a copy probably touched up by his mother, is dated April 18, 1843.

You remind me of the times when we used to sit in the study (where probably you now are and where this note may reach you some two months hence) reading Tacitus: at least you and my grandfather reading it and I looking on.

Alas! I never had much taste for Latin and Greek, or any of the dead languages; and (except that I should have the satisfaction of knowing that my father's money was not so much thrown away) I greatly doubt if my having been a good scholar would give me now so much pleasure as you might imagine. What I do really regret is the little attention I paid to Ancient and especially to Modern History. If half the time spent on the Classics had been devoted to those subjects, the knowledge of them would prove a far more agreeable companion than Horace, Virgil or even Homer. Do not think I underrate those attainments, which alone make a man the perfect gentleman; but I had no taste for them, though ample time and opportunity for all. As it is, I sometimes attempt to rub them up, but I enjoy nothing so much as Hume and Smollett.[1] This mainly arises from the writers' bringing associations, connected with different parts of my native land, and of scenes, though perhaps only scampered through in a Mail Coach, which my memory, very retentive of localities, enables me to revisit, along with the heroes of my Author. A love of poetry is also a sad deficiency in me, for you cannot suppose that I should learn to appreciate it by being crammed with stanzas of Marmion, not amid Castles and Groves, but in a school of 100 boys. Crabbe's Poems are my favorites (laugh at me if you will), because I can go with him everywhere. As for Thomson, 'void of rhyme as well as reason,' he is quite too lackadaisical for me. To the Southward, in bad weather, I used to spend a great deal of time in reading, chiefly books on Scientific subjects, which are of most importance to me now that I have to work for my bread.

Of French he early acquired a working knowledge, improving it greatly in the winter of 1844–5, before his journey to Paris, by dint of lessons and conversation with M. Planchon, his father's assistant at Kew. With German, also, he was conversant enough to tackle German books on botany; but it was a labour to him. Hence the zest of his repartee to Darwin, of whom it is told ('Life,' i. 126): 'When he began German long ago, he boasted of the fact (as he used to tell) to Sir J. Hooker, who replied: "Ah, my dear fellow, that's nothing; I've begun it many times."'

Among his contemporaries he neither courted popularity nor was constitutionally fitted to practise the arts of popularity. Indeed, he suffered from a nervous irritability of the heart which from his school-days brought on palpitation when he stood up to construe in class. And although he tried to overcome this by joining his college debating society and getting up speeches carefully beforehand, success was denied him.

[1] The continuator of Hume's *History of England.*

Even in later life the delivery of an address meant a strain which brought on physical nausea and severe nervous reaction.

As he grew up, he went far afield on his botanical expeditions. On September 2, 1836, Sir William, sending a belated acceptance of an invitation for Joseph to visit his grandfather, writes : 'I only returned from a Highland tour with Dr. Graham, Mr. Wilson [1] and Joseph last Saturday. The latter had been away some weeks with Mr. Wilson amongst the Aberdeenshire mountains, and I could not communicate with him but by ferreting him out in person, which I did, and found him and Wilson at the old hovel at the foot of Ben Lomond, where they were nearly a week.'

On his way to Yarmouth, he stays at Liverpool with Mr. Melly, a collector of beetles, among whose specimens he sees the Goliathi, which he afterwards collected himself in India ; and at Manchester with Mr. Glover,[2] possessor of a less valuable collection ; at each city visiting the Museum and Botanical Gardens. The Manchester Gardens are 'the finest I ever saw ; finer, I think, than Edinburgh, though not, certainly, so good a collection of plants.'

Then at Hull he stays with William Spence, joint author with William Kirby (a Norwich man) of the famous 'Introduction to Entomology,' examining his rich collection and twice going out entomologising with him.

At Yarmouth he works keenly in his grandfather's and Miss Hutchins' herbaria ; and as a result asks his father to re-examine his own specimens of a certain moss (*Bryum triquetrum*) in order to correct what he feels sure is a wrong ascription of a specimen of his grandfather's. So, too, the latter has just received five specimens of the narrow-leaved lungwort

[1] William Wilson (1799–1871) was a botanist who had been attracted to the study during the open-air life necessitated by an early breakdown from overwork. In 1827 he was introduced by Henslow to Sir W. Hooker, and joined him in his annual students' botanical excursion. Through Hooker he devoted himself to the mosses, and described the mosses collected on Ross's Voyage. His great work, the *Bryologia Britannica* (1855), though intended to be a third edition of W. J. Hooker's *Muscologia*, was substantially a new work of the highest merit. Among the new species added to the British Flora by Wilson, his name is preserved in the rose named after him by Borrer, and the Killarney filmy fern (*Hymenophyllum Wilsoni*) by Sir W. Hooker.

[2] Perhaps Stephen Glover (*d.* 1869), known for his *Peak Guide*, 1830, and *History of the County of Derby*, 1831–3.

(*Pulmonaria angustifolia*). Joseph, examining these, concludes that it is one and the same with our common lungwort (*P. officinalis*), but that Linnæus' *P. officinalis* is not a British plant.

From his visit to Yarmouth he returned on November 8, and on the 10th his father writes :

> I need hardly tell you that the boy has enjoyed his visit much and seems really grateful for the privileges he has enjoyed, especially under your roof. He is quite disposed to work at the classes, and set out yesterday morning before breakfast to enter them. He takes Surgery, Chemistry, Materia Medica, Anatomical demonstrations, and occasionally the dissecting-room. He is gone to-day to endeavour to arrange with Mr. Arnott [1] to give him two hours a day at Latin, as you kindly suggested. Thus you see his time will be fully occupied, and he can only reckon on a holiday now and then to allow him to devote some attention to naturalist pursuits.

Next summer we find him geologising, in Arran, with his friend Thomas Thomson.[2] And to go forward a year, on January 9, 1839, Sir William tells Dawson Turner :

[1] George Arnott Walker Arnott (1799–1868), who had given up the law for botany, was a close friend of Sir W. Hooker, with whom he collaborated from 1830–40 in describing the plants of Beechey's voyage, and in 1850 in the sixth edition of the *British Flora*. In 1839 he acted as Sir William's substitute, and from 1845 till his death held the Glasgow chair of Botany.

[2] This Thomas Thomson (1817–78), naturalist and traveller, was the eldest son of Thomas Thomson, Professor of Chemistry at Glasgow from 1817. A schoolfellow of the Hooker boys, he was equally devoted to science, and at the age of seventeen did some remarkable original work in geology, and later, no less original chemical work under Liebig. He graduated M.D. in 1839 with the Hookers, and entered the service of the East India Company as assistant surgeon. He had a perilous adventure during the invasion of Afghanistan, ill-famed for the massacre of the Khoord Kabul, for he was captured by the Afghans at Ghazni, and narrowly escaped being sold into slavery in Bokhara, 1842. Meantime, as later during the Sutlej campaign and his subsequent stay in the Punjaub, he studied Indian and Himalayan botany. As one of the commissioners for marking the boundary between Kashmir and Chinese Tibet in 1847, he travelled into little known regions, embodying his geological and botanical observations in his book, *Travels in the Western Himalayas and Tibet*, in 1852. At the end of 1849 he joined Hooker at Darjeeling, and travelled with him for fifteen months on his later expeditions, especially to the Khasia Mountains. Returning to England in broken health, he spent several years at Kew, working at his collections, and bringing out, in collaboration with Hooker, the first and only volume of the *Flora Indica*. From 1854 to 1861, he was again in India as superintendent of the Calcutta Botanical Gardens in succession to Dr. Falconer, Professor of Botany. Later he lived again for a time at Kew.

> When I went to bed at a late hour last night I left him writing an answer to you, and indeed he may, with a clear conscience, give a good account of himself for the last three or four weeks, especially as relates to his botanical pursuits. He has worked at plants with a degree of steadiness and ardour that has been most gratifying, and it appears that his industry is likely to meet with its reward . . . [i.e. in selection for the Antarctic Expedition].

Three letters of August and September 1838, from the young Hooker to his father, tell how he went with Dr. Graham on a botanising trip in Ireland (August 2–18); to the British Association Meeting at Newcastle (21–30); and then proceeded to visit Dr. Richardson[1] at Haslar (September 1–4), when the latter was to take stock of him, so to say, before recommending him for the Antarctic Expedition.

Details of travelling in those days have a curious interest in comparison with to-day. Thus, leaving Dublin

> at 4 P.M., started in a track-boat for Ballinasloe, where we were met by a Biancini car, which took us to Galway by 8 P.M. on Friday night; the car and track-boat were of the same company, and we went the whole excursion, 140 miles, for 18*s.* each, including a dinner and a breakfast; this, however, was the only cheap travelling experienced.

To get from Newcastle to Portsmouth he was advised

> to take the coach from Newcastle to London at 9 A.M. on Thursday, which I did for £2. I went the whole distance, including coachmen and eating, for £3. I travelled all night, and arrived in London on Friday night, at 8 P.M. A coach was then starting for Portsmouth, in which I took a place, 14*s.*, and arrived here on Saturday at 8 A.M.

[1] Sir John Richardson (1787–1865, and knighted 1846) saw much active service as naval surgeon, 1807–15, then returned to Edinburgh and took his M.D., at the same time studying botany and mineralogy. He was Naturalist to Sir John Franklin on two Arctic expeditions, 1819–22 and 1825–27. For ten years he was head of the Melville Hospital at Chatham, and from 1838 was physician to the Royal Hospital at Haslar, where young naval surgeons awaiting their gazetting to ships were under him. Again, in 1848–9 he led the expedition in search of Franklin. His second wife, *m.* 1833, *d.* 1845, was a niece of Franklin's. In addition to his works on Polar Zoology and Travel, his special subject was Fishes.

A few more passages may be quoted.

Galway is a horrible town with 30,000 inhabitants, filthy in the extreme, without a single good building in it; the whole neighbourhood is limestone, and the fields are all covered with large stones which are turned into walls of the worst description.

Thursday, botanised about Cliffden, rained tremendously all day; went to Mr. D'Arley's at Cliffden Castle. Mr. D. is a very nice gentleman, hospitable in the extreme, who regretted his inability to take in our party of 12. He is tremendously in debt, but no creditor can go to the expense of arresting him, for the Connemara boys, with whom he is a great favorite, will allow no such intruder near Cliffden Castle. The last person who tried was an Inn-keeper here, but the inhabitants, guessing his intention, would not let his servant enter the village, but beat him unmercifully and sent him off. The police force were collected, who took them, and the malefactors are now lying in Galway jail for the next assizes.

True to his careful upbringing, he is ever punctilious in recording his Sunday observances.

[At Galway] we went to Church twice, and I once to the Roman Catholic chapel besides, with which I was much disgusted; the gallery was well filled with respectable persons, but the body of the Church was crammed with inattentive hearers covered with rags or nearly naked. The English services were good, but the congregations wretched. [Next week, at Killery] for some reason or other no service was performed, nor was there a Church nearer than 20 miles.

It was not a very profitable excursion in its results, albeit he is most careful in his expenditure.

I have regretted the expense, just £10, extremely, as except getting a good stock of the above-mentioned plants, nothing has been done but making as many sketches as I could by waiting behind the party; these I have had no time to finish at all. Of plants I have about 3000 specimens, as far as I can count, all dried as well as I could; this I say with conscience, and as I changed the papers every night, when possible, I am sure you will be pleased. . . . Mosses are extremely scarce here; I think one is, however, the *Hymenostoma rutilans*, as far as I can judge without a microscope; if so it be, a good discovery and the only one; it was very sparing in a wood near Galway, at the foot of a tree on the ground; it is very minute and there are only three or four capsules; the other Mosses you will see are some of them very common and only gathered for my own examination.

Now, my dear papa, such is the outline of the excursion which you were kind enough to allow me to join, solely, as it has turned out, for my own gratification. I have enjoyed it extremely, and feel twice as strong as when I left Glasgow; I hope the remainder of it, and especially the interview with Dr. Richardson, will be more profitable to myself. . . .

Excuse this hasty letter, it is now 3 A.M., and we start to-morrow morning. I am very sleepy, the fleas in Connemara keeping me awake the whole night sometimes.

As to the British Association, the Newcastle meeting of 1838 was his first. It was said to outshine in splendour any former meeting; and he confessed to his grandfather that with all its obvious utility as a common meeting-ground, and its encouragement to the non-scientific who were temporarily proud to be seen with a hammer or vasculum, 'the scientific department fell far behind the amusement and eating.' One notes the number of scientific men he either knew already or was introduced to; the quaint appearance of Dr. Richardson in the Natural History section, as he sat on the left of the Chair, and read the report of the previous day's proceedings,—

being fully attired in a Dumfries Tartan of broad check and a shooting coat of the same. . . . There were not above 50 people in the room, and almost no ladies; those few who were there had come in by accident, and I was afterwards much surprised to hear that ladies were precluded from attending this section of Botany and Zoology on account of the nature of some of the papers belonging to the latter division, [for which, in his judgment, there was not the least occasion].

> [On the 24th.] The Medical section was wretched; when I went in Dr. Bowring[1] was reading a violently radical paper condemning Quarantine laws *and* the Government which allows them.

On the 27th, at the Anniversary dinner of the Literary and Philosophical Society of Newcastle,

> the Bishop of Carlisle was in the Chair and proposed several toasts, among others the Universities of Great Britain, with a long speech, which Buckland[2] answered to; but neither of them seemed to remember that there was such a place as Glasgow, or Edinburgh either, which much offended me and T. Thomson; I thought it an especial bad compliment to Dr. Graham, who was sitting at the same table as the speakers.

The botanist in him was also up in arms next day at a public meeting, when it was resolved that a Botanical Garden be established in Newcastle, provided that it be united to a Zoological one; whereupon 'proposed that it should be called a Zoological and Botanical Garden, and agreed to; I wondered why it should not be called the Botanical and Zoological Gardens.'

The minor *agrémens* of the meeting included the usual dinners and fêtes; the botanical excursion headed by Dr.

[1] Sir John Bowring (1792–1872), merchant, linguist, traveller, diplomatist, financial reformer, and man of letters. Among his varied activities he was editor of the *Westminster Review* on its foundation by Jeremy Bentham; M.P. for the Clyde Burghs 1835–7, and for Bolton 1841; an original founder of the Anti-Corn Law League with Cobden, and plenipotentiary in China during the troubled times from 1854. Having newly returned in 1838 from a Government commercial mission through Egypt, Syria, and Turkey, he was fresh from the exasperating methods of quarantine in the East, which took shape in the *Observations on Oriental Plague and Quarantines* which startled the youthful Hooker.

[2] William Buckland (1784–1856), wit, geologist, and divine, who was Professor of Mineralogy, 1813, and Reader in Geology, 1819, at Oxford, President of the Geological Society, 1824, and Dean of Westminster, 1845. His work, which was valuable and suggestive, included the proof that the 'dressed rocks' of this country were the result of planing by glacial ice-sheet; nevertheless orthodoxy, alarmed at the claims of other geologists, smiled upon him, for in his inaugural address he calmed these fears, and in his 'Reliquiae Diluvianae' (1823) he employed his great knowledge and intuition to correlate the cave remains with the deluge. His famous Bridgewater Treatise of 1836 was another buttress of science as applied to contemporary theology. His drollery and quaint stories were famous.

Graham; the descent of a coal-mine, with its breed of horses remarkable for their short and glossy hair like that of a mouse; and visits to a rope-walk, alkali works, and Richardson's Crown Glass factory, which calls forth a reference to one of his encyclopædic sources of general knowledge:

> The most interesting process was the converting the globe of glass into a flat sheet by merely twisting quickly the iron rod to which it was attached; if you remember, the process is well described in one of the late numbers of the *Penny Magazine*.

CHAPTER II

THE ANTARCTIC VOYAGE: PRELIMINARIES

JOSEPH HOOKER had received a unique bringing up in his father's house. He did not so much learn botany as grow up in it. At one-and-twenty he was probably the best-equipped botanist of his years, and he was just finishing his medical course. From his father's position he also received unique opportunity. Sir William enjoyed the friendship of many influential men, scientific and official, who kept him in touch with any scientific projects that were taken up by Government. Two such were afoot in 1838–9: one, Ross's expedition to the Antarctic; the other, Captain H. D. Trotter's [1] to the Niger. Each would require a naturalist. Had Joseph Hooker failed to secure a place with Ross, he would almost certainly have joined the other ill-fated expedition, most of the Europeans on which died of fever.

James Clark Ross, the distinguished Arctic explorer, was already known to Sir William through their common friend, Dr. Richardson of Haslar. He had told Sir William his prospect of leading the Antarctic expedition which only awaited the Government's definite authorisation. Now in the early autumn of 1838 he was paying a visit to the Hookers' close friends and neighbours, the Smiths of Jordan Hill, whose names in

[1] Captain, afterwards Rear-Admiral Henry Dundas Trotter (1802–59), who had already distinguished himself in the suppression of piracy, headed an expedition in 1841 to the west coast of Africa and especially to the Niger to conclude treaties of commerce with the negro kings. Tropical fevers broke up the expedition; two of the three ships were forced to return after three weeks; Trotter himself continued another four weeks before returning, so shattered in health that he was unable to undertake active service for the space of fourteen years.

successive generations will often recur in these pages. James Smith[1] himself was keenly alive to all scientific interests. Knowing what was afoot, he invited Sir William and Joseph to breakfast that the young man might be presented to Ross. It was an unforgettable morning. Sixty years later, writing to Sabina Smith (Mrs. Paisley), Hooker recalled how he had longed to be at the second table, where Ross sat with the young daughters of the house and kept the party lively. His own turn came later. Ross received him very kindly and promised to take him if he would prepare himself for such a duty. One point was that he should first qualify as surgeon. This meant much hard work: as he wrote to Dawson Turner, October 8, 1835:

> Papa has I know told you of the distant prospect there is of my going on expedition to the Antarctic Ocean: I can hardly conceive my being prepared both as a Medical Man and Naturalist; to pass my necessary examinations will be a great push, while again if I do not devote a good part of this winter to Natural History, I had better not go at all. If the expedition does start and I do not go, I shall be dreadfully disappointed, though I am sure I had better not go at all than go ill prepared: the matter will, I hope, stimulate me to exertion.

From a letter of Sir William's to Dawson Turner, dated January 9, 1839, we catch a glimpse of the difficulties to be overcome and the influences set moving to overcome them.

> To-day's post brought me along with your letter one from Dr. Richardson telling me that their Antarctic Expedition had on Saturday received Lord Melbourne's[2] sanction and would sail on the 1st of May. Dr. Richardson fears that Joseph may not be qualified in time, and indeed strictly speaking he cannot be until the 5th of May: but I have

[1] 'Smith of Jordan Hill' (1782–1867) was a lover of literature and the fine arts as well as a considerable geologist, studying especially the changes of level on the coasts of West Scotland and of the Mediterranean, in relation to a glacial period. In another direction his *Voyage and Shipwreck of St. Paul* became a standard authority, thanks to his experience as a practical yachtsman. His son Archibald, the mathematician, and his daughter Sabina (Mrs. Paisley) were contemporaries and friends of Hooker's.

[2] Lord Melbourne was Prime Minister from 1835–41.

written to Edinburgh to endeavour to have that difficulty obviated, and I have asked the Duke of Bedford[1] for a letter to Sir Wm. Burnett[2] (the head of the Medical Navy Board), and I have written to Sir John Barrow[3] and Capt. Ross: and I trust there will be no difficulties in the way. The poor boy is delighted, and I trust it may be in every way for his good.

Joseph joined him in London; on the 18th he reports that the various friends whose aid he had invoked had duly exerted their influence, and Sir W. Burnett

promised to take Joseph into the Navy as soon as he had completed his curriculum [the end of April] and, if I wished, to give him an appointment at Haslar Hospital and a charge in the Museum there with £120 a year. Then he would be employed until the Antarctic Expedition was determined upon, for there are some difficulties in the way of it, and it is doubtful if it will sail before next year.

Joseph has quite won Brown's[4] heart by bringing him

[1] John, sixth Duke of Bedford, 1766–1839, was an enthusiastic naturalist, devoting himself to botany, agriculture, and the fine arts after his retirement from politics in 1807.

[2] Sir William Burnett (1779–1861). After studying medicine at Edinburgh, and seeing much active service as naval surgeon, he had a brilliant career as Inspector of Naval Hospitals. In 1822, Lord Melville appointed him to the Victualling Board, as colleague to Dr. Weir, the chief medical officer of the navy. Then becoming Physician General of the Navy, he introduced valuable reforms, among other things improving the position of assistant surgeons.

[3] Sir John Barrow (1764–1848, Bart. 1835), born of peasant stock in Cumberland, was distinguished from boyhood by his mathematical gift and his adventurous spirit. Thanks to the appreciation of Sir George Staunton, he accompanied Lord Macartney both to China and the Cape, and from 1804–45 was second Secretary to the Admiralty. He was specially interested in Arctic discovery, having had stern experience of the ice as a youngster in a Greenland whaler. A link with the Hookers was his friendship with Dr. Richardson, and the fact that he had studied botany at Kew Gardens before going to the Cape in order to appreciate the natural history of South Africa.

[4] Robert Brown (1773–1858) was called by Humboldt 'facile Botanicorum princeps, Britanniae gloria et ornamentum.' Beginning as surgeon-mate to the Fifeshire regiment of Fencibles, he made a large collection of plants in Ireland where his regiment was quartered, and through his discovery of a rare moss, first made acquaintance with Sir Joseph Banks, by whom he was afterwards offered the post of Naturalist to the *Investigator* under Captain Flinders, 1801–5. The resulting *Prodromus Florae Novae Hollandiae* was a valuable piece of systematic work, and his researches into the reproduction of plants, and especially in the morphology and interrelation of the higher plants, were marked by important discoveries, which carried him as far as the conditions of the time allowed. With these, and the discovery of the nucleus of the vegetable cell, he took a long step towards the development of

some Van Diemen's Land plants which the boy had been studying with considerable attention. We dined yesterday at the Royal Soc. Club and attended the meeting in the evening.

Thus he can add:

My journey has been fully answered in respect to Joseph. . . . Humanly speaking, his way is clear before him for an honourable scientific career.

And on June 18:

Should it please God that Joseph returns safe from his present expedition, and if I have the same friends I have now, it may be in my power to keep this appointment [the Glasgow professorship] in the family by applying to have it made over to Joseph.

As it turned out the preparations took nearly five months longer; part of this time Hooker spent at Haslar, 'a most improving situation under Dr. Richardson's eye,' just as his future friend, Huxley, was to do seven years later, while waiting for his appointment, so long delayed because the discerning Richardson kept him back till a scientific post offered in the *Rattlesnake*. The remainder of the time from the middle of June, Hooker spent as Assistant Surgeon attached to the *Erebus* at Chatham, where the ships were fitting out—Assistant Surgeon and Botanist—for it was in this capacity that he went after all, not Naturalist to the expedition, as he had confidently hoped. For that responsible post Ross finally determined to take a man of longer standing and some established repute, albeit the young Hooker pressed him very shrewdly, as appears from the following descriptions of some official interviews.

physiological as well as systematic botany. In 1810, on the death of Dr. Dryander, he succeeded to his post as librarian to Banks, who, dying in 1820, left Brown his library and herbarium, with reversion to the British Museum, and £200 per annum, with his house in Soho Square. In 1827 he arranged for the library and herbarium to pass immediately to the British Museum, while he was appointed Curator. In this position he had an official influence comparable to the influence of his strong character and intellectual powers among his friends.

Golden Cross, Charing Cross: April 27, 1839.

MY DEAR FATHER,—You will be surprised to hear from me so soon again, and I assure you the unfortunate cause has given me much vexation.

In my last letter I told you that I had not seen Captain Ross; I have since, after much hunting, and the result of the interview has been most unfortunate. The following is a correct statement.

One of the first questions I asked him was in what capacity he was to take me; he told me 'as Asst. Surgeon and Botanist,' adding 'that he had appointed the Surgeon, Dr. or Mr. McCormick,[1] to be Zoologist.' I saw at once that this would completely interfere with all my duties, but I said nothing, desiring first to know whether he would take me in any other capacity; so I asked 'whether he would take a Naturalist with him and give him accommodation, provided Government would sanction or send him.' He put off my question twice, evidently seeing my drift, which I did not wish to conceal; telling me that such a person as a Naturalist must be perfectly well acquainted with every branch of Nat. Hist., and must be well known in the world beforehand, *such a person as Mr. Darwin*;[2] here I interrupted him with 'what was Mr. D. before he went out? he, I daresay, knew his subject better than I now do, but did the world know him? the voyage with FitzRoy was the making of him (as I had hoped this exped. would me).' Captain Ross

[1] Robert McCormick (1800–90) was a Yarmouth man, though of Ulster descent. He studied medicine at Guy's and St. Thomas', and became a naval surgeon in 1823. He had special qualifications for the post of surgeon and naturalist on the *Erebus*, for he had seen Arctic service under Parry in 1827, and when on half pay for four years after thrice invaliding home from his special detestation, the W. Indian station, he had worked at geology and natural history in the study and in the field. Though afterwards he distinguished himself by conducting a boat expedition in search of Franklin (1852), he came to loggerheads with the Admiralty on the question of the promotion he considered due after his exceptional service in the Antarctic, and the end of his career was clouded over with a sense of grievance.

Readers of recent Antarctic exploration will recall his name in the appellation of 'McCormick's Skua,' the Antarctic gull first described by him.

[2] Charles Robert Darwin (1809–82) was the son of Dr. Robert Waring Darwin of Shrewsbury, and grandson of Erasmus Darwin, physician, botanist, and man of letters. His mother was Susannah Wedgwood, daughter of the potter. Hooker took his *Voyage of the Beagle* as a model of what his own Journals of travel should be. The story of their intimate friendship, both before and after the publication of the *Origin of Species* in 1859, is fully told hereafter.

answered, 'Well, perhaps you are right, but at any rate it would never be worth the while of any one to go, who was really capable, as far as mental acquirements are concerned.' Being determined not to be put off, I asked him again 'would he take a Government Naturalist?' He said, 'Certainly, and give him every accommodation,' at the same time adding, what was as much as to say, 'You would never be fit.' I said nothing, but must have looked very sorry and angry, which however he did not see, as he went on, speaking as kindly and almost as affectionately as ever, offering to write me letters of introduction to the surgeon and chief officers of the ship at Chatham, charging them to give me every opportunity of going ashore. I thanked him and left him. Major Sabine was in the room at the same time, and he must have felt for me, after having been so anxious that I should be sent as Naturalist alone. I then went immediately to Mr. Children,[1] who was highly indignant, and said I must not go if I am not to be the only Naturalist, or at least the head Naturalist, for that it is utterly impossible that we should agree, each having an equal claim on going ashore, and he the better right. Mr. Brown and Mr. J. E. Gray [2] both said the same thing, and Mr. Children then offered to go to Sir William Burnett to put off my examination, telling me to meet him afterwards.

This I did, and found Sir William had put off my examination till when I choose, and had strongly disadvised my going except as the only Naturalist in the ship, the more especially as *Dr. McCormick was to be my superior.* Mr. Brown has gone to Capt. Beaufort,[3] Mr.

[1] John George Children (1777–1852), mineralogist, entomologist, and astronomer, held posts at the British Museum from 1816–40, and was one of the secretaries of the Royal Society in 1826–7 and 1830–7. He was a friend of Sir Humphry Davy, who made many experiments in his private laboratory. His personal kindness to the young Hooker was typical of his character.

[2] John Edward Gray (1800–75), began his scientific work as a botanist, and was responsible for the greater part of his father's book, *The Natural Arrangement of British Plants*, the first British Flora arranged on the natural system. A quarrel over scientific personalities diverted him from botany to zoology, and in 1824 he entered the British Museum as assistant to Dr. Children, whom he succeeded as Keeper of the Zoological Department from 1840 till his death. His great work lay in the improvement and organisation of collections, and the scientific descriptions which he wrote.

[3] Sir Francis Beaufort (1774–1857), rear-admiral and K.C.B., retired from active service, severely wounded, in 1812, after a brilliant career of twenty-two years. The excellence of his surveying work led to his appointment as Hydrographer to the Navy in 1829, where he was eminently successful during his twenty-six years' tenure of the post.

Lubbock,[1] and Mr. Forster,[2] to recommend my being sent as Naturalist, but how can I go, when Capt. Ross would be obliged to take me, and at the same time think me unfit? There therefore remain only two ways or situations under which I can go, either as Naturalist to the expedition or as Asst. Surgeon and Naturalist to the *Erebus*, a situation which Sir William Burnett promised me if I liked it. You can, I know, but have the same opinion as Mr. Children and Brown. The more I think of it, the more perplexed do I feel. That Capt. Ross did *not* intend to treat me thus two weeks ago I am sure, from his asking me to tell the quantity of *preserves for animals* required, and his great good nature to me now precludes me from attributing to him any other motive than that he is misguided, and that Dr. McCormick (who, he told me, had been preparing for such an Exped. for three years) has been palmed upon him by someone. Supposing I were to go under these circumstances, all my notes on Molluscs and sea animals will naturally revert, from the Admiralty, to the Zoologist, besides which he will have more time on shore than I can. The most painful part of my duty remains to be done, viz., going to Capt. Ross and respectfully declining his appointment and telling him that I am still trying for the appointment of Naturalist to the Expedition, which all strongly advise me to do. Mr. Children and Brown have been most kind, the former especially; I can never thank him too much; I have invariably made a point of telling them everything without the smallest concealment, and have been glad to find how their opinions coincide with mine. On your account, after all the kindness, trouble, and expense you have put yourself to for my comfort and good, I feel this annoyance very deeply, but you may rest assured that I shall conduct myself well and prudently (doing nothing without the best advice) as far as lies in me. I shall deeply regret it, if I lose the chance of going with

[1] Sir John William Lubbock, Bart. (1803-65), banker by profession, was a distinguished mathematician and astronomer. He was treasurer and vice-president of the Royal Society, 1830–5 and 1838–47, and the first vice-chancellor of the London University (1837–42). His eldest son, Sir John Lubbock, afterwards Lord Avebury, was similarly distinguished in business, science, and politics.

[2] Edward Forster (1765–1849), botanist; vice-president of the Linnean Society, 1828, who used to snatch the early hours of the day for his study, mainly of British plants, before going to work in a city bank. His herbarium was presented to the British Museum.

the Exped., but I should much more deeply regret going against the advice of my friends and losing my time.

Matters straightened themselves out, however. I am appointed from the Admiralty as Asst. Surgeon to the *Erebus*, and Capt. Ross considers me the Botanist to the Expedition and promises me every opportunity of collecting that he can grant.' McCormick, as will be seen, proved anything but exacting during the voyage, and indeed made friends with him at once when he reached Chatham, and looked after him when he met with a slight accident.

A letter of July 13 to his father tells of another official interview, the tone of which he resented and remembered against the Society when it made claims on his work or the disposal of his collection:

At the same time as your letter was brought off one came from Capt. Ross calling me up to town on Tuesday to attend the Commission of the Royal Society for the purpose of giving instructions to the Botanist. Mr. Royle,[1] Dr. Horsfall,[2] Mr. Pereira[3] and Capt. Ross were there. They gave me a long list of advices with little new in them or worth reporting but an order to send seeds to the Bot. Gardens in India; you can guess who wanted this. Pereira

[1] John Forbes Royle (1799–1858). His love of natural history made him throw up his prospect of a commission in the Indian army and enter the Company's medical service, so that he could study Indian botany. In 1823 he became superintendent of the Saharunpore Gardens. He studied and identified many Indian drugs, and with the aid of collectors, gathered vast collections, especially of Himalayan plants, which he brought back to England in 1831. In 1837 he became F.R.S. and Professor of Materia Medica at King's College, London, while at the East India House he organised a department relating to vegetable productions, with a technical museum. In his *Illustrations of the Botany, &c., of the Himalayan Mountains*, 1839, he recommended the introduction of the cinchona plant into India. But it was not till 1853 that Royle, at the invitation of the Governor-General, drew up a report on the subject, which in turn was only carried out in 1860, two years after his death, by Sir Clements Markham.

[2] Possibly meant for Thomas Horsfield (1773–1859), an American doctor and botanist who took service in the Dutch East Indies, but finally joined the English service when the Dutch Malayan colonies were temporarily taken by us in 1811. In 1820 he was appointed Keeper of the E.I.C. Museum in Leadenhall Street, publishing various botanical and zoological papers.

[3] Jonathan Pereira (1804–53), the great authority and lecturer of his day on Materia Medica. In 1839 he had begun to publish his great book, *The Elements of Materia Medica*, and had been appointed examiner in the subject at the London University.

talked a great deal and, without exaggerating, much nonsense, confusing the genera of different localities in an extraordinary manner. None of them seemed cordial to me in the least degree. On leaving the room, no one even wished me a pleasant or successful voyage, except Mr. Robertson,[1] the Secretary, who has always been very kind to me whenever I have occasion to attend at the R.S. rooms.

A few more extracts:

The Gunroom officers are about to petition Ross that I may mess with them; it is extremely kind of them and chiefly McCormick's doing, but I hope Ross will refuse, as I cannot, if they offer, and it will put me to an additional expense of no mean importance.

H.M.S. *Erebus*, Chatham, July 28, 1839.

Mr. McCormick returned last week from Devonshire, and finds that the Government are very loth to make such large grants for the Natural History department, and Sir Wm. Parker [2] says he does not see what Nat. Hist. has to do with the Expedition at all, which has annoyed Capt. Ross exceedingly. Anything that they won't supply my Surgeon will make up from his own pocket; he is very zealous indeed in the cause and offers me every encouragement. . . . In the way of medical duty I have very little to do as far as regards the *Erebus*, but the men of the *Terror* are so much inferior in constitution and morals that there are 5-1 of them ill, to what there are of our men. There are besides a whole swarm of women and children on the lower deck of the hulk, who are a perpetual annoyance.

Sir William paid him a visit at Chatham; and though warmly welcomed by such of his future companions as were there, writes on his return home (August 27, 1839):

I could have wished you had some zealous Natural

[1] Probably Archibald Robertson (1789–1864). Originally a naval surgeon, after 1818 a successful practitioner in Northampton. He wrote on medical subjects, and was elected F.R.S. in 1836.

[2] Sir William Parker (1781–1866) was the famous admiral who was at the Admiralty under Lord Auckland, 1835–41.

History companions to keep up the zest of the thing, and though I think very favourably of most of your companions, I could have wished to have witnessed their conversation taking a more scientific and soberer turn. Above all I should have liked to have seen them pay more respect to the Sabbath. Do you do so, my dear Boy, and carry something of the Sabbath into the week and I am sure you will be a happier man for it.

The days pass in preparation till well on into September.

Our Mess Room [he writes to his grandfather] is fitted up with redwood and painted Birds-eye Maple; it is abundantly lighted from above and calculated to hold ten, half that number is all that will at present occupy it. Each has a small cabin of his own; its dimensions are 6 × 4; it is fitted with a bed-place, a book shelf, a seat, table, etc.; below the bed are very large drawers for our things; it is lighted by a large circular bull's eye on deck; we fit them up as we please; mine is to be painted satinwood, with brass rods and curtains before the door and bed, to be used in hot climates when, with the door shut, they would be far too close; the bull's eye is then removed and a grating replaces it, which ensures a current of air.

He expects his whole outfit, uniform, books, instruments, private stores, to cost £150. His grandfather sends him a travelling thermometer. He had economically waited to buy a new watch until his first expenses were settled; now he was forestalled by his father, who gave him 'a beautiful Chronometer watch.'[1]

It is the admiration of all the officers, so much so, that I expect that it will be taken from me as soon as we get to sea. Of books also I have a good store and some for general reading, all Constable's 'Miscellany,' for instance. The rest are chiefly Botanical with a few on Zoology and Geology. . . . My messmates are all readers and careful of

[1] This watch he used to the end of his life on his travels and at home, wearing it in preference to the watch which Robert Brown left to him. It has been presented to the Royal Geographical Society by Hyacinth, Lady Hooker.

books: they are delighted we have lots of Cook's[1] and Weddell's.[2]

As botanist [he writes in his Journal] my outfit from Government consisted of about twenty-five reams of paper, of three kinds—blotting, cartridge, and brown; also two Botanising vascula and two of Mr. Ward's[3] invaluable cases for bringing home plants alive, through latitudes of different temperatures. I was further, through the kindness of my friends [*i.e.*, his father], equipped with Botanical books, microscopes, etc., to the value of about £50, besides a few volumes of Natural History and general literature.

Thus Natural History came off very badly in the matter of public equipment. Of this and his own work as a volunteer in the neglected department of marine zoology he writes seventy years later to Dr. Bruce of the *Scotia* expedition:

It does not, I think, appear in the Narrative of the Voyage that I was the sole worker of the tow-net, bringing the captures daily to Ross, and helping him with their preservation, as well as drawing a great number of them for him.

Except some drying paper for plants I had not a single instrument or book supplied to me as a naturalist—all were given to me by my father. I had, however, the use of Ross's library, and you may hardly credit it, but it is a fact that not a single glass bottle was supplied for collecting purposes,

[1] James Cook (1728–79). His first great voyage in the *Endeavour* was in 1768–71, when he was accompanied by Sir Joseph Banks; the second, in the *Resolution* and the *Adventure*, in 1772–5, when he was accompanied by a staff of naturalists, etc., headed by the two Forsters; the third, in the *Resolution* and the *Discovery*.

[2] James Weddell (1787–1834) held the record for furthest south before Ross. He was a common sailor of twenty-one when in a lucky hour his bullying skipper handed him over to a man-of-war as a refractory subject. With education he became a very competent officer, but being discharged at the peace in 1816, took command of a Leith ship for a sealing voyage to the newly discovered S. Shetlands. He did much exploration, surveyed the S. Shetlands, and in February 1823, on his second voyage, reached 74° 15′ S. latitude in an ice-free sea.

[3] Nathaniel Bagshaw Ward (1791–1868), medical man and botanist, was the inventor, about 1827, of the Wardian case, in which growing plants can be transported without watering through the extremes of heat or cold. By its means the Chinese banana was taken from Chatsworth to the Pacific Islands; 20,000 tea plants were taken by Robert Fortune from Shanghai to the Himalayas, and the cinchona introduced into India.

empty pickle bottles were all we had, and rum as a preservative from the ship's stores.

The epic days of scientific exploration began when Banks and his men joined Cook on his first voyage. To this epoch still belong the voyages of Darwin in the *Beagle* and of Hooker in the *Erebus*. But the expedition to the Antarctic, which was to give Hooker his first great opportunity, was not intended simply to be a search for new lands nor a mere 'dash to the Pole.' Geographical discovery was subsidiary to its main scientific purpose—that of filling up the wide blanks in the knowledge of terrestrial magnetism in the Southern hemisphere, especially in the higher latitudes.

Much had already been done in the Northern hemisphere since Halley in 1701 drew up the first chart of the variations of the compass, based upon the observations made during a voyage of discovery sent out by the English Government. Finally, thanks to Humboldt,[1] a chain of magnetic observatories had been established in Germany and the Russian Empire in 1827, and extended by the famous physicist Gauss,[2] in 1834, all over Europe, where simultaneous observations were constantly made. It was needful to perfect the charts not only of variation, but of dip and magnetic intensity, elements which were already known to be in a constant state of fluctua-

[1] Baron Alexander von Humboldt (1769–1859) was the leading naturalist and traveller of his day. His books inspired Darwin with the desire to travel. He spent five years in Spanish America from 1799 to 1804; the arrangement and publication of his collections and notes took twenty years, which he spent in Paris, where he had the assistance of Cuvier, Gay-Lussac, and others. Then in 1829 he undertook an expedition through Russian Asia for the Emperor Nicholas, which lasted nine months.

His most famous work was *Cosmos*, a survey of the physical sciences and their interrelation (1845–58). His great interest in geography and exploration of the still unknown tracts of the world, the configuration of the country, climate, the distribution of life, was an interest in which Hooker shared, and which drew them together in Paris in 1845; for though he was then settled at Berlin, he was frequently sent to Paris on political missions.

[2] J. K. F. Gauss (1777–1855), Professor of Mathematics and Director of the Observatory at Göttingen, was a mathematician of singular brilliance, equally distinguished in astronomical research, geodesy, and the problems arising out of the earth's magnetic properties, inventing, among other instruments, the declination needle. He was responsible for the foundation of the Magnetic Association, in connection with whose work Ross's expedition was sent out.

tion, undergoing local and transitory as well as periodical changes. Observations, moreover, must extend over a long period.

The many explorers within the Arctic Circle had recorded much information. Ross himself had found the Northern Magnetic Pole and seen the compass dip vertically to 90°, and Gauss had calculated the Southern Magnetic Pole to lie in 72° 35′ S., 152° 30′ E. But as his materials were imperfect and the position he had calculated for the Northern Pole was 3° wrong, he inferred the Southern Pole to be in 66° S. and 160° E. His inference required verification. Permanent stations should be established at suitable spots in the Southern hemisphere, where simultaneous observations might be maintained in connection with the European stations, while the *Erebus* and *Terror* acted as floating observatories on their voyage. Besides the hourly records of the three variables every day for three years, on the four 'term days' of the European Magnetic Association simultaneous records were to be kept at intervals of not more than five minutes during the twenty-four hours: in fact, on the term day which fell in Tasmania, Ross and his colleagues took these observations at intervals of two and a half minutes.

These considerations took shape in a series of resolutions passed by the British Association for the Advancement of Science in 1838. They were pressed upon Lord Melbourne's Government by an influential Committee and strongly supported by the President and Council of the Royal Society, to whom they were referred as the acknowledged advisers of Government in matters of science. But it was not till the foreign scientific institutions, led by Humboldt himself at Sabine's suggestion, threw their weight into the scale, pleading for national co-operation in magnetic work where private enterprise was out of the question, and urging the superiority of the British Navy and the unequalled experience of its officers in polar work, that the Government early in 1839 agreed to fit out the expedition at a cost of £100,000.

As a result two exploring ships, each with a crew of sixty-four men, were carefully fitted out under the experienced Arctic navigator, James Clark Ross, who had shared in no less than seven Polar expeditions—namely, the *Erebus*, a bomb of 378 tons, 'of strong build and capacious hold,' especially strengthened to bear the pressure and shocks of the ice, and the *Terror*, 340 tons, which had been similarly strengthened for Arctic service in the winter of 1836, when many whalers were reported beset by the ice in Baffin's Bay, and which had been employed the following summer by Sir George Back in his attempt to reach Repulse Bay. 'They possessed every superiority,' writes Hooker, 'except that of sailing qualities for manœuvring amongst ice.' So well found were the ships that they suffered no vital injury from storm or collision, or from frenzied battering by the masses of pack ice in the long-drawn fury of the Antarctic gales: nor, thanks to the precautions taken, did the crews suffer from the dreaded scurvy which cut short the rival cruise of the *Astrolabe* and *Zélée* under D'Urville.[1]

Ross was instructed to land the observers and instruments for fixed magnetic observatories at St. Helena, the Cape, and Van Diemen's Land, finally calling at Sydney, the centre of reference for magnetic determinations. He carried with him portable observatories, and with these he was to make special observations at intermediate oceanic islands (Kerguelen's Land being particularly recommended) simultaneously with the fixed observatories and those in Europe.

Then, after refitting at Van Diemen's Land, he was to begin his southward explorations, first to determine the Magnetic Pole, and incidentally to extend geographical discovery, 'while seeking fresh places on which to plant your observatory in all directions from the Pole.'

The Antarctic afforded more of 'those yet unvisited tracts of geographical research' than the Arctic. It had been visited

[1] Dumont D'Urville (1790–1842), the French navigator and accomplished man of science, whose first claim to fame was the identification and preservation of the Venus of Milo. His exploring voyage in search of La Pérouse, 1826–9, took him to Australasia and the Pacific; in 1837–40, again in the *Astrolabe*, with the *Zélée* as tender, he made two voyages to the Antarctic. Compelled by scurvy to refit at Hobart, he started in January 1840, as Wilkes six weeks before from Sydney, in the very direction in which it was known that Ross was about to sail.

by fewer navigators, and the conditions were less favourable. Cook in 1774, then Bellinghausen the Russian, Weddell with his furthest south of 74°, and Biscoe and Balleny, Messrs. Enderby's sealing captains, all between 1820 and 1839 had passed the Antarctic Circle. Balleny was the immediate predecessor of the French, the American, and the British expeditions in 1840 and the following years. After the lapse of seventy-three years the soundness of his observations has received striking confirmation. In the course of his voyage he obviously saw the ice wall of Côte Clairée, 'discovered' the following year by D'Urville. This, however, he took for an enormous iceberg, and ultimately decided that what seemed to be land behind it was probably a distant fog bank hanging over the ice. Early in 1912 the *Aurora*, belonging to the Mawson expedition, sailed over the position of the supposed land.

This Côte Clairée was a sore point for the French and American expeditions, for Lieutenant Wilkes[1] of the United States Navy 'discovered' it independently a week after D'Urville, and a great contention for priority ensued. With all Ross's admiration for the courage and endurance of both, the reader divines in his plain words a touch of national pride as he records at full length Balleny's superior claim, if land there was, to either: more than this, he must have dimly felt a kind of poetic justice in the event. For although he had been on a friendly footing with Wilkes, in the outfit of whose expedition he had taken much interest, and who later sent him privately a chart of his discoveries before the *Erebus* sailed South from Tasmania, he was somewhat nettled on reaching that island in 1840 to find that both the French and American expeditions, knowing his plans, had endeavoured to forestall them; and he writes ('Voyage' i. 116) that this 'certainly did greatly surprise me. I should have expected their national pride would have caused them rather to have chosen any other path in the wide field before them than one thus pointed out, if no higher considerations had power to prevent such an interference.'

[1] Lieutenant Charles Wilkes commanded the *Vincennes* and its four consorts on the Antarctic exploring expedition sent out by the United States Government in 1838–40.

Acknowledging, however, that they were within their rights in so doing, whatever the results to him, he gave up his original plan. His instructions left him a certain latitude, and, where England had so constantly led, he did not choose to follow. He therefore resolved to start his cruise in search of the Magnetic Pole farther to the east along the meridian of 170° E. His chief reason for choosing this particular meridian 'was its being that upon which Balleny had in the summer of 1839 attained to the latitude of 69° and there found an open sea.' It was not, he adds, because he feared to fail where the American and French had failed to do more than barely cross the Antarctic Circle. Their ships, unlike the *Erebus* and *Terror*, were ill-adapted to battle with the ice. Even in longitude 170°, where Ross met with a belt of pack ice 200 miles wide, they could not have forced their way through. Thus in 1839–40, though D'Urville added Louis Philippe Land to the South Shetlands group—south of Cape Horn—and south of Tasmania traced Adélie Land for about 150 miles before approaching the supposititious Côte Clairée;—though Wilkes followed the same line with its barrier of pack ice another 20° westwards, the ice, impenetrable by their ships, debarred them from so much as reaching latitude 70° S. In signal contrast to their moderate achievements, Ross himself, thus diverted from his original plan, was rewarded with superlative success in the discovery of Victoria Land, with its great volcano Mount Erebus, 13,000 feet high, in 77½° S., and its stupendous ice barrier, which he traced for 250 miles, twice forcing his way beyond the 78th parallel.

Unable to effect a landing so as to visit the southern Magnetic Pole, 150 miles inland, he was able to place it very accurately from abundant observations.

Ross made three expeditions to the South in the *Erebus* and *Terror*—the first, 1840–1, from Tasmania and back to Tasmania again, lasting five months, when he discovered Victoria Land and the Great Ice Barrier; the second, 1841–2, from New Zealand and back to the Falkland Islands, east of Cape Horn, lasting four and a half months, when he revisited the Barrier; the third, 1842–3, from the Falkland Islands and back to the Cape, lasting three and a half months, when he visited Louis Philippe Land and the South Shetlands.

Between the first and second came a stay of three months in Tasmania, a visit to Sydney and a stay of three months in New Zealand. Between the second and third came a stay of, altogether, six months at the Falklands, broken by a seven weeks' expedition to Hermite Island in Tierra del Fuego, and west of Cape Horn.

The original voyage out to Tasmania, which lasted nearly eleven months, followed an unusual course in order to touch at various oceanic islands, to establish observatories there and at the Cape, and to pass certain points of magnetic interest. The journey home from the Cape, however, by way of St. Helena, Ascension, and Rio, occupied only four months. Thus four years had elapsed since leaving England on September 30, 1839, before Ross and his men once more reached English soil on September 4, 1843.

CHAPTER III

THE SOUTH AND ITS SCIENTIFIC SCOPE

THE long preparations at last completed, at the end of September 1839 they set sail on an adventurous voyage for how long they knew not. Its exact scope and length depended on the captain and his undivulged instructions. In the end, as has been said, they reached home within four years; but there had been talk of a fifth year or more. In three successive summers they entered the ice. The first voyage was the most rewarding, the second the most perilous. Ross indeed failed to reach his formal objective. He found a continent instead of open sea: the Magnetic Pole was 150 miles inland. The icy sheet which barred nearer approach to the shore stretched a full twenty miles further to the north than it does now: and for sailing ships at the mercy of winds and tides it was impossible to land here or winter with reasonable prospect of safety.

Geographically, however, they achieved unlooked for triumphs. The experiences of their predecessors offered little or no prospect of new discoveries, but as Captain Scott wrote of that 'wonderful voyage':

> When the extent of our knowledge before and after it is considered, all must concede that it deserves to rank among the most brilliant and famous that have been made. After all the preceding experiences and adventures in the Southern Seas, few things could have looked more hopeless than an attack upon that great ice-bound region which lay within the Antarctic Circle; yet out of this desolate prospect Ross wrested an open sea, a vast mountain region, a smoking

volcano, and a hundred problems of great interest to the geographer; in this unique region he carried out scientific research in every possible department, and by unremitted labour succeeded in collecting material which until quite lately has constituted almost the exclusive source of our knowledge of magnetic conditions in the higher southern latitudes. It might be said that it was James Cook who defined the Antarctic Region, and James Ross who discovered it.

For over half a century the expedition held the record for furthest South '—and it was from the land Ross discovered that Scott, Shackleton, Amundsen, and again Scott set forth on their great Southern journeys. The regions beyond the Antarctic Circle yielded next to nothing to the botanist: they were barren far beyond the barrenness of the Arctic Zone. A seaweed was only once found floating within the Antarctic Circle. At Cockburn Island one sole lichen was found, painting the exposed rocks with red and orange—a lichen, strangely enough, abundant in the Arctic, and next seen by Hooker on desolate summits of the Upper Himalayas, over against the Tibetan Plateau.

The sea, however, had other harvests, and as elsewhere Hooker, unable to botanise, or not wholly engrossed in working at his collections, studied the floating creatures brought in by the tow-net or dredge, establishing for the first time the occurrence of highly developed animal life at a depth of 400 fathoms, so here he determined the presence of abundant infusoria in the icy waters, which provided the ultimate means of subsistence for higher forms. Multitudes of small shrimps fed upon them, and supported abundance of whales: they were, moreover, eaten by the fish; while birds and seals lived upon both and were themselves the prey of the killer-whales.

This zoological interest appears from the very outset of the voyage and continues to the end, though of the third trip to the South he is compelled to write: 'Amongst the animals very little or nothing has been done. I lost all my gauze in the pack from the water being so full of little pieces of ice, and in the clear water it has always been blowing with heavy seas on.'

Dr. Richardson warmly encouraged him in the work; skill with the pencil being a special qualification in dealing with sea creatures which could not be preserved. To add to our knowledge of the structure of animals, he insisted, is the most certain way of attaining a scientific reputation; to be the first to discover or name a new species is a very secondary matter.

But, rich as the collections were that he brought back from the voyage, they were never fully worked out, to the great loss of marine zoology and the disappointment of their zealous collector. The 'might have been' was sharply brought home to him when, sixty years later, he read Dr. Bruce's report of his Antarctic work, 'The Scientific Results of the Voyage of the *Scotia*.'[1]

> There is [he wrote to Dr. Bruce, January 10, 1901] always something painful to me when I come across the scientific reports on Antarctic expeditions, due to the wholesale destruction of the great collections made by Ross and myself of marine and submarine animals of all classes. Ross was an indefatigable collector, who never lost an opportunity, whether on sea or ashore; but except my collection of Diatoms published by Ehrenberg,[2] and discussed in my 'Flora Antarctica,' there is nothing to show of the stores of the pelagic materials obtained with so much zeal and care by Ross and myself. Thereby hangs a tale which, if we two have the pleasure of meeting again, I may unfold to you.

But his enthusiasm was unabated when his forgotten harvest was at last fully garnered. Eight years afterwards Dr. Bruce sent him Vol. V. of the 'Invertebrates of the Scotia Expedition': he replied on February 14, 1909:

> I have again to thank you for a magnificent addition to my Antarctic library. It is really a noble work, and I find

[1] Cp. vol. ii. p. 441.

[2] Christian Gottfried Ehrenberg (1795–1876), Professor of Medicine at Berlin, was the founder and chief representative of the study of microscopic organisms. He was one of Humboldt's companions on his journey to the Ural and Altai mountains.

in the several articles a great deal that interests me very much, especially in the subject of the geographical distribution of the various orders and genera so graphically and scientifically treated. . . .

I well remember the deep sea *Pycnogon* which we dredged up in the *Erebus*, especially the *Amnothea communis*, which astonished the crew. It is much to be desired that zoologists would follow the example of most botanists in giving the geographical range of the species they deal with.

From the moment of starting down Channel the naturalist's eye is alert, whether it be that a wren is observed seven miles out at sea, or sea-water examined for the microscopic cause of its luminosity at night, or the activity of the young of a small crab from the Antilles, harbouring in their thousands on a piece of driftwood, swimming with the last five abdominal segments that in adults are turned in upon the thorax.

Even after Madeira and the Cape de Verdes had furnished some botanical material to work upon, this did not fill up his time, and botany took second place after general naturalist's work.

To his Father

March 17, 1840.

Since leaving St. Helena, my time has been employed exactly as before; the net is constantly overboard, and catching enough to keep me three-quarters of the day employed drawing; the dissections of the little marine animals generally take some time, as they are almost universally microscopic. Though I never intend to make anything but Botany a study, I do not think I can do better than I am doing; it gives me a facility in drawing which I feel comes much much easier to me; it pleases the Captain beyond anything to see me at work, and, further, it is a new field which none but an artist can prosecute at sea; the extent of this branch of Natural History is quite astonishing, the number of species of little winged and footed shells provided with wings, sails, bladders or swimmers appears marvellous. The causes of the luminousness of the sea I refer entirely to animals (living). I never yet saw the water flash without finding sufficient cause without electricity, phosphoric water, dead animal matter, or anything further than living animals (generally *Entomostraca Crustacea* if anybody asks you). These little shrimps are particularly numerous, especially two species of them, thousands of one kind being caught in one night. The library of Natural History that you fitted me out with is to me worth any money. Blainville's *Actinologie* and Edwardes' *Crustaceae* are particularly useful, as by them I can name many old species and detect the wonderful new forms I meet with. My collection amounts to about 200 drawings done from nature under the microscope. . . . As I am learning to use my left eye to the microscope, I do not find my eyesight affected even by candlelight.

His discovery of the Antarctic infusoria is recorded step by step in his Journal. To begin with, he writes on February 15, 1841, in lat. 76° S.:

Much young ice was seen to-day of a light brown colour; when dissolved in water it deposited a very fine sediment, composed of exceedingly minute, transparent, flat quadrangular flakes, each formed of numerous parallel prisms of a perfectly regular form, giving each flake a fluted appearance; numerous circular discs, also transparent, were scattered among them; they were very minutely reticulated, and had often opaque centres. All the young ice was very full of it; when lifted out of the water it did not appear discoloured; many acres were covered with it. I suppose it to be some insoluble salt, whose appearance is probably connected with the volcano.

This facile conclusion impressed itself on the other officers; Ross himself forgot to correct it by Hooker's fuller examination, and (Voyage, I. 243, II. 146; cp. II. 332) records the general belief that the colouring matter consisted of fine ashes from Mount Erebus, eighty miles away, while ascribing the determination of its real nature to Ehrenberg, who examined specimens after their return. But against this note in Hooker's own copy are penned the words: 'I recognised them as diatoms, &c., at the time. J. D. Hooker.'

On the second voyage, the Journal records, December 21, 1841: 'Much of this ice is discoloured, as was the case last year and from the same cause. When melted it gives out a strong animal smell.' And again, off Louis Philippe Land, December 28, 1842—a point repeated in the letter to his father of March 7, 1843, describing the voyage:

> All day the washed pieces of pack ice have been stained with yellow, caused doubtless by the infusoriae in the stomachs of the Salpae, which are washed up against the ice and leave this stain (the same as last year). When the wind was light and the fog thick in the morning, I recognised the animal smell very strong from the pack, precisely similar to that of brash ice, with the Salpoid remains, omitted last year by me, in the cabin.

Letters to Ross after their return (September 1 and 4, 1844) speak of two pamphlets on Antarctic Infusoria received from Ehrenberg—'in hard German,' one containing descriptions, the other 'drawings of *Asteromphalos Humboldtii*, *Cuvierii*, *Rossii*, *Darwinii*, and *Hookerii*. I think, Sir, that we are in good company, though I can give you no more idea of what the species are like further than that the magnified figures resemble the objects at the far end of a kaleidoscope.'

Before this was sent on to Ross, Hooker 'commenced trying, with the German dictionary, to spell out [the] descriptions of our Infusoria.'

> I find Ehrenberg has described 70 new species from the contents of two pill-boxes and three small bottles, and has not yet examined the whole of what I had. As far as I can make out they seem to throw extraordinary light on the subject, and to have been the most important collections ever brought to this country. The amount of species in what you have must be enormous, as my specimens were mere scraps in pill-boxes from the dredge, and a portion of a large bottle you have of condensed *brown Ice*.
>
> The other packets I sent were of dirt from the roots of Cockburn and other Island mosses, which also seem to contain animals. . . . Ehrenberg finds animalculae in all soundings, and I feel quite convinced that those you have

will alone immortalise the Expedition. No person seems to have thought of collecting such things before for scientific purposes.

Happily Hooker's short-sighted eyes stood the strain of the microscopic work fairly well, though he had to turn his unexpectedly good opportunity to account under constant difficulties. This, as the voyage drew towards its close, he describes as follows (March 7, 1843):

> During our now homeward passage I shall have plenty to do with tropical plants and sea animals; the latter I must keep up, for there never was such an opportunity as this ship affords for the study, being a slow sailer and my having such accommodation below for drawing and describing them; not that I care for them at all; somehow with all the time I have devoted to them they have not won my affections, because I feel sure that two studies in Nat. Hist. cannot be well prosecuted together, and though an easier study, marine animals require much more time than plants to investigate fully; the drawings will do me some credit if it be only for the time taken and the novelty of their being often done with the microscope lashed to the table. My eyes are as good as ever they were in strength, but my shortsightedness '*semper idem*' (always worse and worse). The spectacles you were so good as to send me were not half strong enough; however, they are much nicer than are procurable out of England, and I shall get new glasses at the Cape. Between examining mosses and the glare of the Ice and snowy spicules in the wind, my eyes smarted very much during the time the ships were in the pack and watered, but never inflamed. They are all right again now. Your spectacles (green) were a great comfort.

So also with his botanical drawings, done at sea from specimens in his collections. He chooses the best model he can, and if art is deficient, at least he is accurate. Finding a sudden chance to send home his collections from New Zealand, the Aucklands, and Campbell Island, he says (June 6, 1841):

> The notes were all finished in the Ice, where the smooth water enabled me to resume my old post in the Captain's cabin.

As far as I could I imitated Bauer's[1] style of drawing dissections, but as the only sketches on board of that artist are two in Parry's Voyage, I have not much to copy from and I do not expect that they will please you much, and further when the ship gets through a pack she at once meets the troubled waters, and commences rolling about so that I have to lash my portfolio and microscope and to prop myself up. However I get on as well as I expected. Some of the notes are in a very rude state, for the notice of the opportunity was sudden. That they may prove correct is all that I hope for, as I endeavoured to stick to facts. . . . These are . . . both as numerous and as well done as I could.

He did not restrict himself to scientific drawing, however. In the same letter he tells his father:

At present I am attempting a sketch of the ships off the Barrier and burning mountain in 78° South for you, and should I succeed you shall have it; my talent for sketching is, however, far below *par*, and without colours it would be nothing. There is rather a nice print published of Weddell's two ships bearing up in 74° 15′, by Huggins, which would be worth your buying; a few shillings would cover it, and the Icebergs in it give a very fair idea of those floating masses, though they are not flat-topped like the most of those we have seen, nor is the colour at all good, as they should have a blue tinge.

Doubtless his artistic power was improving, for a year earlier (February 3, 1840) he is much more severe upon his general drawing. 'My sketches are characteristic of the different places visited, but miserably done; they are not intended for any person but you to see.' Still, at the end of the voyage, he feels that his execution is not equal to his aims, though many of his sketches were utilised as the basis of

[1] Francis Bauer (1758–1840), the superb botanical draughtsman employed by Banks, who left him a pension that he might continue his work at Kew. His name appears as illustrator on the title-page of Sir W. Hooker's *Genera Filicum* (1838–40); but more than half the plates were drawn by the new draughtsman, Walter Fitch, who was to serve Kew and the Hookers for half a century.

illustration for Ross's 'Account of the Voyage of Discovery and Research.' [1]

To his Aunt (Mary Turner) he writes (April 18, 1843):

> In drawing I do not improve much, though I have made several sketches of the different places we have visited. There is now but one tolerable artist in the Expedition, Mr. Davis [2] of the *Terror*. Dayman [3] (Aunt Ellen's acquaintance), who was the best, is left behind in Van Diemen's Land. Your pencil would be invaluable here, though you [would] have grown heartily tired of Bergs and Ice. Capt. Ross used often to make me sketch coastlines of hills and valleys of snow, which is most miserable work. Could I have coloured, nothing would be so grand as a view of the scenes we have visited, if in fine weather; but let the weather be what it will, an Iceberg is always a treacherous thing at the best.
>
> I am very anxious to know what Fitch [4] is about; he has sent me a very pretty fancy sketch of flowers, for which I am extremely obliged to him; it was very kind of him to think of me; in return I have been making a sketch of a curious Iceberg with a hole in it for him. The berg is fair enough, but the sea will not do. He could copy it and with excellent effect; it was blowing hard and there were some black scudding clouds near the moon, which was reflected on the tips of the waves, close to the edge of the berg. The water should be of an intense cobalt blue, and it should reflect a white glare on the sea. There are no harsh lines on an Iceberg; the shadows should be faint and the lights bright.

This drawing, duly copied by Fitch, was doubtless among those shown to Prince Albert, when Sir William was summoned to Buckingham Palace in the spring of 1842 to give some account of the progress of the Expedition.

[1] See the list, p. 86, footnote.

[2] J. E. Davis was second master of the *Terror*.

[3] Joseph Dayman was mate on the *Erebus*, and afterwards lieutenant on the *Rattlesnake*, in which Huxley was naturalist. In 1840–1, while Ross made his first cruise to the South, Dayman was one of the three officers who remained in charge of the magnetic observatory in Tasmania.

[4] Walter Fitch (1817–92) was originally a pattern-drawer in a calico printing factory. He entered Sir W. Hooker's service in 1834, and for half a century continued as the official draughtsman for the Kew botanical publications.

Landscape drawing was by no means one of the lighter occupations banned by Sir William. Like his father-in-law, Dawson Turner, the friend and connexion of Cotman, he cared for art beyond his own botanical draughtsmanship. 'I rejoice that you make drawings of scenery. They will be invaluable.' And in the same strain his shipmate Dayman writes on August 27, 1841, from Tasmania to Hooker in New Zealand:

> I am particularly happy that you have found the drawings you made on the passage out to be of more value than you expected—if it be only as an encouragement to make more, for upon my word without flattery (which you know by this time I am incapable of) if you do not something of the kind, I do not know who will. As far as poor McC[ormick] is concerned, one of the main objects of the Expedition has already failed.

Valuable as his zoological researches were, both in satisfying his restless intellectual interests and in giving him fuller understanding of living Nature, his father—strict botanist of the older school—mistrusted any swerving from the closest allegiance to botany. He took alarm at the remark (February 3, 1840), 'My time has been so completely occupied with sea animals that I have little time for other drawing.' When he showed his son's first collections to Robert Brown he diplomatically abstained from mentioning these zoological dissipations, for 'Brown's idea is that without neglecting such things, your time even at sea ought to be *mainly* devoted to studying the plants you have collected,' a thing that proved easier to do in the calm of the pack-ice than on the unquiet expanse of the Southern Ocean.

Nor was this his only stricture. To try too much is to become ineffectual. He urges his son to stick to botanical work exclusively—to avoid wasting his time in unnecessary entertainments; counsel indeed scarcely needed for one who cared so little for the ordinary attractions of society. But Sir William's definition of frivolity is strangely wide.

The first halting-place of the expedition was the beautiful island of Madeira, lovely with semi-tropical vegetation, and twofold lovely as the first relief after a tedious sea voyage. Several hospitable friends of the family lived here, and Hooker rejoiced to explore the wonders and beauties of the island so familiar to him from books. He and his fellow officers had long planned an excursion to the valley of an ancient crater in the mountainous heart of the island, and he sent home a lively description of the jaunt. This gallop up to the Curral is one of the 'unnecessary entertainments.' True, Joseph did not fail to collect all the plants he could find both here and in the Cape de Verde Islands and St. Helena, where also he roamed afield; but the season was too late—everything was burnt up: not to add that he was unpractised in making a large collection. Worse still, an old hand, Cuming,[1] visited St. Helena a week or two after he was there, and in one strenuous day made a much more brilliant collection. Sir William accordingly admits his excuses as to drought ashore, damp and ill accommodation afloat, but confesses to considerable disappointment. Robert Brown, his botanic idol, likes Joseph's sketches and notes; but as to the collection, merely sends suggestions for better preservation of the specimens, such as the use of brown paper in the tropics, instead of blotting-paper, which ferments.

And Sir William, repeating that he ought in future to secure, if possible, an assistant collector to leave him free for the mental work of describing and drawing, adds, it is too much for a man to *collect* well and to *note* well. Assuredly he is well employed but is not specialising enough. Great opportunities lie before him. No botanist has been to Southern New Zealand since Menzies [2] and Vancouver.[3] In Tasmania

[1] Hugh Cuming (1791–1865), conchologist and botanist, who was long settled at Valparaiso. He spent 1835–9 in exploring the Philippines. It was on his way back to England *via* the Cape that he visited St. Helena.

[2] Archibald Menzies (1754–1842) began his botanical career as a gardener in the Edinburgh Botanic Garden; was encouraged by Hope, the Professor, to qualify as a surgeon, and completed his reputation as naturalist and surgeon on Vancouver's voyage in the *Discovery*, 1790–5. He was elected to the Linnean Society in 1790.

[3] George Vancouver (1758–98) sailed as a seaman in Cook's second voyage, and rose to be a captain in the navy. After the Nootka Sound dispute with Spain, he was sent to take over the district again and explore the coast from lat. 30° northwards. On the way out (1791–5) he explored much of Australia, New Zealand, and Tahiti, returning by Cape Horn.

he should visit some of the high mountains, 'which *everywhere* afford what I consider *by far* the most interesting plants.' The Algae in the high south latitudes are particularly worth collecting, and indeed should be collected everywhere if no phaenogamic plants be available, even if they be known species, in order to determine their distribution.

Throughout, it may be noted, Sir William is the systematist, the collector, and describer, urging his son to look for more plants and especially those missed by the latest travellers, such as Wright[1] in the Falklands, and to get his friends to collect specimens 'in *quantities* not in driblets' at all stages, so as to have ample material for Floras of all the places he visits, and the mistakes he corrects in his letters are those of identification tested by extant accounts. On the same principle, just as Robert Brown bade him 'collect everything,' so Hooker sagely acknowledges, 'such scraps as are useless for other purposes may yet, so long as they exhibit the Natural Order to which they belong, prove of service in illustrating the geography of plants.'

But later collections were more satisfactory. No extenuating circumstances needed to be invoked when, at last, in June 1842, there arrived the plants and notes from Kerguelen's Land, the Aucklands, and Tasmania, which rumour had sent to the bottom along with the ship that carried them. Among these notes Lady Hooker reports 150 drawings, 'with highly magnified dissections, some almost worthy, my husband says, of Bauer's pencil.' Sir William, after looking through the collection with Robert Brown, writes enthusiastically: 'Believe me, dear Boy, they have given me *infinite* pleasure, for they prove that you must have been diligent, and consequently successful.' And again (July 7, 1842) of the drawings and notes: 'I expected much of you; but these have far

[1] William Wright (1735–1819), a naval surgeon who, being unemployed, took up private practice in Jamaica (1764–77), finally becoming honorary surgeon-general of the island. He corresponded with Banks and others, discovering especially a native species of cinchona in Jamaica. After botanical study in England and military adventures abroad, he finally settled in Edinburgh in 1798. Among his friends was Sir W. Hooker, to whom he presented a collection made in Iceland to replace Sir William's that had been burned.

exceeded my expectations and do you credit. . . . And Brown is charmed with what you have done.'

The long stay at Kerguelen's Land, Tasmania, Hermite Island, and the Falklands, the travel through New Zealand, the short stay at the Cape and Sydney, and flying raids on Lord Auckland Island and Campbell Island, provided suggestive material for his works on the Floras of the Southern lands and the Antarctic regions: works which afforded not merely a thorough list and account of the plants and the conditions under which he saw them existing, but discussed the comparison of South and North, the questions of distribution, the problem of the oceanic islands and the former connection of the Southern continents, leading slowly but inevitably on to the evolutionary theory in which he was to be Darwin's confidant, critic, and supporter. Darwin's own 'Voyage of the *Beagle*,' indeed, was the most recent of the various travel books that inspired him. It was in the press while he was approaching his M.D. examinations, and the old friend of his family, and of Darwin himself, Mr. Lyell of Kinnordy, sent him a set of proofs that had come from Darwin. Time was short: Hooker slept with the proofs under his pillow, and devoured them eagerly the moment he woke in the mornings. Before he sailed Mr. Lyell sent him a copy of the book, a gift most gratefully and enthusiastically acknowledged. As the voyage continues he tells Mr. Lyell, 'Your kind present is indeed now a well-thumbed book, for all the officers send to me for it.' [1]

If Darwin's was the last of the travel books that inspired him, Cook's voyage was the first. As has been noted already, it fired him at a far earlier age than Darwin himself was stirred by Humboldt's 'Personal Narrative,' a fact on which he dwells again when writing to James Hamilton, his old college friend, after he had sat on the very spot in Kerguelen's Land from which the view of the Arch Rock was taken, and the picture of the men killing penguins.

[1] Thus J. E. Davis, second master of the *Terror*, later thanking Hooker for the 'young library' sent to him, writes: 'I like Darwin's Journal much: he has accomplished what Old Johnson said of Goldsmith when he heard he was going to write a Natural History: "he will make it as interesting as a Persian tale."' (See also the letter to Lady Hooker, p. 136.)

Such pictures once visualised were ineffaceable. It was the same elsewhere. In his letters he repeatedly brings a view home to his father by recalling an illustration or description in some familiar book of travels—as in Madeira and at Teneriffe, Webb and Berthelot, or at the Cape, Burchell's Travels. In describing a plant fresh from its native ground, his strong visual memory is ready to prompt some detailed comparison with a dried specimen once studied in his father's herbarium.

As to his duties on the *Erebus*, he gives a detailed description in his letters to his grandfather. There was little sickness on board: on his professional visits each morning to the sick bay, he seldom found much to do: indeed, as has been noted already, during his stay at Chatham before the ship sailed he remarked the superiority in conduct and health on the *Erebus's* crew over the *Terror's*, albeit during the voyage the *Terror's* officers prided themselves on keeping the stricter discipline on board.

He was fortunate in his captain and fellow officers. Ross was a friend of his father, and respected by him both for his religious feeling and for his scientific aptitudes. Sir William, it will be remembered (II. 12), coming down to visit his son at Chatham, found the junior officers, in the rôle of Jack ashore, lacking in scientific seriousness of conversation, and—what was worse in his eyes—respect for the Sabbath. Nevertheless, they were good fellows; and interested in science when not, like the surgeon and those trained in magnetic work, professionally concerned. The *Erebus* was, and they were proud of it, a discovery ship, not a surveying vessel; and they had been chosen as suitable for a voyage of this kind, although it came to be generally recognised that Ross chose for his executive officers men who were never likely to rival the brilliancy of his own career. They were not, like the lieutenants of the *Rattlesnake*, hostile to use of the tow-net as 'messing the decks': on the contrary, scientific observations went on every day; and every day if possible soundings were taken to test the ocean temperature at various depths, and the tow-net used.

Hooker was uncertain at first with regard to McCormick;

the surgeon and nominal naturalist to the *Erebus*, under whom he was to serve, for technically his collections, other than botanical, were liable to be merged in his senior's; but on the high seas, where botany gave insufficient occupation, Hooker slipped into the position he had first desired, of Naturalist *de facto* to the Expedition. As he writes (February 3, 1840):

> McCormick has collected nothing but geological specimens, and pays no attention to the sea animals brought up in the towing nets, and they are therefore brought to me at once. . . .
>
> (March 17, 1840, at the Cape.) McCormick and I are exceedingly good friends, and no jealousy exists between us regarding my taking most of his department; indeed he seems to care too little about Natural History altogether to dream of anything of the kind; for my part I am rather glad to have an opportunity of doing more than is expected from my department. . . . He takes no interest but in bird shooting and rock collecting; as of the former he has hitherto made no collection, I am, *nolens volens*, the Naturalist, for which I enjoy no other advantage than the Captain's cabin, and I think myself amply repaid.

Most of his work, however, was done under Ross's wing, whose special branch of science lay in terrestrial magnetism; but he was keenly interested in Natural History and, adds Hooker to his father (February 3, 1840), 'he knows a good deal of the lower orders of Animals, and between him and the invaluable books you gave me, I am picking up a knowledge of them.' No doubt he would not have been so gracious to a mere assistant surgeon who was not the son of his distinguished friend, and indeed in all Hooker's early undertakings when he had to deal with officials, he was greatly helped, and knew that he was helped, by the social and scientific prestige at his back, and the introductions he received to notable persons who could help him.

> My time during this sea life has not been, I hope, so uselessly employed as I expected it might have been. Capt. Ross, as soon as he heard that I was very anxious to work, gave me a cabinet for my plants in his cabin; one

of the tables under the stern windows is mine wholly; also a drawer for my microscope, a locker for my papers, etc. To me he is most kind and attentive,—forestalling my wishes in many respects. One day he finds a 'box that will do nicely for Hooker,' then a seat at his cabin table, and a place always clear for me to sit down, when tired of standing at the drawing-table. Two towing nets are constantly overboard for sea animals. . . . Almost every day I draw, sometimes all day long and till two and three in the morning, the Captain directing me; he sits on one side of the table, writing and figuring at night, and I on the other, drawing. Every now and then he breaks off and comes to my side, to see what I am after. . . .

I have now drawings of nearly 100 Marine Crustacea and Mollusca, almost all microscopic; some of them are very badly done, but I think that practice is improving me, and as I go on, I hope that some will be useful on my return. Were it not for drawing, my sea life would not be half so pleasant to me as it is. In the Cabin, with every comfort around me, I can imagine myself at home. Other duties are given me to do; indeed, on finding how idle I was to be I asked the Captain if I could not in any way be useful to him, when he gave me the Hygrometer to take four times a day, at 9, 12, 3, and 9; and for two days in the week at 3 A.M., after the registering there is to draw out tables for different Meteorological purposes. The Captain has a compound microscope exactly like your large one, which I use whenever I require it, indeed he has made everything in his cabin my own. He has expressed himself much pleased with my Botanical collections, from which I judge that he never saw a really good collection, for I never look back upon a day in which I should not have done more than has been done, though at the time I hardly well knew how to carry what I had got. . . . It would have amused you to have come into the cabin and seen the Captain and myself with our sleeves tucked up picking seaweed roots, and depositing the treasures to be drawn, in salt water, in basins, quietly popping the others into spirits. Some of the seaweeds he lays out for himself, often sitting at one end of the table laying them out with infinite pains, whilst I am drawing at the other end till 12 and 1 in the morning, at which times he is very agreeable and my hours pass quickly and pleasantly.

The years pass; but the same note is continued in a letter of April 20, 1843. Community of intellectual interests, no doubt, minimised the inevitable little rubs of months of close quarters in a sailing-ship, frankly acknowledged by the young assistant surgeon.

> Our Captain is still always to *me* most kind and attentive, indeed his whole conduct to me, ever since we left, has been quite uniform, and I have an immense deal to thank him for; as you may suppose, we have had one or two little tiffs, neither of us perhaps being helped by the best of tempers; but nothing can exceed the liberality with which he has thrown open his cabin to me and made it my work room at no little inconvenience to himself. He is quite now the same to me as ever he was, and will be I doubt not to the end of the Expedition, so that my situation is most comfortable, nor would I change with any ship in the service.

But whatever his equitable claim in such circumstances he would not lay himself open to the charge of grasping at more than his due.

> Whenever the seine was shot I attended on the return of the boat, to pick out the fish that were wanted; a very few I kept for myself and Richardson[1] should he not get them, but my duties of course precluded the possibility of my making any notes or a large private collection. Captain Ross often feels himself jammed between me and McCormick, when the latter wants to keep a nice thing for his government collection, and I of course want to put it with ours, for he makes no general collection of anything but rocks and birds, and as I take the drudgery of collecting all the other branches of Nat. Hist. with the Captain's assistance, it would not be fair that I should be refused the credit of bottling down the more scarce and beautiful. Whenever there is the slightest difficulty I always give up, remembering the proverb against 'those who wrestle with sweeps.'

[1] I.e. Sir John Richardson of Haslar.

Botanical work on board ship was done under difficulties of its own, especially at the outset. As has been seen, the early collections found small favour in the sight of his scientific friends at home, who, as his father said, looked to the actual results apart from inexperience and the extenuating circumstances of drought ashore and wet on board, when in the tropics the specimens pressed in the ordinary blotting-paper fermented, and the presence of the passengers for the Cape left no room for dealing properly with the plants. When they left, the sick bay was available for the naturalists,

> and a great comfort it is [he writes on March 28, 1840], as it is spacious, and hitherto I have been very much at a loss where to lay out my plants, not liking to take advantage of the Captain's cabin for so extensive a job, and our berth being too full during the day to grant me room enough. Hitherto I have always laid them out and changed them after my messmates have turned in, which often kept me up very late after my excursions; further, until the Captain had reduced his cabin into order I had no place to put my collections, and they used to get sadly kicked about the lower deck; now, however, I have a nice cabinet in the cabin, where there is nothing to fear but the universal dampness of the ship, and a few cockroaches which did me some little damage, eating out the stems of some plants, and leaving the leaves.

He accepted his father's criticisms as a stimulus to better work. The conditions being what they were, this criticism was perhaps rather uncompromising, considering that when he sent his collections of some 200 species home from St. Helena (February 3, 1840) he did not himself think he had much to show for his labour:

> Some are good specimens, others are only sent as mementoes. I can hardly expect you to be much pleased with them, though I assure you I never spent an idle day ashore; nevertheless I never came off at night, without being convinced that I might have done much more than was done. Capt. Ross wished me to delay sending them till we arrive at the Cape. . . . I do not care that my

collections should be mentioned in the public journals (like McCormick's) should they even be worth it, which I doubt—as all I care for is to please you. I grow every day more selfish and totally indifferent to public opinion; I still scorn the Royal Society's commission in botany, and if I only hear that the present collection does not go to you, my next *first set* shall be a different one, but you shall not be the sufferer. The Royal Society ordered me to send them a first set, and when they have a right to order me, I will; as it is, I am so sure that this set is for you, that I make it a tolerable one. Good as a set it may be; but I fear you will not think it so as a collection.

Letters were very slow in reaching the exploring ship: sometimes they pursued her vainly half over the globe: and thus it was not till two and a half years later (November 25, 1842) that he could speak of being reassured as to his later work.

The dissatisfaction my first plants gave has weighed on my mind until the receipt of your last letters, and all along made me fear that I was physically incapacitated for the high trust reposed in me, which the longer I remain in the Expedition the more honourable do I feel it. My services now are not those of a day, although but a few days have been spent in collecting.

Botany at sea meant for the most part collecting on lonely islands and examining the collections afloat when weather permitted. A significant note in a letter to Robert Brown (November 28, 1843) explains:

In a few days we start again for the Ice, and as soon as we reach smooth water and the pack, I shall begin finishing my notes on the vegetation of the Falklands and Hermite Island.

Botany at sea also meant collecting floating seaweeds and examining them and the animal life upon them.

Till within a few days [he writes from the Cape on March 17, 1840] no floating seaweeds have been seen, when they suddenly appeared whilst cruising off St. Helen's Bay about sixty miles north of the Cape, whilst we were beating to the Southward; they certainly (though only of one kind) gave a most exalted notion of a submarine forest, with its accompaniment of a parasitic vegetation; with fish for birds, corals for Lichens, and shells for insects. Whilst going six or seven knots through the water, we, stationed in the quarter boats, harpooned these weeds as we passed, and very good fun for botanising it was; the largest brought on board had a short thick branching root from which sprang four great stems, the longest 24 feet. . . . It belongs to the genus *Laminaria;* the old stems are brown, with flat white corals on them, and some parasitic seaweeds; the matted roots contain numerous other seaweeds, shells, Crustacea, corals, Molluscae, Actineae and red-blooded worms. The leaves are infested with Patellas, Sertularias, and Flustrae. From one specimen I took four seaweeds and upwards of thirty animals, by carefully pulling the root to pieces. Nor were these large seaweeds; many were seen twice as large if not larger. What extraordinary power can have torn them up by the roots I cannot conceive, for, from their length, they must grow far below low water mark.[1]

Nevertheless, however engrossing the twofold interest of these occupations, the old spell of botanising ashore always gripped him anew with irresistible attractions. The same letter tells:

I have heard naturalists complain of the tedium of a sea voyage; such cannot be naturalists or must be sea-sick (which I have never been for an hour). I do not mean to say I would not be better employed and happier perhaps studying Botany ashore, with more comforts around me, but I assure you my weeks fly, though from my slow working I have not much to show, and, unaccountable as it may appear to you, when we draw near shore I feel quite thrown out of my usual routine of employment. I must own,

[1] Writing to his father on May 3, 1842, from the Falklands, he gives an explanation with which some observant naval officers supplied him:

'The officers of the *Arrow* are very nice fellows. One of them told me that as the *Macrocystis* grows large, it finally weighs up the stone, which was its moorings, and then the whole plant goes off to sea, which fully explains the reason for our finding so much of it alive at sea.'

however, whenever my foot has touched *terra firma*, there is a sort of magic in the place that makes me grievously loth to quit it again. There are also peculiar emotions attending the seeing new countries for the first time, which are quite indescribable. I never felt as I did on drawing near Madeira and probably never shall again. Every knot that the ship approached called up new subjects of enquiry, and so it is with every new land or even every barren rock. It was the same on approaching the Cape and viewing Table Mountain; I could have, and did, sit for hours wondering whether this knoll was covered with heaths or *Rutaceae*, whether this rill produced the *Wardia*, or that rock the *Andraea*, where was Ludwigsberg, Wynberg, the tree fern and all the spots which the mind associates with our mutual pursuits, our friends, or our home. Selfish as I doubtless am and proved myself to be at home, there is one idea, the prosecution of which I often dream of, and that is, to tell, of all other persons, my father, mother, and brother of what I have seen; I never view a new scene but I think what pleasure it will give me to view it over again with you all, to map to you the places where my specimens were gathered, to paint the views to my mother and to spin to William the yarns of incidents that befell my excursions, while grandpapa and my sisters will look upon me as 'the Monkey that has seen the world.'

As his field of study becomes more suggestive we see his work passing from the collector's individual notes to the wider questions of geographical distribution, so attractive to the range of his mind. The details become the tissue of his generalisations.

The earliest botanical *impressions de voyage* for instance, at Madeira, overflow with his delight at finding the rich plant life, known heretofore only from books and dried specimens, now flourishing in semi-tropical exuberance. The experimental cultivation of the tea plant appeals instantly to the practical instinct which did so much for commercial botany in the years to come. So too the 'cabbage' of Kerguelen's Land, an excellent food for sailors, and the Tussac, or Tussock, grass of the Falklands, with its prospect of acclimatisation in the Western Highlands for pasturage; to both of which he makes constant reference, alike scientific and practical. He sends five sets of his St. Helena specimens home for various recipients; he takes some 300 specimens away with him from the Cape on his first short visit there (March 17–April 6, 1840) for examination at sea.

By the time he has visited Kerguelen's Land (May 12–July 20, 1840) his researches begin to take definite shape, both in subject and in outlook, foreshadowing what was to appear in his Flora Antarctica. Here emerges his serious interest in the problems of distribution thrust upon him ever more forcibly by the plants, living and fossil, so far removed from any parent continent, and by the nature of Antarctic vegetation in general. He found the Kerguelen flora in form peculiarly S. American, with some plants common to the Auckland group and more to the Falklands. Later in the voyage he is enabled to write under date November 25, 1842, 'My regions are different both in climate and forms from any other.' At Kerguelen's Land above all, his favourite cryptogams, so much less known than the flowering plants, and here relatively abundant, invited his study. 'You direct my attention,' he writes to his father (September 7, 1840), 'particularly to Cryptogamia; believe me that I have at Kerguelen's Land strained every nerve to add to its scanty Flora in that particular.'

The Journal contains a very full description of this lonely, rugged, storm-swept island, for

> though two months there, to the last day I went botanising, and as far as I know I have left no hole unexamined or stone unturned. . . . You cannot conceive the delight which the new discoveries afforded as they slowly revealed themselves, though in many cases it was all I could do to collect from the frozen ground as much as would serve to identify a species.

Indeed the very first day he landed,

> arriving on board, I found that I had ascertained the existence of at least thirty species of plants in one day, and within

two miles of the harbour, thus proving that Mr. Anderson[1] was either not *ingenious* or not *ingenuous.*

During the two months of his stay here, while the portable observatory was set up for a long series of magnetic observations, not only did he enlarge the list of local species from 18 to 150, especially among the Cryptogams, but, by analysis of his material here and elsewhere, he was able to show the relative increase among the lower forms of Antarctic vegetation,[2] the peculiarities of plant life in the lonely Oceanic islands; the relation of the island floras to each other and to those of the Southern Continents and of the Arctic regions.

His Journal records a curious discovery in the two small lakes between Christmas Harbour and Northwest Bay.

> In these lakes there occurs a most remarkable plant, which resembles *Sabularia aquatica,* forming green patches a foot or two below the surface of the water on a loose muddy bottom; here it flowers, the close imbrication of the *calicine segments* and those of the Corolla protecting the stamens from the influence of the water. Each *germen* contains a small bubble of air, generated, of course, within the ovary. Winter seems to be its flowering season, and I found it in flower after a long search, under a coating of 2 inches of ice; as far as I have hitherto examined it seems to differ from the characters of any Natural Order.

The 'Cabbage' (*Pringlea antiscorbutica*), as has been said, comes in for a good deal of notice, along with other useful plants on the island. He writes in his Journal:

> Even in this remote corner of the globe, and scanty though the vegetation be, it has more than an ordinary interest, from the utility of two of its products. The

[1] William Anderson, at first surgeon's mate, afterwards naturalist, on the *Resolution* under Captain Cook. In the account of Cook's voyages, he is referred to as 'the ingenious Mr. Anderson.' He wrote a full account of the Kerguelen Cabbage aforesaid (*Pringlea antiscorbutica*).

[2] Dicotyledons to Monocotyledons as 1 : 2; grasses as 1 : 2·6 of the whole.

destruction of its former forests has produced abundance of good coal.[1] Cook mentions the remarkable cabbage, which, to a crew long on salt meat, is an invaluable anti-scorbutic, and to many, a most agreeable dish; unlike other pot-herbs, it possesses after boiling so much of its essential oil, as entirely to neutralise or destroy any symptoms of heart-burn or flatulence; nothing can be more wholesome than it is. The root eats like horse-radish and the young hearts like coarse mustard and cress; the seeds are the food of the numerous ducks on the island; growing as it does near the sea, on a spot upwards of 1000 miles from any land where fresh vegetables can be obtained, it seems planted by Nature's hand for the poor mariner, when suffering under his own peculiar malady.

This curious plant was one of Cook's discoveries; Hooker had been specially urged by his father and Robert Brown to investigate it on the spot, and it recurs again and again in the letters on either side. From seed he brought back with him, young plants were raised in Tasmania, though it seems without success in establishing the plant as a staple of food. Sir William at first failed to raise it at Kew; his son writes:

I do not understand your not getting the Kerguelen's Land Cabbage to grow. I have had fifty plants of it from seed. I had it growing in a bottle! (hanging to the after rigging), on a tuft of Leptostomum during all our second cruise in the Ice, and brought it alive to Falklands. It was sprouting before the Cape Horn plants went home, from seeds I scattered under the little trees. We used to amuse ourselves planting it here and there where we go. I shall fill a Ward's case with Lyall[2] (it is the *Terror's* second case) at St. Helena, with native plants, and sow the seeds among it. Try it again in a cool place very wet and shaded, in a black vegetable mould like peat. Do not bury it but lay

[1] 'If I could get a piece,' responds Sir William enthusiastically, 'I would have it framed and glazed.'

[2] David Lyall (1817–95) was assistant-surgeon on the *Terror* and a useful botanist.

it on the surface. Depend upon it they will grow if cool and damp enough.[1]

Some points in its development quite baffled him; he writes (July 6, 1841):

The examination of the Cabbage was made on the Island and several times since, and I send it in despair of understanding its organisation. You will remark that the radicle is pointing away from the funiculus and is on the upper side of the seed as it hangs, and how it gets there, supposing the foramen of the ovule to be where Lindley[2] describes it should be, I cannot conceive, for in its turning it must go ¾ round the seed. I suppose Brown understands it all; the flowers I nowhere saw, but he has them in the museum from Anderson.

Brown, it may be remembered, was the inheritor of the collections of Sir Joseph Banks, who had sailed with Cook.

Two grasses form most rich and nutritious fodder for cattle, as we proved by some sheep being let loose on the Island, who soon ran wild, and though they were landed hungry and lean, they very soon fattened and thrived. Goats, pigs, rabbits, sheep, and perhaps small cattle, would

[1] After his return, however, he had to confess to Ross (Sept. 14, 1845) that the seed he himself brought back to Kew 'never vegetated, though we sowed all and in all manner of situations.' He wished to name the plant *Rossia kerguelensis*, but 'our friend Brown had already applied the MS. name, given both because of the anti-scorbutic nature of the plant and because Pringle wrote upon *scurvy*, which has not much to do with the matter, it must be confessed.' (To Ross, September 1, 1845.)

[2] John Lindley (1799–1865). Like Brown and Bentham, Lindley, a hard worker and man of versatile powers, took a conspicuous part in building up the natural system of classification set forth by Jussieu as against the artificial system of Linnæus; the convenience of which was merely for identifying plants. Through the friendship of Sir W. J. Hooker (for he was an East Anglian) he became assistant librarian to Sir Joseph Banks: then Assistant Secretary and Secretary to the Royal Horticultural Society, 1822–60; Professor of Botany at University College, London, from 1828; editor of the *Gardener's Chronicle*, 1841, till his death. He was mainly responsible for Kew Gardens being preserved and made over to the nation as the headquarters of botanical science, though knowing full well that his opposition to officialdom would exclude him from receiving any appointment. His chief works were *The Theory and Practice of Horticulture*, 1840; *The Vegetable Kingdom*, 1846; the editing of *Botanical Register*, 1829–47, and various works on the Orchids. In his views of species he has been described as an evolutionist without knowing it.

> all thrive well on the Island, and would be no ordinary boon to the whalers. The little *Ranunculus* is the only acrid plant I have found near the harbour, so I suppose it must have been this that Cook's party ate for cress; it appeared to me anything but wholesome.
>
> Among the seaweeds many are doubtless edible; on one occasion I found our gunner seated on a rock with his feet in the surf passing down what he called dulse; it certainly was eatable raw; I need not add my friend was a Scotchman. The Lichens are all much too tough to afford any hopes of rivalling the Iceland Moss. Some of the *Musci* might be used by the Laplanders as they do their own, as swaddling clothes for their babies.

Strange that this was an island in S. latitude corresponding to that of Jersey in the northern hemisphere.

To the last hour of his stay at Kerguelen's Land he was absorbed in the strange interests of the place, and writing from Tasmania, November 1840, with the prospect of visiting another oceanic solitude, Campbell Island, he speaks of it as

> another edition of Kerguelen's Land, I suppose. I know I shall be happy there, for I was sorry at leaving Christmas Harbour; by finding food for the mind one may grow attached to the most wretched spots on the globe, yet hitherto I fear I have rather played with Botany than done any good at it.

The long stay at the Falkland Islands in 1842 gave time for generalising upon the botanical material collected in the South. The main lines of his thought begin to stand out clearly in his letters of this date. To his father he writes on November 25, 1842:

> The Cryptogamiae are far more numerous. I am not aware of having omitted any species of any Nat. Order which came under my notice; this perhaps prevented my getting any better specimens of some Phaenogamic plants that were in flower, but anybody can collect them, and no botanists *will* attend to the Cryptogamic. I am further anxious to know the proportions that the Nat. Orders bear to themselves at different Antarctic Longitudes and to

themselves in each locality, as an object of primary importance to the elucidation of Bot. Geog. and the effects of climate upon the Vegetable Kingdom. Several of the tabular results I have drawn out show a delightful accordance, nor do I know of any result of this Expedition which gave me such pleasure as to find how beautifully the grasses rose in the scale of importance, beating even Brown's published ideas, and yet they are not the only plants by whose abundance or want the botanical nature of a country may be judged of. As we go South, Fungi disappear, Lichens increase, Pleurocarpi[1] diminish, in proportion to Acrocarpi,[1] as do the proportion of Pleurocarpi which fruit to the barren ones. Cyperaceae decreases, and Dicotyledons bear a smaller proportion to Monocotyledons. Nothing so satisfies me, that I have observed carefully in any Island, as to find these laws to hold good in the collections made long ago and when it is too late to remedy any defects, to look for more grasses or to wonder if I have not made too many species of my Cyperaceae etc.

And to Dr. Boott[2] four days later he enlarges on the proportion of the Rush tribe to the Grasses occurring in this region.

The descending scale for the Southern regions is beautiful and in perfect accordance with what was to be expected from the climate and position of the several islands.

Australia, 0·7 : 1.
Campbell's Island, 1 : 5.
New Zealand, 1 : 1.
Auckland Island, 1 : 1·9.
Falklands, 1 : 2·5, and
Kerguelen's Land, 0 : 5.

[1] Two divisions of the Mosses.

[2] Francis Boott, M.D. (1792–1863). Born in Boston of British parents and maintaining friendships in both countries, he took up the study of medicine in 1820 (M.D. Edin.) and practised successfully in London 1825–32, with ideas on fresh air in advance of his times. Another innovation was to discard the traditional black coat and knee breeches of the physician for the ordinary dress of the day—blue coat with brass buttons and yellow waistcoat. But with characteristic fidelity he changed no more with the fashion, and his endeavour to avoid singularity in 1830 ended by making him more singular than ever in 1860. Inheriting a competency, he devoted himself to botany, specialising on the genus Carex, his *Illustrations* of which appeared 1858–67. He contributed a monograph of 158 species to Sir W. J. Hooker's *Flora Boreali-Americana*; his collection he bequeathed to Kew. He became a member of the Linnean Society in 1819; secretary 1832–9, and treasurer 1856–61.

These results, however, I must beg you to keep to yourself, as we are not permitted to communicate *Botanical Information* (does it deserve the name?) except through the Lords Commissioners!

He perceives also that the distribution and abundance of vegetation in this region depends not on the height of the mean temperature, but on the amount of moisture in the air and the equable level of heat and cold, free from extremes.

To establish this accurately would prevent critics from repeating that 'nothing of importance had been done towards investigating the causes of difference in Geographical distribution since the publication of Humboldt's work.'

To his Father

March 7, 1843.

I long to see your new work on Ferns; perhaps you will do something to their Geographical distribution, which seems most dependent on a uniform and moist temperature such as Islands enjoy. All the Magellan species that inhabit the Falklands, there become harsh and coriaceous, from the vicissitudes of temperature, and of the hygrometric state of the air to which they are exposed. . . . The Hygrometer I consider of more importance than the Barometer in all ordinary cases, that is, where the Islands are not large and the mountains not high. . . . I have lately been examining some of my hygrometer observations and find that the difference between the vegetations of the Falklands and the Fuegia may be well accounted for. When the results are placed in a tabular form it is quite surprising to see to what vicissitudes of temperature and moisture the Falkland plants are exposed. Now the mean temperature of the Falklands is the highest, but its plants are exposed to dry winds, great heat of the sun's rays unimpeded by any vapour when it is calm, and great cold at night, whilst those of Fuegia are not so, and enjoy perpetual moisture, and are very sensitive to extremes of temperature, as also to dryness.

His original intention had been to write a Flora Antarctica, where his work would be on a fairly little exploited field. As he reached the Cape on the outward voyage he was already planning the book.

March 1 and March 17, 1840.

I am now beginning to consider what are to be the limits of my Antarctic flora; if I confine it to 23° North of the S. Pole it will consist of one species, I suppose, and that the *Protococcus nivalis,* nor would this be a fair limit to poor Flora, as she is guided by climate, not parallels which man has laid down and called latitude. My idea is, to be guided very much by the temperature of the Islands and the nature of the plants they contain. It will be, however, difficult to draw the line; the Straits of Magellan must, I suppose, come in with the Falkland Islands, whilst the Southern Island of New Zealand, Van Diemen's Land, and the Cape will be excluded. The mean annual temperature of the Antarctic Ocean is said to be nearly that of the Arctic; if this is the case there must be some unknown reason for the comparative barrenness of the Islands of the two seas.

It was a different matter when, later, his father suggested that he should undertake complete Floras of some of the places he had visited. His answer (November 25, 1842) shows a natural diffidence at the thought of embarking on so much more complex a task.

In proposing me to publish Floras of New Zealand and V.D.L., I fear you overrate my Botanical powers, for I am very ignorant of any plants but those I have seen. My strict Flora Antarctica will always begin where the Pines cease, and I should like it to contain the most of the country S. of Magelhaens (but Darwin[1] will give me good limits there) provided I can gain access to the proper materials. Auckland and Campbell Islands, Kerguelen's Land, and the Falklands will be the only other stations except what few you have from Macquarie Islands. Do tell me in your next what the things are which Frazer[2] sent you: and ask Brown whether any things have ever been collected in

[1] As having visited the country on the voyage of the *Beagle.*

[2] Probably Louis Fraser, 1810–66, who was on the Niger Expedition of 1841–2 and afterwards took charge of Lord Derby's zoological collections at Knowsley.

Prince Edward's, the Crozets, Royal Companies Islands, Emerald Island, and whether Webster's Deception Island or Cook's South Georgian plants are in the Museum. Tristan D'Acunha and St. Paul's and Amsterdam, though in such low latitudes, have an Antarctic Botany, but I have seen none of them.

However, he set to work on his own plants and his books during the next six months with this end in view. One more botanical letter to his father may be quoted to illustrate his work on the Cryptogams, with its tendency to simplify classification and its relation to his Herbarium work. After the third visit to the ice he writes on the way from the Antarctic Circle to the Cape:

March 7, 1843.

During the past voyage I have re-examined all my Antarctic Mosses. . . . The Andraeae puzzled me exceedingly and occupied me very many days, for I had to examine many hundred specimens. I do hope they are scrupulously accurate, for I always compared the present examination with what I made on the spot, and consider most of the mosses to have had three examinations; where there is so much novelty I may have made varieties into species, but in a field so new some allowance must be made. . . .

There are hardly any new genera, nor have I any wish to get a notoriety by having '*Hook.*' tagged on to the end of a string of barbarous names. I should be far more proud of placing a well-known plant in its true position and relation to others than naming another and leaving others to squeeze it in between what he may think its congeners.

All other mosses are divisible into *Acro* and *Pleurocarpi*; there are five groups I consider quite natural, and the three first of them abnormal; these are what McLeay's[1] quinary system acknowledges, but you must not think that I am led away by any system, for I formed this system before I saw McLeay's and before I understood his views. When we met we never broached the subject of his system, for I felt myself too ignorant of the subject; I cannot,

[1] William Macleay, of Sydney, son of the Colonial Secretary, was a naturalist of some note, inventor of a now forgotten system of classification which posited the number 5 as the basis for the structure and grouping of all living things.

however, forget a remark he made, saying 'he was glad I paid so much attention to the minute Orders and to Cryptogamic Botany, *for in them would be found the foundation of a truly natural system.*' Now, though I do not put any faith in the quinary arrangement, I believe that 5 *happens to be* the number of groups into which mosses most naturally divide themselves, and I am convinced of the truth of the circular system. Fries [1] first developed it in the Fungi, as Brown knows, for he pointed it out to McLeay, who wrote a paper on it (Fries's work); again Berkeley [2] takes it up in the 'Annals,' vol. i, and quotes Montagne [3] in strong confirmation. Until, however, Lindley took it up I do not know any other steps taken towards arranging the groups of plants on a fixed plan. Amongst mosses there are many beautiful analogies in the groups, but how to characterise the genera is quite a puzzle to me. *Gymnostonum* must be split up, for there is hardly a genus of Acrocarpi to which each of its species is not far more allied than to its congeners in the present arrangement.

The other drawings are attempts and nothing more, for they are the first Lichens I ever drew, and I am no hand at

[1] Elias Fries (1794–1878), a Swedish botanist, successively Professor (1834), Director of the Botanic Gardens (1859), and Rector of the University (1853) at Upsala. He was an especial authority on the Cryptogams.

[2] Miles Joseph Berkeley (1803–89), the great mycologist, was directed to Natural History by the influence of Henslow at Cambridge, finally devoting himself to the Cryptogams and especially to Fungi. In 1828 he first came into touch with Sir W. J. Hooker, for whom he described all the fungi in the volumes supplementary to *The English Flora* of J. E. Smith. For half a century all the exotic fungi received at Kew passed through his hands, and over 400 papers on fungi stand under his name, apart from those at which he worked in collaboration. His *Introduction to Cryptogamic Botany* (1857) remained for many years the standard book on the subject, while he was one of the pioneers of Plant pathology, popularly remembered as the investigator of the potato murrain in 1846.

[3] Jean François Camille Montagne (1784–1866), botanist, was left fatherless very young, entered the French navy at 14, and took part in the expedition to Egypt. On his return to France in 1802 he studied medicine, and in 1804 was attached as surgeon to a military hospital at Boulogne. He became chief surgeon to Murat's army in 1815 and again in 1819, and in 1830 was head of the military hospital at Sedan. He left the army in 1832 and devoted himself to the study of cryptogams. Elected to the Académie des Sciences in 1853, and to other Societies, and received the cross of the Legion of Honour 1858. He contributed many papers to the *Archives de Botanique* and to the *Annales des Sciences naturelles*, besides working out the Plantae Cellulares for Webb and Berthelot's *Phytographia Canariensis*, Dumont d'Urville's *Voyage au Pôle Sud*, Gay's *Historia fisica de Chile*, etc., etc.

colour. I have descriptions in full of them, but I can make no hand of the genera of Lichens, there seems to me a sad want of tangible characters except amongst the larger.

I have also done a little towards the Flora of the Falklands, and a good deal of an introductory paper on the Geographical distribution of the Antarctic plants, their relations to the Arctic, and the analogies between the Antarctic, Polynesian, and American floras.

From the Cape I intend to carry on drawing up to England and studying what Cape and Rio plants I can pick up, that I may know something of the more common Tropical Nat. Ords., of which at present I am totally ignorant. You will indeed be surprised when you will find at what a loss I shall be to give you the names of the most common garden plant, but I have not seen a rose since leaving New Zealand or any other flowers but Antarctic.

CHAPTER IV

THE VOYAGE OF THE *EREBUS* AND *TERROR*: PASSING IMPRESSIONS

FOR reconstructing the history of the four years' voyage, abundant materials exist. The official account is Ross's book in two volumes, 'A Voyage of Discovery and Research in the Southern and Antarctic Regions, during the Years 1839–43' (John Murray, 1847).[1]

This abounds in good matter; not even the full-dress style of the period, very conscious of its epaulets, can mask the essential interest of these visits to the young colonies of the South, to the solitary fastnesses of oceanic life, and the unimagined wonders of an ice-world in a 'furthest south' four degrees beyond any previous record.

Next comes Hooker's MS. Journal, upon which he drew for some of the material of his letters home. These letters, or

[1] To this Hooker contributed (from his *Flora Antarctica*) botanical accounts of Kerguelen's Land, I. iv. pp. 83–7; Auckland Island, I. vi. pp. 144–8; Campbell Island, I. vi. pp. 158–63; the Falklands, II. ix. pp. 261–77, including an account of the Tussac Grass, p. 261; Hermite Island (Fuegia), 'the great botanical centre of the Antarctic Ocean,' II. x. pp. 288–302; and Cockburn Island (in the South Shetlands), II. xii. pp. 335–42, together with a description of the Fossil wood in Van Diemen's Land, II. i. pp. 5–11, and of hunting wild cattle in the Falklands, II. ix. pp. 245–53.

Most of the illustrations are by J. E. Davis, Second Master of the *Terror*; nine are from Hooker's drawings, some signed, some marked in his own hand in the copy of the book given him by Ross: these are Mount Minto and Mount Adam, I. chap. vi.; Cape Crozier and Mount Terror I. viii. (unsigned); Panorama of the Great Barrier, I. Appendix (unsigned); Seal Hunting on the Ice, II. ii. (the engraved signature is queried in pencil); Catching the Great Penguins, II. iv. (the central figure in the black hat is pencilled Bates); Mode of Pushing through the Pack during a Fog, II. iv. (unsigned); Tussac Grass of Falkland Islands, II. viii.; Hunting Wild Cattle in the Falkland Islands, II. ix.; 'Balsam-Bog' Plant (*Bolax Glebaria*), Falkland Islands, II. xi.

copies of them, are faithfully preserved, bound in a large quarto volume. His letters home were generally transcribed by the willing hand of his mother—who frequently Johnsonised the style to her own liking—for distribution among friends and relations, official news being of the scantiest, while letters, to these others, were regularly sent to her to copy. This solidly bound volume contains fifty-two autograph letters, ranging from four to twenty-seven closely written quarto sheets in a minute hand, twenty-nine in copy only, and twenty-seven duplicates which had returned in course of time to Kew. A still larger companion volume contains 234 letters received by him during this period.

So much of this abundant material may be cited as will suffice to show the impression made upon his mind by new scenes and new ideas, his occasional jaunts, more and more coloured by his scientific objects, a few sketches of the people with whom he came in contact, a passage or two to show his sensitiveness to Nature, and his power of describing what he saw.

At Madeira, as ever and again on his travels, his eye is instantly caught by any likeness to his beloved Highlands, whose beauty had sunk deep into his mind from his earliest days. Equally he recalls the pictures of the same scenes in the books of travel so well known to himself and to his father.

> On first nearing Madeira, I was strongly reminded of some of the islands on the West of Argyllshire, only the volcanic rocks are much redder, and clothed here and there with low brushwood; the tops of the hills are often capped with pines.
>
> The ravines are quite like Scotch ones, but more sparingly wooded, and the faces of the very deep ravines are most admirably like the view in Webb and Berthelot, full of vertical perpendicular lines which are dotted with trees. These views came into my mind directly I saw the realities.

With the botanist's eye he notes for his father the botanist, the belt of chestnuts running halfway up the mountains: 'the tops of the Mts. more sub-divided into conical peaks than the Scotch hills and covered with grass': the mingled tropical and temperate fruits growing in the island: the joy of the crews on arrival when 'all hands were busy spreading Bananas on our bread instead of butter and relishing grapes more than tea': though he found little in his diligent search for Alpines on the extremely dry and barren rocks of the Currâl, for 'Neither the season nor place were favourable to botanising.'

Here he received the warmest of Scotch welcomes from a Mr. Muir, formerly a Glasgow merchant, and a great friend of his grandfather, 'who had charged me particularly to call upon him,' finding his house by the help of a passing Englishman, after his enquiries, couched in Dog-Latin with Portuguese terminations, had produced no effect on the natives.

Though unable to accept Mr. Muir's instant invitation to stay at his Quinta as long as the ships lay off Funchal, he was constantly there, and notes with special pleasure, in the little parties got up to meet him, the absence of ceremony among the British families living there. Indeed there were so many Scotch and Glasgow acquaintances dining one night with another friend, that 'the conversation was wholly upon Glasgow or Britain, and Mr. Shortridge had a long discussion with me concerning the respective merits of Mr. Almond and Mr. Montgomery [two Glasgow ministers]; distance lent energy to the cause, and I supported the former with much more warmth than I should have done at home perhaps.'

A party from the ships now carried out a long cherished plan of visiting the famous mountain glen known as the Currâl. On the way, Hooker's unceasing interest in the practical side of economic botany, already stirred by the discovery that the coffee served him at dinner was home grown, made him pay special attention to the 'Jardine,' a tea plantation among the chestnut woods some 2000 feet above the sea, belonging to the late British Consul, Mr. Veitch. In this temperate region, with a soil composed of a fine vegetable mould over volcanic detritus, he notes that 'neither bananas, coffee, nor dates will grow here, but the climate seems peculiarly well adapted to the cultivation of Chinese plants; Camellias flourish, including the rare *C. oleifera* which produces the oil used in China.' Mr. Veitch was hoping to grow tea regularly and cut into the monopoly of the East India Company. To Hooker he confided his plans and methods, 'telling me that it was his duty to impart his knowledge to me as Botanist of the Expedition, and only hoped I would not use it to his disadvantage on the Island.' His visitor was allowed to take specimens of the plants, but our time was too short to allow of our waiting and tasting Mr. Veitch's tea. The owner very naturally praises his tea, as equal to the true Chinese herb. Mr. Muir informed us that it was execrable, and pronounced so by every one that had tasted it.' On the other hand Lieutenant Bird testified to its excellence, while Captain Crozier, commander of the *Terror*, reconciled these opposite views, '—— tells me he has often drunk Mr. Veitch's tea, and that formerly it used to be so bad that bare civility could hardly tempt him to swallow it and *not do the other thing*, but that which he tasted this time was very fair tea indeed.'.

The lonely waste, where hardly any animal life was to be seen, was tenanted by strange human beings.

> After leaving the Jardine we continued ascending through the forest, the trees gradually dwindled away and nothing remained but a short herbage with numerous bushes of a Cytisus with which the hillsides seemed spotted. On emerging at the top of the valley, about 3500 feet, we were suddenly attacked by a party of pseudo Highlanders male and female, chiefly children, ragged, dirty Portuguese, each armed with a long pole, iron shodded (*sic*) for climbing, with which they assailed our ponies, causing them to spring over the rough ground at a rate which nearly rendered my seat untenable. This was done apparently for effect, for we came suddenly upon one of the most slpendid views I ever beheld. We stood upon the brink of a tremendous precipice which formed one side of a gully about 2000 feet deep and ¾ of a mile across. On looking over nothing was seen but the tops of a few projecting trees, and at the bottom a small stream that dashed along and was all but invisible. The opposite precipice was steeper and more bare than that on which we stood.

The whole scene very much reminded me of a view among the Grampians of Forfarshire, where you come suddenly upon the Glen of the Dale; Glen Dhu stretches away on one hand, and on the other you look down into the broad valley of Clova; the present, however, was infinitely grander, and the numerous laurel trees gave it a different aspect. The river dashing at the bottom, which looked like a mere burn, brought Scotland forcibly to my mind; it foamed away with a murmur which from the distance we could scarcely catch.

The ragged Highlanders, for I can call them by no other name, were most troublesome, begging and offering us their climbing poles. . . . On seeing me scrambling among the rocks they paid me particular attention.

. . . On reascending I found my companions seated among some rocks, surrounded by a brood of the most extraordinary ragged urchins I ever beheld, of all ages from five to twelve, dressed in tatters with high peaked carabooshes, their long hair streaming over their faces, which were of a most determined Portuguese cast. They excited our compassion by kneeling round us and begging by holding up their hands with the palms together like Catholics invoking the Virgin. Some of them were really pretty, though [with] very coarse features; among them was a very old woman whose husband had been lost among the cliffs or rather killed. They had large black eyes and seemed remarkably healthy, though they live in the most wretched holes and feed upon chestnuts, scarcely ever touching other foods. Even the little babies were sucking chestnuts. A few dogs were spectral animals.

. . . On a grass bank, where we had left our horses, there was spread for us a famous cold luncheon prepared for us by Mr. Muir. Dr. Lippold[1] had joined us just before reaching the Jardine, and he certainly amused us not a little during dinner. The young half savages clustered around us whilst eating, forming a ring, which gradually approached and hemmed us in. Now the little German abhors the Portuguese beyond any other nation, and he could not brook these unfortunate urchins drawing near

[1] Dr. Lippold had been sent to Madeira to collect plants and seeds, partly for Kew, partly for the Duke of Bedford.

> us. He used accordingly, every now and then, to start up, take his stick, shout, hooroosh, shake his coat-tails at and scare the poor little snips out of their senses, who would run up the hills with amazing agility, their scanty clothing tripping and causing them to tumble over and over as they scrambled along on all fours, almost to our table-cloth.

An unfortunate result of this excursion was a sharp attack of rheumatic fever, caused by lying on the damp grass at lunch when overheated. Hooker was laid up in the ship for a week, and could scarcely go ashore to make his farewells. The report of this from friends in Madeira made his parents very anxious, for it was many months before they received his letters reporting himself perfectly well. In later life, it is true, his heart was not strong; but through all the following years of strenuous travel and unceasing work, the minor troubles which persisted indicated no serious weakness.

At Teneriffe there was no time to travel the twenty-eight miles to Orotava in order to see the famous Dragon's-blood tree. The brief afternoon ashore gave opportunity of very little collecting. Nor was Hooker able, much as he wished, to see the two English Jacks taken when Nelson made his unsuccessful attack on Sta. Cruz. The church where they hung high out of reach, since an English middy had audaciously carried off the third, was too far away. However, 'I was much amused by the little urchins grinning and repeating the words "English flag" when asked where the *Parochia* was.' So in the town itself 'the only remarkable thing I saw was the camel used as a beast of burden.'

Their next point was the Cape Verde Islands, 'not that we knew we were going there, for everything regarding our destinations has been kept a profound secret until we cast anchor in the harbours!' It strikes an old-time note indeed to be told that:

> On our arrival (November 11) a slaving schooner was lying in the Bay, and I understood that a more cautious one had made sail on discovering us heaving in sight. The present one remained some days, and when taking her departure her drunken skipper saluted us, and mocking,

> told us he was going nigger hunting to the Coast. We had no commission to catch slavers or to do mischief further than resenting personal injuries.

If Madeira afforded the first vision of real tropical verdure, the Cape de Verdes intensified it with the unimagined grace and beauty of a cocoanut grove, the one redeeming feature of the prevailing Saharan desolation near the coast. The fertile interior was twelve miles away from Porto Praya; still, in a week here, during the bad season, Hooker managed to collect 110 species in a tolerable state and saw perhaps 100 more in a useless state—a very fair proportion of the 300 brought home by a previous collector. Of the famous Baobab tree he remarks that neither to himself nor to Captain Ross did it give the impression of being such a slow growing and ancient tree as was reported by those who had seen one cut down.

Distance was not the only obstacle confronting the botanist. Returning from their first day's outing they found that 'the Consul had very thoughtfully left word for us to prepare ourselves for the coast fever (or yellow fever), which was certain to lay hold of all Europeans who should expose themselves as we had done.' Nevertheless they went not once again, but twice, further afield to the beautiful valley of St. Domingo in the interior, the first time entirely, the second half way, on foot.

> The Consul persuaded us to ride, assuring us that a walk of twelve miles there and twelve back would assuredly be followed by fever. We therefore hired two ponies, the only two we could procure, and the very worst I ever saw, and a Jackass for which we drew lots. Mr. McCormick and I soon relinquished our beasts, and sent them back before leaving the Town, and the Jackass, having performed the feat of unassing Mr. Hallett and running through the Town with our poor purser hanging to his neck, we determined to walk.

After the Saharan desolation of the lower country, where under the tropical sun the soil of black volcanic slag and ashes scorched the feet in walking, the picture changed suddenly.

> So enchanting is the scenery of these glens, and so suddenly do they start up beneath the feet, that one almost feels persuaded that the author of 'Rasselas' was there before him, or that the scenes of the Arabian Nights were not all laid in the East.

Evening fell cool and refreshing as they descended this valley, and 'one little bird sang so like a robin that we all exclaimed at once we were in England.'

To give his father a notion of the fantastic peaks and pinnacles of the surrounding mountains, he employs his frequent method of reference to their common knowledge of the literature of travel. 'They reminded me of the Organ Mountains of Rio de Janeiro, only these were much sharper.'

Hospitality was freely offered by a Portuguese of some position in Porto Praya, but educated in France. In this remote valley he lived with his wife and several little slaves; his property surrounding his house being cultivated with tropical fruits and plants.

> During dinner our hostess arranged three little slaves round the table; they were very clean and neatly dressed, quite young and jet black. After dinner they each received an embrace from their mistress and came to us for the same (which I assure you [he tells his sisters] was not withheld because of the swarthiness of their complexions, and was accompanied with a donation of fruit). Our host said he treated them as his children, and would not part with one for anything. On taking our departure we gave our kind host all our shot and I my powder flask, as the only recompense he would take.

So delightful had the excursion been, that on the Monday (17th) he repeated it, in company with Wilmot[1] and Lefroy. This time they left early, and managed to ride across the first six uninteresting miles, when 'Mr. Wilmot was the first to find out how to make a Porto Praya pony gallop (if it ever can).

[1] Lieutenant Eardley Wilmot was an engineer officer. A close friend of Lefroy (see ii. 343) he had joined in his effort to improve the training of officers at Woolwich. With Lefroy also he was selected for magnetic work on Ross's expedition, his destination being the Cape Observatory.

It is accomplished by exaggerating the motion of galloping yourself on the saddle, kicking your heels into the animal's flanks, and personifying a flying postboy.'

This day there was time to botanise; and after dinner with the friendly Frenchman they ascended a peak immediately behind his house, shaped like a steep cone with a pinnacle on the top of it, amid prophecies that they would break their necks.

The ascent culminated in an arduous climb, and a descent which seemingly could not be worse and was at least fresh, on the further side. Swinging down from ledge to ledge, while an agitated group of little niggers far below shouted and gesticulated unintelligibly,

> I was well rewarded by finding, when about half way down, a lovely fern with beautiful soft green foliage growing like our *Cystopteris* out of the crevices of the rocks; it grew with lots of the Campanula and Umbellifer (found on the way up) which so put me in mind of old Scottish forms of plants, that I only wanted a companion who had botanised over Ben Lawers to share my joys with me. [Before returning,] I emptied my pockets into my travelling portfolio, which I may mention here is the only good way of preserving plants in the tropics, and were it not for the weight, ought to be looked upon as an indispensable addition to the vasculum. The poor withered herbs that I gathered on my previous excursions used on my return to be more crumpled still from the fiery heat of the sun beating on the vasculum, and sorry specimens they have made, though invariably put into paper immediately on my return.

No time was left for geologising, though the relation of the limestone and the volcanic rocks was an inviting problem. But the whole scene left a deep impression, and the Journal records:

> Man always looks back with pleasure to such spots as this, where disinterested kindness has been shown him; when to this is added a new country and the charms of a scenery half tropical and half—what is dearer still to me—Scottish, both as to scenery and general features of a scanty

vegetation, his happiness to whom the works of Nature have charms, is, for the time, complete.

Three more Oceanic islands were visited before the Cape, the unusual course west to St. Paul's Rocks, then south to Trinidad off the Brazilian coast, then east to St. Helena, being followed in order to fix certain magnetic determinants.

On the eight or ten detached rocks of St. Paul, some sixty feet high, 'a wretched cluster about as big as all the houses in the Crescent put together,' Hooker did not set foot. Landing in the tremendous surf was so dangerous that Captain Ross gave up the second visit, on which he had intended to take Hooker. Botanically, however, this was little loss. Not even a lichen grew on the rocks, and his shipmates brought him back specimens of the only seaweed which grew there, serving to make a rude rest for the Noddy, interwoven with a few feathers.

Trinidad was a shade less inhospitable, its valleys possessing a little vegetation. Among its mountain crags

> we easily pictured to ourselves the figures of gigantic Turks, bishops, &c., on the summits: there was no wood but a very remarkable tree on the top of the highest hills (2000 feet ?)—it struck me that it was a tree fern. All over the coast there are remains of barked white trees lying on their sides, but no live ones. They lay in different directions, and except the introduction of goats has, by eating up all the young trees and leaving the old ones to perish, destroyed the vegetation, as was the case at St. Helena (see Darwin), I am at a loss to conceive how they have so universally disappeared.

The one accessible beach on the lee side, where a landing was effected in the morning, was stony and barren and hemmed in by precipices; in the afternoon the surf on the windward side seemed hopeless. However:

> When about to give up the attempt one of the party espied a small cove to the N. of the Nine Pin rock, and there we landed with great difficulty. A narrow platform of rock afforded us a footing. When within 100 yards

of the shore, a grapnel was dropped and the boat was then backed to the rocks, a bowman carefully paying out the rope; then taking advantage of a lull another seaman with a lead line jumped ashore and made it fast; a third was stationed at this line in the boat, then, as the surf rose, the grapnel line was held tight and the lead line paid out, thus preventing the boat from being cast ashore; when the reflux came the contrary was done. In the intervals we jumped ashore and the instruments were handed out after us. To gain the beach from this we had to walk along a ledge of rock up to our middles in water, carrying the instruments by turns, both men and officers. . . .

After ascending about 600 feet of a shelving debris we found ourselves at the foot of a continuous precipice, that shut us in completely. The rocks were in most places perpendicular and smooth, without a sign of vegetation but a few lichens; in other places the rocks were broken up into quadrangular blocks, which when moved came tumbling down and bringing others with them, which continued their course till they reached the Captain's instruments on the beach where he was conducting his [magnetic] experiments. These were materially affected by the iron in the rocks.

As bearing on the problem of distribution, the population of this lonely island is carefully noted. Besides the sea-birds, Noddy and Tern, whose eggs were sought by the Grapsus crab, 'of insects I saw a Hemerobius, a small fly, cockroaches from the wreck of a vessel, common house-fly, and some spiders.' The land crab was as much in evidence then as to more recent visitors to the island—'a very short, strong, thick-set animal,' with 'an enormous mouth and large savage black eyes. When threatened he takes up his post under a stone, and commences opening his claws, and putting them to his mouth in a menacing attitude, evidently expressing a desire to eat you, opening his formidable mandibles at the same time.'

Arrival at St. Helena was the more welcome because of the slowness of the voyage.

> The *Terror* has been a sad drawback to us, having every now and then to shorten sail for her. I cannot tell you how delighted we were to get here (St. Helena), having been upon salt Junk for 74 days, with hard biscuit for vegetables. . . . The weather has been during the voyage very fine indeed, though very hot at times, so much so that sleeping upon deck is quite delightful. . . .

St. Helena as a colonised island was very different from the others. Appealed to as a fount of botanical culture he pokes fun at himself as a practical gardener. Strawberries and similar European plants refused to fruit in the absence of a regular summer and winter season. He suggested on theoretical grounds two alternative methods of checking their 'running to leaf'; 'between these two methods I hope I have hit a gardener's plan, or what will look like one; if the more orthodox plan succeeds my suggestion will, I hope, be looked upon as the invention of a fertile brain instead of the guess of an ignoramus.'

But 'the plant that pleased him more than any other' was a fine Araucaria (monkey puzzle). Few specimens then existed in Britain, and this, as a new species from Brazil, is described in full detail. The fruit, it was asserted, never ripened; but his keen eye noted several seedlings which the owner of the garden had never observed. He has a boyish delight in climbing the spiny tree and knocking off some cones, because travellers declared the tree unscalable, and at sea he writes, 'even now I look at the cones slung up in my cabin by a true lover's knot with great satisfaction.'

But here also he is confronted by his favourite problems of geographical distribution, of the interaction of imported animals and plants on the old flora. The climate differs on the wet side of Diana's Peak; so do the plants. He perceives a striking phase of what was afterwards to be called the 'struggle for existence' bluntly revealed in the action of animals on plants, plants on each other, and plants again on animals, owing to the introduction of new forms of life into the island.

So, he writes in his Journal from his passing notes—time forbidding fuller observations:

At that particular elevation (about 700 feet, 1000 feet being the average elevation of the interior of the island) there is hardly a trace of the original plants in the soil, they having been completely destroyed by tne introduction of pigs and goats into the Island, which eat up all the young trees, leaving the old ones, which are invariably succulent Compositae, to perish, or else tearing off their bark which is soft and loose. In addition, the soil and climate is so well adapted to the growth of forest trees, which when once they have formed a shelter sow themselves, that there remains no opportunity for the native trees to recover the soil, which is now dry and not adapted to their habits, the rich vegetable mould which they formed being swept by torrents into the valleys subsequent to their destruction. On the northern slope of Diana's Peak I have seen a broad belting of trees put a stop to the descent of the Cabbage trees (a name given to the six or eight species of native arborescent compositae) which cannot exist along with any other vegetation that overtops them, nor can they grow singly. Another tree is said to be completely extirpated—the Ebony. Large masses of the wood are still found in some of the valleys, though I was unable to procure any specimens.

Though the introduced trees have adapted themselves to this soil and climate, the Animal Kingdom and other indigenous vegetation are not to be found under their shelter. The insects and birds which I observed among the native trees were not to be found in these plantations; of the birds in particular I observed this. It is also the case with the Lichens and Insects, two species of Usnea and another Lichen being found on the firs and oaks only, whilst only one species of plant, *Rubus pinnatus* (an indigenous species), grows indifferently on open banks and in the wood—never in *native wood.*

Longwood, with its associations of fallen grandeur, was less to him than the wonders of nature; nevertheless, he writes in his Journal on February 6:

So very much is talked about Napoleon's tomb, that though I felt very little interest in seeing it, I was determined to be no more called a Goth, which name I had earned from my previous indifference, and to go to this more hackneyed spot than Richmond or Kensington Gardens.

His fears were justified when he reached the tomb.

It is situated at the head of this valley, guarded by a sentinel who duns you about the mighty dead, and gives you water that the Emperor drank; on turning your heel upon him, numerous children assail you with flowers, Geraniums, that the Emperor was fond of. On turning into a pretty cottage to get some ale at 2*s*. a bottle, the cork was no sooner drawn than out came the Emperor with it; it was the Emperor this, that, and the other thing; our hostess's daughter came in with the Emperor on her lips; his ubiquity certainly astonished me. As a last resource I commenced gathering Lichens; surely the hero of Marengo could have nothing to do with Lichens on a stone wall, when another disinterested stranger came to inform me that the Emperor had from it marked out the position of his tomb, and that the Emperor was fond of the wild plants I had in my hand. I fairly took to my heels, heartily wishing that for my own sake as well as for the good cause of humanity, the Emperor had had his wish of living and dying in some remote corner of Britain.

The Cape was reached on March 17, and left on April 6, 1840. There is little to note during this brief stay. Hooker's impressions of the Cape date from his second and longer visit. This time he collected, as has been said, some 300 species of Cape plants to study on the voyage. A long five weeks of sailing brought the ships to Kerguelen's Land, where Ross's prolonged magnetic observations kept them from May 12 to July 20.

Though this lodestone of Hooker's childish imaginations deserved all too well its other name of Desolation Island, its fascination for him was reinforced, as we have seen, by a still stronger spell, the charm of discoveries leading on to luminous generalisations. The letter to his father from Hobart (August 16, 1840) describing the place deserves fairly full quotation.[1]

[1] The passages enclosed in square brackets are from the Journal.

We proceeded to Kerguelen's Land, and after twice being blown off in a gale we at last, on May 12, anchored in Christmas Harbour. During the passage there were few sea-animals, so I studied Cape plants with Harvey, Endlicher,[1] and De Candolle.[2]

From a distance the Island looks like terraces of black rocks; on which the snow lies, causing it to look striped in horizontal bands. On the melting of the snow, the flats appear covered with green grass and the hills with brown and yellow tufts of vegetation. The shores are almost everywhere bounded by high, steep precipices, some of frightful height, above which the land rises in ledges to the tops of the hills. The varied colour in the vegetation gave me hopes that the country might be rich in mosses, &c. [nor could anything the *ingenious* Mr. Anderson in 'Cook's Voyages' said persuade me to the contrary. . . . Surely, I thought, this cannot be such a land of desolation as Cook has painted it, containing only eighteen species of plants].

Christmas Harbour is well described and figured by Cook, indeed the accuracy with which he made a running survey of the coast is quite marvellous, and shows how talented a man he was. I cannot say so much of his Surgeon and Botanist, 'The ingenious Mr. Anderson,' as our copy calls him. Had Cook been here in winter he would have found it a different place to lie in from what it is in summer; the winds blow into it from the N.W. with the most incredible fury, preventing sometimes for days any intercourse with the shore. We have the chain cables of a 28 gun-ship, and yet we drove with 3 anchors and 150 fathoms of chain on the best-bower, 60 on the small, and a third anchor under foot, *the Sheet*. Such a thing was never heard of before!

[1] Stephen Ladislas Endlicher (1804–49), a Hungarian, Professor of Botany in Vienna from 1840, and author of a Genera Plantarum.

[2] Augustin Pyrame De Candolle (1778–1841), a Genevese whose most important work was done in France between 1796 and 1816, when he returned to Geneva. He used his immense knowledge of botany to become the leading systematist of his period. (For the adoption of his system by Bentham and Hooker in the *Gen. Pl.*, see ii. 19 *seq.*, 22, 415.) Beginning to work out his great system on too large a scale (1818–21) he continued it in the more manageable *Prodromus Systematis Naturalis Regni Vegetabilis*, in seventeen volumes, 1824–73, ten of which were the work of his son and successor, Alphonse. The latter, like Hooker, was strongly interested in distribution and economic botany, writing a *Géographie Botanique* in 1855 and *Origine des Plantes Cultivées* in 1883.

During our stay I devoted all my time to collect everything in the botanical way, and I hope you will not be disappointed with the fruits of my poor exertions. You say you hope I shall double the Flora and I have done so.[1] I was much surprised at finding the plants in a good state of flower and fruit (all but two).

My time was my own to leave the ship when I liked, for the Captain took off all restrictions to my going where I liked. My rambles were generally solitary, through the wildest country I ever saw. The hill tops are always covered with snow and frost, and many of my best little Lichens were gathered by hammering out the tufts or sitting on them till they thawed. The days were so short and the country so high, snowy, and bad that I never could get far from the harbour, though I several times tried by starting before light. As far as I went the vegetation did not differ from that of the bays. . . .

I went several boating excursions in the neighbourhood, and in one was dismasted and nearly swamped. So Captain Ross would send no more, and I am promised to be of a longer and better party on the next opportunity. Two Lycopodia, one splendid one, and a Fern were all Mr. McCormick added to my collection. He brought numerous splendid quartz crystals and zeolites, &c., together with lots of coal and fossil wood. The latter we had long before found, and I first detected it lying in immense trunks in the solid basaltic rock; its existence here is wonderful in the extreme; I have plenty of specimens.

[In the absence of trees, the coloured patches of Lichens on the hillsides, the heaving belt of seaweed girdling the shores, took the place of forest green or autumnal tints.]

The Lichens appear here to form a greater comparative portion of the vegetable world than in any other portion of the globe, especially when it is considered that from the want of large trees there can be no parasitical species. The rocks from the water's edge to the summit of the hills are apparently painted with them, their fronds adhering so closely to the stone that they are with difficulty detached; in other

[1] Sir William had written: 'I wish I could have a day's botanising with you in Kerguelen's Land. I think we could at least double the Flora. Look well to the Cryptogamia and see how far south the Algae extend and what are the species.'

cases they seem to form part of the rock which, from its excessive toughness and hardness, almost defies any attempt to procure specimens that can be satisfactory. But it is at the tops of the hills that they assume the appearance of a miniature forest on the flat rocks, and nothing can be prettier than the large species with broad black apothecia that covers all the stones at an elevation of from 1000–1500 feet. A smaller species like a little oak-tree grows in spreading tufts also upon stones, and is of a delicate lilac color. Near the sea they are generally more coriaceous, especially a yellow one that then forms bright yellow patches on the cliffs. In the caves, also near the sea, a light red one is so abundant as to tinge such situations with that color, and many other species inhabit the rocks and their crevices.

Seaweeds are in immense profusion, especially two large species, the *Macrocystis pyrifera*[1] and the *Laminaria radiata*?; the former of these forms a broad green belt to the whole Island (as far as seen) of 8–20 yds. across within 20 feet or so from the shore. Here its branches are so entangled that it is sometimes impossible to pull a boat through it, and should any accident occur outside of it, its presence would prove an insurmountable obstacle to the best swimmers reaching land. On the beach the effect of the surf beating it up and down is very pretty, but not so striking as the view from a little elevation, of a bay, with this olive green band running round it. The sea birds, etc., when on the water, always fly over or dive under to reappear on its other side. The *Laminaria* hangs down from every rock within reach of the tide, perpetually in motion from the lashing of the surf, and yet from its shininess and strength always unhurt. I think I may safely affirm that no other species in the vegetable kingdom has so secure a rooting as this seaweed has on the bare rock. I have often sat upon the cliff overhanging the sea at the N.W. bay during a gale of wind, and watched the surf break with terrific violence on the rocks, which are often themselves detached and alternately brought backwards and forwards by the swell and reflux with a deafening roar; still the coriaceous fronds of this weed are with impunity

[1] This 'is the only strictly Antarctic plant of the island, which floats alive in the water and increases there like the Sargasso weed: hundreds of miles from any land 64° South is the highest latitude in which I have seen it.' (To Bentham, April 27, 1842.)

washed backwards and forwards, then form attachments defying the power of the sea. . . . [The only use in Nature I can assign to it is the shelter it affords to a species of Patella from the attacks of the gulls, which prowl about during low water and secure as their prey any other unfortunate shell-fish which is exposed. The weight of the fronds of the *Laminaria* hanging down over the dry rocks forms an insurmountable obstacle to the birds.]

The birds, unused to man, were devoid of fear. In the shallow bay next to the Arch Point, were myriads of the beautiful Sheathbill as the sailors called it (a *Chionis*), so tame that it allows you to come quite close to it. It was something like a pigeon, black legs (not webbed), beak and eyes; it ran with great agility among the rocks [like ptarmigan, helping itself by the first joint of the wings, which is provided with two callous extremities admirably adapted for this purpose] and came close to examine me; its plumage is of a spotless white, with a slight pink tinge on the primaries of the wings; the bill was a sheath common to the two nostrils. On one occasion I thoughtfully sat down on a stone and commenced whistling a tune when, on turning my head, I found I had unwittingly been performing an Orpheus's part, for upwards of twenty of these beautiful birds had gathered about me, and were gradually approaching, declining their heads and narrowly watching my motions, and would even perch on my foot, rocking their heads on one side in the most interesting manner. Among them were some penguins, peering over the rocks . . . so tame that they allowed me to take them by the beaks.

Among the stones were feathers in amazing quantities and

many skeletons, especially Penquins', which are, I suspect, destroyed by a very large gull, whose bill is like that of a hawk, and its webbed feet terminated by hooked claws of great strength. The penguins' food is, I suspect, fish, at least the stomach of a common one was full of such matter; and the white birds are omnivorous, eating flesh, seaweeds, and insects. One that we kept on board used to run about the decks after the sailors, and at their dinner used to help itself from their dishes, eating meat boiled or raw, raisins, rice, salt meat, and would drink water, limejuice, and grog! Its tameness and gentleness rendered it a general favourite, but its spotless plumage soon turned gray, and then black.

So too the common Jack penguins were easily tamed.

At first we had about a dozen on board, running wild over the decks following a leader; they cannot climb over any obstacle two or three inches high, so we thought them safe, until one day, the leader finding the hawse hole empty, immediately made his exit, and was followed by the rest, each giving a valedictory croak as he made his escape.

[As food, the sheathbills] are tolerable eating, rather tough though, and they have a rank flavour and smell when newly killed, and require soaking before cooking, when they eat well in pies and mulligatawny.

[The penguins'] flesh is black and very rich, and was much relished at first for stews, pies, curries, etc.; after a day or two we found it too rich, with a disagreeable flavour, whence partly from prejudice I believe, they were dropped, except in the shape of soup, which is certainly the richest I ever ate, much more so than hare soup, which it much resembles.

Certain annotations in the presentation copy of Ross's Voyage deserve passing mention. They unmask two pieces of unconscious humour on the part of Dr. McCormick, one a mistake, the other the fruit of a well-laid practical joke. In the scientific appendices, McCormick (II. 409) describes the Kiwi or Apteryx, that wingless bird, as seeking 'larvae and seeds of a rush (*Astelia Banksii*), its favourite food.' On the margin is pencilled 'grows on *high* trees only.' And on p. 414 he describes the nest of the albatross, which 'only lays one egg. In one instance only I found two eggs in the same nest (both of the full size, and one of them unusually elongated in its longest diameter), although I must have examined at least a hundred nests.' Indeed a puzzle, anxiously detailed; but we smile at the accusing pencil, 'placed there by Oakeley,' the mate of the *Erebus*.

CHAPTER V

TASMANIA AND THE ANTARCTIC

FROM August 16 to November 12 they stayed at Tasmania. The dominant person in the island was the Governor, Sir John Franklin,[1] who, seconded by Lady Franklin, gave all aid and welcome with the enthusiasm of an old Arctic explorer, indeed volunteering to take a share himself in the long term day observations, which reminded him, he declared, of old times in the North.[2] Nor, later, did he forget Hooker. Lieutenant

[1] Sir John Franklin (1786–1847). Though he fought at Copenhagen and Trafalgar, it was as an explorer that Franklin won chief distinction and became the friend of the elder Hooker. From 1800 he had spent three years with Flinders in the *Investigator* surveying the coasts of Australia. In 1818 he first joined in the search for the North-West Passage, for the discovery of which he ultimately paid with his life. Sailing eastwards from Spitsbergen, the expedition had to turn back; but Franklin, commanding the *Trent*, under Buchan in the *Dorothea*, revealed himself as a great commander and a scientific investigator and was elected F.R.S. in 1822. In 1819–23 he led an exploring party along the Saskatchewan and the Coppermine rivers and eastward along the coast; in 1825–7 he descended the Mackenzie river and followed the coast west, trying to meet Beechey, who was pushing east from Behring Strait. From 1837–43 he was Lieutenant-Governor of Tasmania, where, as will be seen, he welcomed Joseph Hooker; in 1845 he set out on his last voyage in the *Erebus* and *Terror*, Ross's ships in the Antarctic, accompanied by Ross's second in command, Captain Crozier, and was heard of no more. Between 1847 and 1857 no less than thirty-nine search-parties were sent out from England and America. Piece by piece the mystery was solved. Franklin was one of those who died while the ships were hopelessly beset by ice for eighteen months; Captain Crozier and the rest, 105 in number, perished as they tried to march homewards.

[2] Ross's 'devotion to his beloved pendulum' was the dominant note. In the primitive room whose floor was Mother Earth, for lack of timber, 'the officers relieve one another in regular watches, and I never met with such devotees to science. You would be delighted to see Captain R.'s little hammock swinging close to his darling Pendulum, and a large hole in his thin partition, that he may see it at any moment, and Captain Crozier's hammock is close alongside of it.'

Dayman, who was left in charge of the magnetic observatory, writes, 'Sir J. Franklin expressed his regret that he had not seen more of you while you were here.' Others had occupied his attention.

Lady Franklin had established a Natural History Society, or rather Soirées, that met every fortnight, on Monday evenings at Government House, and Hooker was elected an honorary member. Lady Franklin herself was, it seems, somewhat imperious, and to the young man incomprehensible in her autocratic ways. Hence he writes (November 9, 1840):

> Lady Franklin . . . would like to show me every kindness, but does not understand how, and I hate dancing attendance at Government House. I have dined there five or six times. . . . She very kindly invited me to go to Port Arthur in their yacht, to botanise; we were three days away,—two of them at sea, and the third, a Sunday, it rained furiously. I got about 500 specimens on Monday, and a few after service on Sunday, though Lady F. did not like it, and very properly, but I thought it excusable as being my only chance of gathering *Anopterus glandulosus.* Do not think this is my habit. Captain Ross is too strict, were there no other reasons.

His own disinclination to spend his time in meaningless amusements can be gathered from letters of the period. Herein he was fortified by a letter from his Glasgow friend, the botanist Arnott, who warns him to collect, not to dance or amuse himself: 'H.M. does not pay for this.' He quotes the example of Lacy and Collie, who were not employed to play the fiddle on Beechey's voyage, yet that seemed the principal part of their occupation!

His main concern from April to July 1841 was botanising work that afterwards bore fruit in his 'Flora of Tasmania.' He has an eye, however, for human affairs. Among the trees charred by the natives' bush-fires from ancient times, he marks some few hollowed out by fire to form their houses: a meagre record of the thousands of native Tasmanians, for of them all 'only three remain, all males, and they consist of an old and a middle-aged man and child. They are very savage, but seldom seen—only once lately, and then near the lakes in the interior.'

As to the better society in Tasmania, the last of the Convict settlements and acquainted with bushrangers, it 'is perfectly English,' a commendation bestowed on the most comfortable houses he enters in any Colony, and

> there is a marked line drawn between the children of convicts or ex-convicts and those of honester, *even if less capable*, folk. Wealth is accumulating fast: and the banks allow 10 per cent. on deposits.
>
> Literature, however, is at a low ebb, and except a few English families, there are none who take the better periodicals, or would comprehend them if they did.
>
> There are lots of splendid Pianos and Harps, and few who can use them. Three hundred copies of Gould's most extravagant book[1] are purchased by these colonists, solely for the pleasure of seeing the show of it on their tables.

Looking back after a couple of months' absence he exclaims, altogether Van Diemen's Land was quite a home to us and a most attractive place.' His remembrance is of his personal entertainers, and the best is of those who could provide him with the music he loved:

> There is really so much good society, wealth, and splendour in the private houses: music is much cultivated, and all the new operas, &c., are procured as soon as published. Many of those pretty Strauss Waltzes you used to play I have heard here. At Government House there is always excellent music, and the military band is one of the best in the lines.

So little had he gone into society at Hobart that on the eve of departure he winds up:

> You would hardly believe it, but Mr. and Mrs. Gunn[2] are

[1] Either the *Synopsis of the Birds of Australia and the Adjacent Islands*, 1837–8, 72 plates, or the first of the seven folio volumes of *The Birds of Australia*, 1840–8, 601 plates.

[2] Ronald Campbell Gunn (1808–81) emigrated to Tasmania in 1839, becoming superintendent of convict prisons and a police magistrate. A keen naturalist, he opened a correspondence with Sir W. Hooker and Lindley, exchanging plants for books and scientific apparatus, and sending zoological collections to J. E. Gray at the British Museum. [The *D.N.B.* wrongly names him Robert.]

the only persons I have had to take leave of in Hobart Town. Except the officers of the 51st I know no other persons here, and they appreciate me much more than if I had been gay, they are a set of excellent fellows,—the best regiment I ever saw.

In the same letter comes a reference to one Jorgen Jorgensen, about whom Sir William had bidden him make enquiry. Jorgensen's special connection with the Hookers began with the fact that on the way back from a famous journey to Iceland, an account of which is given later, he had saved Sir William from perishing on a burning ship.

Jorgen Jorgensen had nearly slipped my mind. I have seen him once or twice, but he is quite incorrigible; his drunken wife has died and left a more drunken widower; he was always in that state when I saw him, and used to *cry* about you. I have consulted several persons, who have shown him kindness, about him, and have offered money and everything, but he is irreclaimable; telling the truth with him is quite an effort. When once openly employed by his friends against some bush-rangers, he was at the same time betraying his employers. He wrote to me asking me to lend him your 'Tour in Iceland'; Mr. Gunn was luckily present and told me that he had had a copy lent him many months ago and still not returned. He lives entirely at the Tap, where he picks up a livelihood by practising as a sort of Hedge lawyer, drawing out petitions, etc.

It would be unpardonable to withhold an account of this meteoric personage, which is to be found in Appendix A.

All were sorry to leave Hobart Town, where, as Hooker tells his cousin, Mrs. Fleming (Jane Palgrave),

we were treated with the utmost kindness by the inhabitants, who received us like brothers and gave us balls and parties innumerable; indeed nothing could exceed the attentions paid to us; they rivalled one another in loading us with their favours. The Governor's house was open to us, and he gave all the ship's company vegetables from his garden every day, with fruit for the officers. . . . All this was, however, too good to last, and when the time came to leave, there were many bitter regrets, especially when we thought that the Yankees and French had made fine discoveries to the Southward a few months before, and that we were looked up to as about to eclipse all other nations, and that it remained to be proved whether we deserved their kind treatment or not; this was, however, a spur to us all, and we sailed down the Derwent bent upon doing our utmost.

The first voyage lasted, as has been said, from November 12, 1840, to April 6, 1841. The three weeks from November 20 to December 12 were spent on the Lord Auckland Islands, where the long term-day magnetic observations were made and Hooker reaped a rich botanical harvest, as also at Campbell Islands, December 13–17, while New Year's Day brought the first sight of the ice. This time they got through the pack ice, a stretch of 200 miles, in four days, more fortunate than in the next season further to the east, when the pack stretched 800 miles and held them forty-seven days. As a rule, the great expanse of ice quieted the waves, and Hooker welcomed these periods of comparative calm for his microscopic work or drawing; but a hurricane in the pack, hurling the masses of ice about like huge missiles, such as lasted for three days on the second voyage, smashed bowsprits and rudders and would have sent any other ships to the bottom. The weather was nearly always bad; the reader of Ross's voyage counts eleven storms punctuating the incessant chronicle of thick weather, fog, snow squalls, high winds and seas, after two months of which February 18 brings the grateful record of the first night on which stars were visible.

Of this journey he writes to his father after returning to Tasmania, on April 8, and August 24, 1841.

Hobart Town, Van Diemen's Land: April 8, 1841.

MY DEAR FATHER,—Yesterday at 4 P.M. we anchored at our old station opposite the Paddock, and accordingly I hasten to have this letter ready to send you by the first opportunity, which will be in a few days. We have indeed had a most glorious and successful cruise to the southward, and seen many wonders hitherto quite unexpected, though it has been very unprolific. We reached 78° 3′ S. Latitude and approached as near to the S. Magnetic Pole as was possible, within 150 miles, having laid down its position with perfect accuracy from observations made to the N.W. and S.W. of its position. We have run along and roughly surveyed an enormous tract of land extending from 72° to 79° S. Latitude; every part of it further south than any hitherto discovered land, and our progress was finally arrested by a stupendous barrier of ice running 300 miles E. and W. I shall, however, give you a list of our positions every day at noon since leaving V.D. Land, last, that Maria may lay it down in your S. Polar chart and I shall add a small chart of the coast we have seen. (P.S. I have too much to say to leave room.)

And now as regards the object of the expedition, it is certainly a failure, our intention having been to have made observations on the actual site of the S. Magnetic Pole, and also to have wintered within the Antarctic Circle, that we might have made a series of experiments with such instruments as must be used on land—from the first object we were deterred by the Pole's lying inland, among a stupendous range of mountains covered from their tops to the sea beach with everlasting snow and ice. Nor can we anywhere approach the mainland as the sea is covered with streams of ice and sometimes extending in one continued line for many miles. In approaching such a coast the danger arises from the chances of a shift of wind, or a gale which would prevent our working off, when all the ice would set down on us and jam us; or, what is quite as bad, we might be becalmed and frozen in, for the sun here has no power to melt the ice even in the height of summer; wintering in such a Latitude Captain Ross pronounced as totally impracticable, as we should be frozen in, and only get out when a current should take the pack, which would imbed us, north, and melt it in warmer water.[1]

[1] As he further explains to his father (Nov. 25, 1842) who had been told by the Admiralty that they were then to winter in the ice, perhaps in order to keep some term days in the South Shetlands:—'We cannot remain in the pack except under sail, for the S.W. wind would gradually blow us out of it, . . . and it is idle to suppose that an accessible harbour could be found where the ice and snow are perennial. There is no great winter cold to shut us in safely, in a few days, or summer's heat to thaw it.'

All the polar voyagers were astonished beyond measure at the stupendous masses of ice, and their singularly regular figure; they are all square or oblong squares generally about 60 to 100 feet out of the water, and of course seven times that below, its $\frac{7}{8}$ being always under water, they are all formed along the coast and drifted north from it,—84 have at one time been counted from the mast head, of all sizes, from $\frac{1}{4}$ mile to 6 miles long; this was in about 70° South. The whole of the land surveyed from 72° to 79° presented the appearance of range upon range of peaked mountains, covered everywhere with snow, except where the precipices were too perpendicular for it to lie, and these are exposed to constant disintegration from the masses of snow rolling from above down their faces, and sweeping huge masses on to the Icebergs below, which when they are removed from the coast by a gale, transport these erratic boulders. All the coast of one of the Islands we landed on, is lined with masses of ice covered more or less with sand, stones and rocks. In such situations it is impossible for plants to grow, and I add that during the whole time that we were within the Circle, the Thermometer never rose above 32° and very rarely so high, you will not be surprised at this; on board the ship its average range was 18°–24°, never lower than 12°, of course ashore it must be much colder. The sun is very powerless here; at 75° North the sun in summer raises the mercury in a black bulb Therm. to 100° and upwards, but here only to 42°. The sea is equally unproductive, its temperature 29°, and 28° is the freezing point of sea water. When near the shore, I have always been looking for some trace of vegetation in the sea, but now I am perfectly convinced that in this longitude vegetation does not enter the Circle. Emerald Island, off which we passed some seaweed, is probably the Southern limit.

The success of the Expedition in Geographical discovery is really wonderful, and only shows what a little perseverance will do, for we have been in no dangerous predicaments, and have suffered no hardships whatever; there has been a sort of freemasonry among Polar voyagers to keep up the credit they have acquired as having done wonders, and accordingly, such of us as were new to the Ice, made up our minds for frost bites, and attached a most undue importance to the simple operation of boring packs, &c., which have now vanished, though I am not going to tell everybody so ; I do not here refer to travellers who do indeed undergo unheard of hardships, but to voyagers who have a snug ship, a little knowledge of the Ice, and due caution is all that is required. At one time we thought we were really going on to the true South Pole, when we were brought up by the land turning from S. to E., where there was a fine Volcano spouting fire and smoke in 79° S., covered all over with eternal snow, except just round the crater where the heat had melted it off. I can give you no idea of the glorious views we have here, they are stupendous and imposing, especially when there was any fine weather, with the sun never setting, among huge bergs, the water and sky both as blue, or rather more intensely blue than I hăve ever seen it in the Tropics, and all the coast one mass of beautiful peaks of snow, and when the sun gets low they reflect the most brilliant tints of gold and yellow and scarlet, and then to see the dark cloud of smoke tinged with flame rising from the Volcano in one column, one side jet black and the other reflecting the colors of the sun, turning off at a right angle by some current of wind and extending many miles to leeward ; it is a sight far exceeding anything I could imagine and which is very much heightened by the idea that we have penetrated far farther than was once thought practicable, and there is a sort of awe that steals over us all in considering our own total insignificance and helplessness. Everything beyond what we see is enveloped in a mystery reserved for future voyagers to fathom.

But you are all this time wondering what are the fruits of this Expedition to me especially. During our stay at Lord Auckland's group I made a collection of plants with which I hope you will be pleased, among them were two tree ferns, and many new species. I have accompanied them with as full notes as I could, especially relating to geographical position ; there are some most remarkable new genera, and I think a new Nat. Ord. among them. . . .

All my time when we have had fine weather to the S. has been taken up in examining them, and I fully think that Mr. Brown will be much pleased with the notes and drawings, which are numerous ; they must, however, be judged very leniently. I have endeavoured to be careful, and when the motion of the ship is such that my things have to be lashed to the table and I have to balance myself to examine anything under the microscope I fear many errors have crept in. . . .

To his Father

August 24, 1841.

Much do I wish that I had opportunity to devote myself entirely to collecting plants and studying them, but I want you to know how I am situated, that we are comparatively seldom off the sea, and then in the most unpropitious seasons for travelling or collecting. This is my main reason for devoting my time to the Crustaceae, &c., a study to which I am not attached, and which I have no intention of sticking to. My other reasons are that there is no one else to study what there will be no other opportunity in all probability of seeing alive, and the ready use of the pencil is indispensable to the subject. Again, the discoveries we have hitherto made are not only beautiful but most wonderful, curious and novel. The collection is almost all of my own making and Capt. Ross's (altogether indeed). No other vessel or collector can ever enjoy the opportunities of constant sounding and dredging and the use of the Towing-net that we do, nor is it probable that any future collector will have a Captain so devoted to the cause of Marine zoology, and so constantly on the alert to snatch the most trifling opportunities of adding to the collection, and lastly, it is my only means of improving the expedition much to my own advantage (as far as fame goes) or to the public, for whom I am bound to use my best endeavours. I again repeat that I have no intention of prosecuting the study further than I think myself in duty bound. In harbour I only collect them with Seaweeds, and never draw or do anything but stow them away ; and as for [when I am] at sea, I hope the notes and drawings I sent home will show that I do not neglect Botany, nay, that I have spent as much time, as the heavy seas and bad weather of 70° S., would allow me to plants and mosses. All this renders me most anxious to see the termination of the voyage, for I have no wish but to continue at Plants. Not that I am anything but extremely comfortable here both in my mess, the cabin and the ship. My only regret is that the necessarily altered course and prospects of the voyage stand so much in the way of Botany. The utter desolation of 70° South could never have been expected, and Capt. Ross as fully expected to winter, and collect plants in spring and leave the ice for good and all as I did, as also that we should be able anywhere to land and collect as in the North. It cannot be helped now, we must again return to the Southward, and I shall be again employed alternately collecting sea animals, examining plants and sketching coast views. I shall, however, never regret having gone the voyage, for I doubt not we shall enjoy the thanks and praise of our countrymen for what we have done. No pains has been spared to render the voyage serviceable, we have done our best, and Capt. Ross's perseverance has been put to the most severe test in penetrating as far as he has, and for my own part I am willing to work night and day, as I have done, to make accurate sketches of the products of our labors. To me it will be always a satisfaction to know that I have done according to my poor abilities, and if I cannot please Botanists I am not therefore to be idle when I may do some good to zoology. Could I with honor leave the expedition here, I would at once and send home my plants for sale as I collected them, but now my hope and earnest wish is to be able on my return home to devote my time solely to Botany and to that end the sooner we get back the better for me. My habits are not expensive, but should I not be able to live at home with you, I would have no objection to follow Gardner's[1] steps and gain an honorable livelihood by the sale of specimens.

It is well worth setting down another and quite unlooked for impression of these scenes, for some of the most curiously

[1] George Gardner (1812–49) was a Glasgow man who studied under Sir W. J. Hooker. His botanical journey to Brazil in 1836 was made possible through Sir William, who helped him to secure a number of subscribers, including the Duke of Bedford, for the plants he might collect. He returned in 1841 with a vast collection, an enumeration of which he published, as well as accounts of new species, and a paper on the connection of Climate and Vegetation. His full account of his travels appeared in 1846. In 1844 he was appointed Superintendent of the Ceylon Botanical Garden, where his active career was cut short by apoplexy, March 10, 1849. The vacant post was offered to Hooker, but refused by him.

effective descriptions of moving incidents come from simple, unlettered souls. They do not reflect upon the nice choice of words. The occasion makes the artist. They feel strongly if they feel at all, and their feeling bursts out in the first natural expressions of a forcible if limited and ungrammatical vocabulary. Such an 'inglorious Milton' was the blacksmith of the *Erebus*, a lively Irishman named Cornelius Sullivan. He first wrote down an account of their joint adventures on the second voyage from the dictation of his friend, James Savage, a seaman who had joined the ship at Tasmania. But this half story was obviously inadequate. He was moved to add the wonders of the first voyage.

> My friend James [his exordium runs], before i begin to give you anything Like a correct acct. of our dangers and discoveries, it is but justice to this My first voyage to the South, to give you an acct. of our Discoveries, before you joined the Expedition—this is the most Sublime but not the most dangerous.

With a sailor's eye on the weather and a poet's eye on its pictorial effects he tells us:

> Janry the 11th at two oclock on Monday Morning, we discoverd Victoria Land the Morning was beautiful and clear. at 7 oclock in the afternoon we were under the Lee of the land, sounded in 250 fathoms of water—not a cloud to be seen in the firmament, but what lingered on the mountains —Large floating Islands of ice in all directions. Hills vallies and Low Land all covered with snow. The snow topd. mountains Majestically Rising above the Clouds. The pinguins Gamboling in the water the reflection of the Sun and the Brilliancy of the firmament Made the Rare Sight an interesting view.
>
> That night we Stood out from the land, we did not Loose sight of it for the Sun was high above the Horizon at midnight as it would be in England on a christmas day.
>
> While we were in these distant Regions we had no night I mean dark.
>
> 12th Do. Captn. Ross went on Shore he took possession of the Land without opposition In the name of Queen

> Victoria—hoisted the British Colours Gave the Boats Crew an allowance of Grog with three hearty cheers for Old England.

The set phrase for taking possession is delightful where the only opposition could have come from the curious crowds of penguins, martial in looks but mild in behaviour, for :

> The Species of Penguins amphibious Little Creatures were so thick the Captn. Could not enumerate them, But the beach was Literally coverd. with them.
>
> At 12 oclock the Captain Come on Board we made all Sail Running by the Land to the Eastward Blowing very hard and Still Keeping out to Sea to avoid Danger.
>
> On the 13th we made Mount Sabrina [a poetic lapse for Mt. Sabine. Doubtless he had no more acquaintance with 'Comus' than with the learned and gallant Secretary of the Royal Society, but at all events the shipping list gave him the name, for the *Sabrina* was one of Weddell's little boats when he made the previous record for 'farthest South'] here is a Phenomena. This splendid mountain Rising Gradually from the Sea Shore to the Enormous height of Sixteen thousand Eight hundred and ninety feet high. I Could compare it to nothing Else but the Speir of a church drawn out to a regular taper point. Protruding through the Clouds. But beyond this as far as the Eyes Could Carry the object Seemed more Interesting.
>
> My friend if i could only view and Study the Sublimety of nature—But Lo i had to pull the brails.

The prose of life has a most unhappy way of obtruding itself, especially on board a sailing-ship in dangerous waters. But though his interrupted musings could never be wholly satisfied, he picked up scraps of knowledge as he went along and moreover added reflections of his own.

> This noble battery of Ice which fortifyd. the Land two hundred feet high. And floating islands in all directions this Strange Scenery was Remarkably Striking and Grand. The bold masses of Ice that walld. in the Land, the romantic gulf of the mountains as they glitter in the Sun Rendered this Scene Quite Enchanting. this Mountain is most perpendicular

mountain in the world—we have Seen it at night a hundred and fifty miles Distant.

We shapd. our Coarce a Long the land to the South East a Distance of two hundred miles farther. On the 28th we discoverd. Mount Erebus this splendid Burning Mountain Was truly an imposing Sight.

The height of this mountain Six thousand feet hight with a gradual ascent from the Sea Shore. From the Summit of this mountain issues Continually Vast Clouds of Smoke when Scatterd. about with the wind forms a Cloudy Surface of Smoke a long the Surface of the mountain.

At the west End of Mount Erebus it plainly appears there has been a Desperate Eruption from the Craggy appearance —it is Sufficient to Convince an accute observer.

The south side of this Splendid mountain was Lost to our view, Land and Ice obstructed the Scene. We did not land here nor did we deem it Safe to Land neither; we could not see fire nor matter, the Sun Shone so brilliant on the Ice and Snow it completely Dazzled our Eyes. Yet it is my firm belief that this must be an imposing sight in the dark of winter.

As to the Barrier, 'or as I should call it nature's handiwork,' what could be more impressive than the artless record of how the plain sailors were struck dumb by the wonderful sight, and, if we read between the lines, felt that they need not be ashamed of their emotion when their experienced Captain himself was equally touched? Thinking from the masthead view that they would 'run down' the barrier by midnight, they set sail in the evening.

But as far and as fast as we run the Barrier apperd. the Same Shape and form as it did when we left the mountain. We pursued a South Easterly Cource for the distance of three hundred miles But the Barrier appeard. the Same as when we Left the Land. On the first of Febry. we stood away from the Barrier For five or six days and came up to it again farther East, on the morning of the eight Do. we found our Selves Enclosed in a beautiful bay of the barrier.

All hands when they Came on Deck to view this the most rare and magnificent Sight that Ever the human eye witnessed Since the world was created actually Stood Motionless for Several Seconds before he Could Speak to the man next to him.

Beholding with Silent Surprize the great and wonderful works of nature in this position we had an opportunity to discern the barrier in its Splendid position. Then i wishd. i was an artist or a draughtsman instead of a blacksmith and Armourer We Set a Side all thoughts of mount Erebus And Victoria's Land to bear in mind the more Imaginative thoughts of this rare phenomena that was lost to human view

In Gone by Ages.

When Captn. Ross Came on deck he was Equally Surprizd. to See the Beautiful Sight Though being in the north Arctic Regions one half of his life he never see any ice in Arctic Seas to be Compard. to the Barrier. So that the South Pole must be degrees colder than the North pole is evident from the Enormous thickness of the ice. An Ice island floats on the water with $\frac{7}{8}$ under water. consequently the ice islands we have Seen two hundred feet hight above the Surface of the water must be Sixteen hundred feet high. That is exactly four times than the Cross of St. Paul's Cathedral in London. To view an iceberg when the Sun shines clear on it for any time is very injurious to the Eyes for the Avalanches in the Ice presents a deep blue and greenish hue. From a concussion of air that generally casts a dimness on the Sight and leaves the object the greatest Source of wonder and admiration. It would take a man of Talents to describe this unequal Sight For no imaginative Power can convey an adequate idea of the Resplendant Sublimity of the Antarctic Ice wall. It is quite Certain and out of Doubt that from the seventy eight Degree to pole must be one Solid continent of Ice and Snow. The Fragments as i call the floating Islands though Large Enough to build London on their Summit must through a Long Succession of years have parted from the Barrier they could never accumulate to such Enormous hight otherwise. Some bergs from one mile Sqre to ten miles and Some Larger but i could not ascertain the sqre of them.

A lighter scene emerges on the return to Hobart Town, April 7 to July 7, 1841. While the ships were cleaned and refitted, all the officers' time was not devoted to science. Hobart redoubled its welcome to the successful explorers; 'our arrival was hailed with delight by the inhabitants. Invitations of all sorts were poured in upon us for riding, hunting, and shooting. The Theatre invented a Melodrama, and a Panorama showed us all off on the ice.'

In return June 1 saw a grand ball given on board. The *Erebus* and *Terror* were lashed together, the decks roofed in, a covered way run to the shore over a bridge of boats to meet a direct road cut through the woods for 300 yards by Sir John Franklin, decorations and supper on a lavish scale, the whole paid for—and it cost a pretty penny—by a contribution of so many days' pay by each officer. Mrs. Fleming, in a letter written a few months later, receives some description of the frolics, which were kept up till 8 next morning, when the hosts were

> left to the misery of seeing the broken supper, the lamps taken down, and the horrid contrast which twelve hours always produces on such scenes.
>
> The lower deck was shut up and the Captain's cabin fitted with mirrors, brushes, and combs, &c., &c., and all the little nick-nacks you ladies use at toilet, and maid servants from Govt. House to match. Parties of us were stationed at the gangways to show the ladies below, and it was great fun to wait for a lady and gentleman coming along the passage and the moment she emerged into the blaze of light, offer an arm which she of course accepts, and lead her to where the maid servants are, through the crowd, while her poor husband, brother or father stared about him and asks for his partner.
>
> . . . We were lionized beyond anything, and the glorious First of June is to be noted hereafter in the Van Diemen's Land Almanacks as the day on which the most splendid entertainment the Tasmanians ever witnessed was given.

It may be imagined that, as a consequence, many hearts were lost to the ladies of Hobart; indeed, 'two of our officers are engaged in the colony and shall return thither, as soon as we are paid off, to fulfil the contract, or as we tell them, *victimize themselves.* (Don't you look black now.)'

And Lieutenant Dayman, who remained here to manage the magnetic observatory, writes Hooker at Sydney a good deal of chaff about their shipmates, who had had the field to themselves before H.M.S. *Favourite* arrived: 'The Favourites say, if they speak to a girl, they are told she is engaged to one of the "diskivery officers."' But he has no shaft to let fly at his friend; he cannot recall any 'particular admiration' of his to give news of; 'I suppose you are something like myself, a general admirer of the fair sex.'

From Tasmania a short visit was paid to Sydney in connection with the magnetic observatory, lasting from July 7 to August 5, 1841. Sydney in those days, only one year since the importation of convicts had ceased, could boast no shops finer than the Hobart Town ones; round the beautiful harbour stood a few fine houses, in particular the new Government House, still uninhabited, built in the Elizabethan style, the new Custom House and Mr. McLeay's house with its garden full of interesting plants.

The town itself lay in a hollow; its long streets ending at The Cove in dirty wharves where Hooker was nearly drowned in the pitchy darkness one night. It showed some fine buildings of a reddish sandstone; but more were dirty and insignificant, public-houses predominating. George Street was disfigured by the dead wall round the large barracks; the architecture of the churches displayed a sad lack of taste. The streets were lighted by gas and patrolled by abundance of constables at night to keep the peace; but though broad they were ill paved and muddy in the rain. Between the actual town and a wildness as of the far west there were hardly any houses; not even a public-house, such as abounded within; it was a city without suburbs. A few gentlemen's houses were scattered up and down the bay, but no snug cottagers' or farmers' dwellings were to be seen, nor smiling cornfields. An ill-kept Irish hovel on the north shore had no parallel in Tasmania.

Colonial unconventionality is measured by the use of tobacco: 'smoking along the street seems very much practised, to such an extent that notices are often to be seen prohibiting the practice in places where no one in England would think of using the weed.'

The newly arrived emigrants had visionary ideas of their future, as Hooker had occasion to learn when returning one day from a botanising walk.

> About half past five it began to pour with rain, and with a load of plants we were glad to take refuge in the New York tavern, the parlour of which was filled with lady emigrants (from the ship *Queen Victoria*). While drinking our beer we were much amused listening to their conversation. They were apparently of the middle class of English farmers, Yorkshire from their speech. In their delight at being emancipated from the ship, they dreamed of nothing but comforts to await them up the country, and seemed to think that their hardships were over; one talked of having a nice house, with a verandah, on a hill near the water, with a garden, &c.; and really her husband must provide her such a one. Little did she think that she will perhaps have to spend two years in a mud hovel, with a marsh before the door and the bush for a verandah. Another congratulated herself on the prospect of making herself useful by knitting mosquito nets for her father; if in three months' time she is making onion nets, or seines for a neighbouring lagoon, it will be perhaps the highest part of her daily toil. Generally speaking, the young men were smoking cigars and drinking hot or cold grog; one talked of going to a billiard table and another of the theatre, after having spent the day going about to milliners' shops with their consorts. What this colony holds out for a settler I do not know, but to me these seemed a most mistaken set of people in their ideas of future comfort or happiness. . . .
>
> It soon ceased raining and we started off through the town and government domain for the ships, splashing through the mud at every step, while the little urchins compared us carrying our grass trees to Moses among the bulrushes.

The Mr. McLeay here mentioned had lately been Colonial Secretary and was soon afterwards knighted (see p. 9); and his son William (already referred to, p. 84) was a naturalist of some mark. To them Hooker had an introduction from his father, and received a warm welcome. Twice the naturalist came on board the *Erebus* and spent all day looking over the Southern collections. 'He is delighted,' Hooker writes to his father on July 18, 'with my drawings of sea animals, of which many are entirely new ; I must, however, redouble my efforts on that head, little as I care about them, as I hear that the Americans have done much during their voyage to them, and that, McLeay says, is the only thing they have done.'

On the way to Sydney 'the tow-net produced some new and good things for the pencil, and we actually brought up several live animals from a depth of 400 fathoms ! Lat. 33° 32′ and long. 167° 40′, but no trace of vegetable life.'

The presence of living corals at such great depths was pronounced very remarkable.[1] Some of the shells Captain King recognised as South American, especially the small yellow bivalve from the *Macrocystis* (the seaweed found floating far to the south, thousands of miles from the American coast).

Among the Auckland Island sea animals, he marked 'a Galathea very like an Arctic one,' while 'a curious animal from Kerguelen's Land approaches more nearly to the fossil Trilobites than any hitherto discovered, the antennae being apparently wanting, and the eyes are as in the fossil *Entomostraca*.'

McLeay was full of stories of Dr. Buckland and his blue bag ; but only one is recorded in the Journal. 'Dr. Buckland could tell the age of a skull by the taste, which he proved by producing that of an old woman buried a few years before, which tasted greasy, &c. &c.'

A long visit to McLeay's garden proved it to be a botanist's paradise. 'My surprise was unbounded at the natural beauties of the spot, the inimitable taste with which the grounds were laid out, and the number and rarity of the plants which were collected together.'

[1] On Sept. 1, 1845, Hooker writes to Ross : 'I read in the *Ann. Nat. Hist.* a notice of Goodsir's labours with Sir J. Franklin. He seems to be doing remarkably well, as the notice said that 300 fms. was greater dredgings than had ever been obtained before. I wrote an answer to the Editor, saying we had repeatedly dredged at that and at greater depths, giving a few general remarks as proofs.'

The interior of the house, a striking specimen of Colonial architecture, the individual trees and creepers, flowers and shrubs, the revival of nature when the rain ceased and 'a few insects came out, the Diamond birds flitted from tree to tree and the large Sea Eagle or Osprey left his lonely lair and commenced wheeling over the calm waters of the bay,' and beyond the bay 'a rocky precipice christened Sunium, on which it is the intention to build a temple'—all this is fully set forth in the Journal with one very homely touch as to 'Mr. William's workshop':

> The smell of camphor and specimens, so well known to me at home, reminded me strongly of olden times, especially as I found everything in the inimitable mixture of confusion and order in which Mr. Brown's shop at the Museum and his rooms in Deane Street are wont to be.

> (To his Father, August 25, 1842.)—McLeay has promised to collect for me in New Holland, and knowing him as we do, when one thinks that hardly a dozen mosses have been described from that vast country, there can be no bounds to the novelties he may fall in with. He was quite delighted when I showed him the *Sclotheimia Brownii* growing on rocks near his house, and the Dawsonia amongst some roots he had brought from the forests of the interior. He seemed rather cautious about broaching his Quinary system, and I was rather anxious to hear how he thought it would apply to the higher orders of plants. The circular system no doubt holds among the Cryptogamiae, Fries having proved it with regard to Fungi, and Berkeley seems to incline the same way.[1]

The record of the visit ends with the entry for August 5: 'At 11 A.M. sailing down Port Jackson along the cold-looking sandstone cliffs, leaving Sydney with few regrets but leaving Mr. McLeay's fine establishment where there was much to see.'

[1] 'As to McLeay's theory, I fairly worked myself out of that error by the mosses, which I first arranged to please McLeay himself.' (To Harvey, June 8, 1845. Cp. p. 84.)

CHAPTER VI

SOUTH AGAIN: NEW ZEALAND AND THE CAPE

In ten days they made the Three Kings' Islands, and on August 16 entered the Bay of Islands, New Zealand. Here the ships stayed till November 17. New Zealand was still regarded by many who had spent years there as hopeless for colonisation. 'Colonists,' wrote Dr. Sinclair sweepingly, 'had nothing to do except they put themselves on a par with the natives and breed pigs, cultivate potatoes on the sides of hills and perhaps turn savages.' To a botanist, however, it was fascinating. Hooker, under the guidance of Mr. Colenso,[1] the printer to the missionary establishment, and himself a keen botanist, made a number of excursions into the country, though it was all too swampy to go far, collecting many specimens, especially of the Cryptogams, for the Bay of Islands was otherwise a comparatively well-known centre.

From New Zealand, on November 23, 1841, the ships set out for their second voyage to the South, sailing on a more easterly meridian in order to reach the Great Ice Barrier at the point where they had been compelled to turn back the

[1] William Colenso (1811–99). He was born at Penzance, and was a cousin to the late Bishop Colenso of Natal. As a youth he was apprenticed to a printer of Penzance, and later was employed by the British and Foreign Bible Society in the same capacity. The Society sent him to New Zealand in 1834 with the first printing press established there. In 1844 he became a missionary, and after training at St. John's Coll., Auckland, was ordained to a church in Napier, where he lived till his death. His botanical writings, though numerous, are fragmentary and are chiefly contributions to the Tasmanian *Journal of Natural Science* and of the New Zealand Institute, &c. For sixty years he collected information regarding the language, customs, songs, &c. of the Maori. F.L.S. 1865 and F.R.S. in 1886. Sir Joseph named the genus *Colensoa* after him.

previous season. Turning south at long. 146° W., where little ice had been met by previous navigators, they found the line followed by Cook in 1774 and entered the pack on December 18. But the experience of one year is not that of another. The pack ice extended 800 miles. For forty-six days they struggled with the ice before getting clear of it. The weather was much worse than on the former voyage. On January 19 a terrific storm dashed them about in the ice for twenty-eight hours. Huge waves hurled masses of ice against the ships like battering-rams. The *Erebus's* rudder was damaged. But so well were the ships strengthened against the ice, so closely were their holds stowed, making the hulls a solid mass from side to side, that to Ross's delight and surprise they suffered no further damage. Repairs were difficult, the workers being drenched for hours by the icy water; but within four days the crippled ships were repaired, Captain Ross permitting this work of necessity to be performed on the Sabbath day, as indeed he did again after the collision in the following March.

Escape from the pack was as perilous as remaining in it. On the evening of February 1, clear sea came in sight, but the long westerly swell raised 'a fearful line of foaming breakers' on the pack edge, menacing them through the gathering darkness, an equal danger whether the wind fell or increased to a storm as it threatened to do. The only course was to take the immediate hazard. Two hours' battling with the waves, shotted, as it were, with blocks of ice, brought them into safety, with the loss of part of the *Erebus's* stem. It was worse on board the *Terror*, for there fire had broken out, some blocks of wood having been left too near the hot air stove, and it was only extinguished by flooding the hold two feet deep.

After these dangers, the troubles arising from the looser floating ice were of less account, until, more than a fortnight later, the floes were dispersed by a couple of storms. Then on February 23 the Great Barrier was reached, six miles further south and ten further east than the previous year. From this point it trended N.E. as they followed it for twenty-four hours, till compelled by the approach of winter to turn north and then east again, through the endless floes, making for the Falkland Islands, which lie to the east of Cape Horn.

But this was not the last of their adventures. They had recrossed the Antarctic Circle and hoped to have got clear of ice, when at midday on March 12, 1842, in the midst of a fierce storm, a great berg appeared ahead, and in trying to weather it the *Terror* collided with the *Erebus*, carrying away her bowsprit and foretop-mast. For nearly ten minutes the two ships lay interlocked, drifting down upon the berg and the breakers, each ship, as it rose on the great waves, threatening to send the other to the bottom. Breaking at last from this disastrous embrace, the *Terror* was seen to run before the wind and disappear beyond the lee end of the berg. The *Erebus*, disabled by fallen spars, was drifted down on the berg. For three-quarters of an hour she lay among the breakers, striking her masts against the berg as she rolled, and lashed by the spray falling back from the ice cliffs. But perfect discipline was maintained. At last the hamper was cleared, the mainsails were loosed, and the ship slowly crept from her perilous position by the desperate expedient of a 'sternboard,' i.e. sailing stern foremost down wind, her yardarms scraping along the berg, from which she was only held off by the strength of the undertow. Clearing the berg, they found themselves running upon another, the passage between being but thrice the ship's breadth. It took all the Captain's skill and all the crew's steadiness to get the ship's head round into the channel. Once through, however, they were safe in smooth water under the lee of the berg, and there, to the great relief of all, found the *Terror* awaiting them in anxious suspense.

Next morning, viewing the long line of bergs that showed this sole passage of escape, Captain Ross was inclined to regard the collision as a blessing of Providence, albeit somewhat rudely administered. It had turned them sharply off their original course, which would have spelt worse disaster, to the only practicable place of escape. The sailors were indefatigable. In three days, as they ran before the wind, repairs were effected, and the Falklands, between 2000 and 3000 miles away, were reached on April 6, 1842, 'the first land of any description that has greeted our eyes now for 135 days,' the more grateful because here at length they were told 'that our late success (the first visit to the ice) caused an immense sensation of triumph in England! These are the first flattering words we have received from home; nor can you conceive how welcome is the news, having penetrated beyond even our former *Ultima Thule of Latitude*.'

His own views as to the nature of the Barrier, and of the pack ice of the Antarctic, especially as bearing on the prospects of the third voyage to the South, appear in a letter to his father, dated November 25, 1842.

> All the Ice in the Antarctic Ocean is formed by the gradual accumulation of Snow, on small pieces of Ice which only dissolve by being drifted to warmer latitudes. The Icebergs are probably the accumulation of centuries. These bergs are stranded all along the coast. The Barrier is probably only a large solid pack filling up a broad shallow bight, like that of Benin or S. Australia. Some unusual severe winter, ages ago, first filled it with a sheet of Ice, and as the snow fell it sunk deeper and deeper every year till it stranded; the sun has no power on it now, and so every snow shower must add to its height. What atmospheric changes the revolutions of centuries may produce we cannot know; but whilst the climate of the South is so equable and the removal of the ice by drifting probably proportioned to its slow drifting accumulation to the South of the Packs, these vast phenomena must remain comparatively unchanged. The Barrier, the bergs several hundred feet high and 1–6 miles long, and the Mts. of the great Antarctic continent, are too grand to be imagined, and almost too stupendous to be carried in the memory. With regard to the prospects of this coming cruise, I am anything but sanguine of great success. The past winter has been a very bad one indeed, and further we know that though the sea was clear of ice when Weddell went down, there was ice when the two French and the Yankee expeditions attempted this Longitude; whether they tried to get through it boldly

or no is not to the purpose; there is no doubt it existed. My opinion is that the Packs shift slowly, and that a place open for one season may be shut for many successive ones. I have heard that an English Lieut. called Rea, or Wray, went down in a sealer, and met the Pack in 60°. Now, though I sincerely hope to make the Pack and get through it, rather even than meet no ice, still we twice have been entirely successful, and it is humanly possible that ships can always penetrate at whatever point they take the pack. A little more ice last year would infallibly have stopped us had it detained us a few weeks more. I would give up all my pay to be sure of gaining 78° again, for the French and Yankees will surely laugh if we are foiled in any one attempt. Should we find much ice we shall be a long time in it doing our endeavours to get South: they are fine times for me, as the smooth water sailing is quite delightful, and it is a great comfort to know that, if we cannot get on, we can always go back with the S.W. winds and the drift of the ice. Should we fail we shall all feel it deeply and almost wish to be allowed to try again. It shall not, however, be our faults if we do fail, it may be our misfortune and a very sad one. None of us despair of success in beating the French and Yankees; but it is *ourselves* we want to beat, and thus we are our own enemies.

At the Falklands they stayed five months (April 6 to September 8) and later another month, November 13 to December 17, before the third and last trip to the ice, the intervening two months being spent in a visit to Hermite Island, to the west of Cape Horn.

A long series of magnetic observations was carried out; for Hooker, exploration, hunting, arrangement of collections and letter writing filled up the time. Delighted though he was to 'be fast by the nose again' at Port Louis in the wet and mist of a storm that rose just too late to prevent their entrance into the Sound, first impressions of the Falklands were dismal. 'Kerguelen's Land is a paradise to it. Desolation stares in our faces, except a few houses at the settlement, where there are about sixty souls, including *His Excellency the Governor* (a Lieut. of Engineers) and some Sappers and Miners.'

The purser went ashore after nightfall in search of fresh provisions. Eager to bring Hooker some new botanical specimens, he grappled in the dark with some wayside plant; it turned out to be Shepherd's Purse! 'To-morrow I shall do something better,' is the sanguine comment.

Beef there was in plenty, and horse-flesh at need, for cattle and horses ran wild on the island, for hunting which the Governor offered the use of horses and dogs, and there were wild geese and ducks and rabits for the shooting; but no flour was to be had, nor any green thing but some turnips.

Lieutenant Moody appears to have been somewhat autocratic and not always wise as an administrator; but with natural good sense, Hooker remained on good terms with him, and avoided being drawn into other people's disputes. Moody was greatly pleased with his report on the Tussock grass, the one product of the island with commercial possibilities in it, and sent it to England as a paper to be read before the Geographical Society (November 1842). So that Sir William writes gaily of the interest in the Expedition,

> excited by some little matter which Col. Moody and I laid before the Geo. Soc. from our sons, relating to the Falkland Islands. *You* are considered (how correctly I won't say) the fortunate *discoverer* of the most wonderful Grass in the Falkland Islands, that is to make the fortune of all Highland or Irish Lairds who have bogs, for bogs—'pates' [peats] they will have it, are the proper soil for the plant. And said Bogs for hundreds of miles, where nothing has yet grown, will be clothed with such luxuriant grass as all the cattle in the world cannot keep down. You have no idea of the quantity of letters I have from strangers in all quarters, from the South coast of Kent to John o' Groat's, and from the East of Fife to the West coast of Connaught, humbly begging me, the happy Father of so renowned a son, to give them but the tythe of a fibre

of the root, or one seed; or in default of them a piece of a leaf![1]

But the disagreement of Captain and Governor had other consequences at last, as told in a letter to Sir W. Hooker (April 29, 1843):

> The Governor of the Falklands was very kind indeed to me and we were great chums; but he and our Ross quarrelled most grievously, so that I was often unpleasantly situated; but told them both, that I had nothing to do with their affairs. The worst of it was that Moody let us go to sea for the South without fresh beef, so Smith and I went and shot a bull calf and a horse, which were very good eating; we caught another horse, having run it down with the dogs, quite a little thing, and tried to keep it as a pet on board; but the little thing, which was quite fond of me, died before we got to the ice. However, keep all this to yourself, for I am going to have nothing to do with their rows.

[1] The wonderful Tussock grass, when at last raised, 'has thriven marvellously both in the Orkneys and Hebrides, having seeded abundantly and sown itself (1847),' but did not practically fulfil these glowing anticipations in the Northern hemisphere. Moreover, the first sowings of seed sent home by the Expedition baffled the botanists. This is the key to Hooker's belated satisfaction when writing to Ross in November 1844:

'I am delighted to hear that some of the old Tussac vegetated, as everyone has said that our Expedition seed *all* failed: it is quite a triumph to me, I assure you, as now the Expedition *was* the first to introduce the grass. I have eleven plants in my bedroom, growing very slowly, and there are a great many in the Garden.'

Even then it was not all plain sailing, as a subsequent note to Ross (Sept. 1, 1845) records:

'Your excellent brother's plant of Tussac flowered with us, and turned out the British *Dactylis glomerata*, to our shame and confusion at Kew, for we were sufficiently positive of its being the right thing. The fact is that we have only lately procured young plants and raised seeds of the *true* Tussac, many other things flowered before with various people but none the right. It grows exceedingly slowly and is a rigid wiry grass in its young state and will not (apparently) flower for a long time yet. Pray do not laugh immoderately at us for all this bungling, for all kinds of people, botanists, gardeners, and agriculturists have been deceived with what springs up in the pots. What we now have young plants of and raised seeds of, is not like what I should have expected Tussac to be, but as ten plants were watched sprouting from the seeds themselves and it totally differs from all other grasses, resembling the young plants received from the Falklands, we are pretty sure it will *become* the true Tussac. Enclosed are seeds which will surely germinate, but they must be watched, as lots of other things spring up in the pots. I can give you a young plant if you will tell me where to leave it in Town.'

From the botanist's point of view, the Falklands turned out better than was expected. The mosses took first place for interest; then the monocotyledons, of which he had about forty species, and he found a good many plants undescribed in De Candolle after the publication of D'Urville's lists.

He was grateful for having the run of the Governor's library.

> I often spend a day there and afterwards take on board with me any of his books that please me. Those I have been lately reading are—Pope's Homer's Iliad, Mrs. Hemans' Poems, Daniell's Chemical Philosophy and Pugin's Christian Architecture, a very miscellaneous selection, but even from the last; with all his faults and bigoted Roman Catholicism, I have gained much good. Keith's Evidence (of Prophecy) and Pollock's Course of Time I had read long before without appreciating them as I do now,—Stephens's Travels in the East pleased me much and Milner's Church History, what I have seen of it, for it is too much for me to get through here. (To Lady Hooker, August 24, 1842.)

As regards botanical books, however, he tells his father (August 25, 1842):

> It was very foolish in me to have brought so few books on Cryptogamic plants, having nothing but Loudon's[1] Encyclopædia and the miserable Sprengel[2] to help me. From knowing something of the mosses before, I can get on with them and examine them very minutely, but with the Algae and Lichens I am sadly puzzled. Your parcel to me, when it comes! will be a great catch, if it is only for the Journal, to which Berkeley no doubt still contributes.

It was better when a packet arrived from Sir William:

[1] John Claudius Loudon (1783–1843) was a famous traveller, landscape gardener, agriculturist, and horticultural writer; Fellow of the Linnean Society, 1806. His energy, despite ill health, is illustrated by the fact that at one time he was editing five monthly periodicals, from the *Gardeners' Magazine* to the *Arboretum et Fruticetum Britannicum.*

[2] Kurt Sprengel (1766–1833) was Professor of Medicine and later of Botany at Halle. His investigations greatly stimulated the microscopic anatomy of plants, though his own results, owing to inadequate means of investigation, were not always trustworthy.

Falklands: November 25, 1842.

The books you send out are capital. Lindley's Elements seems a most valuable work to me and the very one I wanted, for I have a very high opinion of him as a Nat. Order man—though he makes too many it is impossible not to admire the thorough knowledge he has of the subject; and now that a linear arrangement will never do, and Fries's Motto 'omnis ord. nat. circulum per se clausum exhibet' is daily gaining proof, Lindley's groups and alliances of plants which, like sects, are more like one another than anything else, must be invaluable. I am no judge of the goodness of this arrangement of the groups, but it is the throwing the Nat. Orders into groups and showing the dependence of one group on another which impresses me; his theory of the mosses is an eyesore to me and shows the folly of theory without practice. . . .

As to his occupations on the treeless, wind-swept island, he tells his father (May 3, 1842):

On this Island my time has been entirely devoted to Botany. . . . Every day adds something new to my collection, especially among the lower tribes. During my late excursion, I found the *Ballia Brunonii*, which I have now gathered all round the world. . . . Altogether this place is better for Botany than I expected, and but for Lichens, &c., it beats Kerguelen's Land, [though] collecting here is no sinecure, for the days are very short and the nights long.

Later he tells his mother (August 28, 1842):

The weather and state of the country, now swamped, prevents my making any excursions to a distance, though I enjoy the short walks about the bay very much and seldom go out without picking up some novelty. At present my time ashore is wholly taken up with seaweeds and marine animals, for which purpose I wander along the beach at low water with long boots on, collecting; but the wind is so cutting and the water so cold, that I often wonder whether my hands spend most of the time in the water or my pockets, whither they are wont to stray, as in days of yore.

As spring approached, even the Falklands put on a brighter face. The forthcoming visit to Hermite Island offered an attractive prospect, despite the fact that, with the equinoctial gales coming on, a long and uncomfortable passage might be expected. There is at least this consolation: 'We know from now *long* experience, that no sea can hurt such vessels as ours, which rise like tubs on the water and tumble about in the waves.'

Already he is beginning to think of the Fuegian Fagi, &c., as described in his father's 'Journal of Botany'; and correcting Webster's confusions in his account of Captain Foster's[1] voyage:

> It is, however, among the Mosses and other Cryptogams that I shall hope for novelty in the S. extremity of the American Continent. . . . You will not wonder that after spending so long a time in the Antarctic regions, I should be most anxious to complete the Botany of this desolate part of the world, by going even to the Horn, and that any new Moss or Lichen from such latitudes appears of infinitely more value to me than a new *Palm* or *Rafflesia* would to you, nor can you well conceive my delight on finding the three curious Halorageous, Portulaceous, and Crassulaceous weeds of Kerguelen's Land at the Aucklands, then Campbell's Island, and again on the Falklands—three curious forms of small Natural Orders, as strictly Antarctic as Parrya or Sieversia is Arctic.
>
> Amongst the lower orders I find it takes all my eyes to get up a tolerably complete collection, for in such dreary

[1] Henry Foster (1796–1831), navigator and surveyor. His most important voyages were with Captain Clavering and Sabine in the *Griper*, to the coasts of Greenland and Norway, after which he was elected to the Royal Society; as astronomer with Parry in his Polar expeditions of 1824–5 and 1827, when his astronomical and magnetic observations won him the Copley medal; and from 1828, when he was sent out in command of the *Chanticleer* to the South Seas to determine the ellipticity of the earth by pendulum experiments at various places, as well as to make magnetic and other observations. His work took him to the South Shetlands, and thence to St. Martin's Cove, behind Cape Horn, a spot afterwards visited by Hooker. Here he met Captain King in the *Adventure*, who was surveying the neighbouring islands. He was accidentally drowned in the Chagres River just after he had at last succeeded in measuring the difference in longitude across the Isthmus of Panama by means of rockets. The account of the voyage was written from the journal of Webster, surgeon of the *Chanticleer*.

climates, where vegetation itself is scarce, I find that everything, in however bad a state, must be taken at once and looked for, in fruit or flower, afterwards. Indeed I often wonder what can be done with the barren specimens I am forced to be content with. [From a letter to his Father, August 25, 1842.]

To his Mother

December 6, 1842.

September 8th we weighed and made sail down the Sound as I was writing a letter to Bessy. On the following day we were greeted as we expected by a stout S.W. gale, which blew almost without intermission until the 16th, during all of which time we were hove to and battened down, most delightful as you may suppose after four months in harbor. On the 16th we were eighty miles to leeward of the Falklands! when, after a short calm, Easterly wind sprang up, and as the sea went down, we ran on rapidly to the Horn. Fair winds took us on to the land; on the 19th we made it early in the morning, consisting of ranges of snowy peaks, and soon after saw the far-famed Horn. The day was beautiful and so we passed in the afternoon right under the cliff, which is quite a fine one,—very steep and precipitous to the Southward. Jagged and peaked at the top, covered with very stunted brushwood of the crumpled or deciduous leaved beech, which was brown as the leaves were not expanded yet. The cliff is of a black color and about 600 ft. high with plenty of Albatross, Cape pigeons, and other sea birds wheeling about it, indeed we were so close that we could see them sitting on the face of it. A little cairn of stones raised by the officers of the *Beagle* is on the top of all.

After rounding (or doubling) the Cape, the Bay of St. Francis opens out and the view is very fine. This bay was supposed to be in Hermite Island until that Island was found to be made up of many enclosing this sheet of water. Horn Island is the most Westerly and, as its name owns, boasts of the Cape. Hermite Island is the Easternmost and Cape Spencer, its most Southern point, is very similar to and abreast of Cape Horn (some two or three miles further North). We beat up the Bay and at night anchored in very deep water under a bluff precipice off the mouth of the Cove. When it came on dark, it was a very curious place, for we were under high black-looking mountains rising at once from the water, and we could just see their white tops glimmering through the darkness.

When the moon got up the view was beautiful, and a more extraordinary anchorage for wildness and sublimity we never lay at. In the morning the quietness of the spot and the green woods, which we had not cast eyes upon for twelve good months, was most refreshing. The little cove was so foreshortened lying amongst hills so high all round that we could hardly suppose it would afford shelter, which it did however, when we were warped about 1¾ miles up towards its head, opposite a few wigwams of the natives. The island is so narrow that we could always hear the hollow roar of the surf on its weather shores, and after one of the hard gales which were common there would be a slight swell in the cove, whose beaches were so steep as sometimes to prevent landing. All along the N. side of the Bay the Mts. are quite precipitous, with a great deal of snow on their ridges. On the South side they rise at an angle of 45 degrees up from the water, with a few cliffs here and there so straight that though the cove is very narrow the top of Kater's Peak, 1700 ft. high, is seen from the ships when in the centre. The head of the cove runs up in a broad densely wooded valley to another ridge of hills which complete the amphitheatre of mountains. Altogether the place reminded me very much of the Trossachs or the head of Loch Long contracted.[1]

The foliage being much like that of the Birch, and the steep mountain torrents keeping up a continual roar which often put me in mind of many a night spent in the Highlands. Nothing is so soothing as the sound of rushing water, and it was very delightful to lie at night in bed with the door and hatch open and hear the little cataracts roaring, however, I soon found sleep much more delightful and forgot the romance,—finally its effects were quite mesmeric (Is that the new name ?). The weather for the first few days was most beautiful, and we began to think the Horn a sadly abused and traduced place. Spring came on rapidly, the

[1] 'In grandeur, perhaps, St. Martin's Cove was little behind that favourite spot. Many things were, however, wanted to complete the picture as Scotch; perhaps, like Glen Croe, it was wild without being really beautiful, and only assumed the latter appearance to us because for eleven months we had not seen a tree.' (To Rev. James Hamilton, November 28, 1842.)

Berberry flowered with bright golden blossoms, the tufts of Misodendrons on the beeches grew quite brilliant, and the crumply leaved beech burst at every twig, emitting a delicious resinous smell. Nature was evidently taking every advantage of the fine days, and I began to think that seed-time and harvest would all be over together in one month, and could not conceive what the poor plants were to have to do during all the summer if spring was so fine. My Father's class song of Spring, all I remember of which is, 'The Larch hangs all its tassels forth,' was nothing to this. I certainly never saw anything like the sudden bound vegetation took in ten or twelve days. We arrived in winter and it was summer already. A few days more, however, changed the face of nature, and after all the Snow had disappeared, two or three hours covered everything with a white mantle and the weather continued very changeable during our whole stay. Clouds and fogs, rain and snow justified all Darwin's accurate descriptions of a dreary Fuegian summer. Indeed all Darwin's remarks are so true and so graphic wherever we go that Mr. Lyell's kind present is not only indispensable but a delightful companion and guide.

The Westerly winds which prevail seldom affect the waters of the cove, but when they are strong and gales set in with drifting clouds, snow and rain, the whole land appears savage to a degree. The force of the wind and its effects are not to be compared to Kerguelen's Land, where the steady torrents of wind came rushing down in one impetuous stream through the valley at the head of Christmas Harbour; here they dash down from the narrow gorges of the mountains, deflected from their course, and burst on the ship with a clap like thunder, tear the water up and are gone in an instant; two will sometimes meet from opposite quarters, and unfelt a few yards off, whisk up a cloud of spray and continue struggling down the Cove until, perhaps, they split and run along in two divaricating lines of foam, as far as the eye can trace them. The gusts were in no instance stronger than at Kerguelen's Land, and from their short duration do not bring a strain on the cable or cause us to drift from our moorings, but from their suddenness they were more remarkable. It was very interesting to walk the deck with hat tied on and watch these freaks of Æolus, or to see a squall or Williewaw, as they are called, strike the *Terror*, heel her over for a minute, and rush on till it met the steady gale outside, of which we felt nothing. On the hills its effects were also very remarkable, especially high up near the Gorges, where the trees which met it in its first burst would be all shattered, and lay in every direction for an acre perhaps ; these, too, are sturdy, tough, stag-headed little obstinate trees whose splintered trunks, though only a few inches (8–14) in diameter, show that their mettle is good.

The poor Fuegians of course attracted our attention before anything else, and surely they are the most degraded savages that I ever set eyes upon. They are considered as the lowest in the stage of civilisation of all nations under the sun,—the Tasmanians, now banished from that Island, alone excepted. They inhabit various scattered parts of the coast in separate tribes, said to be at war with one another. Those we saw amount to about twenty and are said to be confined to Hermite Island. They have wigwams made of nothing but a few branches arranged in the form of a beehive in the woods close to the sea,—there are two or three of them in almost every bay of the Islands, and they wander either across the hills or in their canoes from one to another. These canoes are the most useful articles they possess, though very clumsily made of the Bark of trees sewn together over a framework. The bottom is plastered with white clay, of which a supply is always kept on board to stop a leak—they take great care of their boats, and whenever they haul them up, which is the women's duty, they make a sort of road of smooth pebbles up the beach, and then cut quantities of seaweed over which they drag the boat up high and dry. Little baskets made of rushes woven together, and a drinking cup cut out of the root of a Laminaria, are the only domestic utensils,—wood ashes and clay used as a pigment and a few shells strung on seal sinews their only ornament, whilst their only weapons are a long sling and a very long spear of wood with a bone head so fitted on to the shaft that on striking a seal or penguin the shaft falls out and remains attached to the head by a piece of sinew, and thus encumbers the animal by floating. These Fuegians wear no clothing whatever either in Winter or Summer except such as are given them by us,—more apparently for ornament than comfort. The men do little or nothing except a seal or such like comes in their way, whilst the women are employed collecting limpets and mussels, which are eaten raw or half-cooked and form the largest proportion of their food; to do this the poor things have to go every day often up to their middles in water,—snow falling heavily at times, and with a young child slung to their backs. Their manners are little above the brutes, filthy and squalid to a degree, and they will eat anything but salt meat that we offered them. They are all great thieves and excellent imitators both of language and action, though they have never improved themselves permanently from their intercourse with Europeans. Their language is a most horrible, guttural concatenation of sounds and unlike the New Zealanders, whose tongue is harmonious and beautiful to the ear,—they, as I said before, imitate a sentence of any language readily, whilst few of the N. Zealanders can pronounce $\frac{1}{3}$ of the English words.

Our walks were of course confined to the Island, and there was not much of general interest to attract attention. Beginning a walk was the worst part, as one must tear through the dense wood and force a passage up the hills, —the ridges are generally bare of wood and easily walked over to some distance, but whenever the valley comes wood is sure to be packed into it. Of Mosses, Lichens, &c., there are a profusion, and the collecting them kept me constantly at work. Above the wood, however, the rocks are very bare, from the frequent heavy snow storms, which often overtook us on the hills and made the walk back very unpleasant, the wind clogging it on our persons. Nothing, however, but personal weakness, or too sudden a change, would have made Sir J. Banks feel their effects so much, for we thought nothing of it, and were it necessary, even without a fire, a shelter might be made, which with the warmth of two or three persons close together, might have defied death by cold.

Writing to Mrs. Boott, November 28, 1842, he insists further on this point.

This part of the world (Fuegia) has always borne the character of being eminently rigorous and inhospitable,—

very much because poor Sir Joseph Banks and Dr. Solander, after being accustomed to tropical heat and that hottest of harbors, Rio Janeiro, were rather suddenly cooled down here in the height of summer. The climate in winter is, however, as mild in proportion as the summers are chilly; the annual temperature is assuredly low, but the averages of that of each season are remarkably close.

Sir William was delighted with the living plants sent home from Hermite Island, and writes on March 14, 1843:

> So valuable a consignment has not been received at the Garden since we came here. The two new kinds of Beech, and these the most Southern trees in the world, are invaluable, and the Winter's Bark Tree (of the latter only one specimen was in the kingdom before) are growing beautifully.

Of the third voyage to the South Hooker wrote later to his father (March 7, 1843):

> Now that the voyage is over we are very proud of it (pride in poverty, you will remark), for we have got nothing easily. This cruise was not so hazardous as the last, being less in *Bergy* seas, nor have we been in any so extreme danger, but then as the ships cannot last for ever it becomes daily more uncomfortable on the philosophical principle of the 'Pitcher going 99 times to the well.'

Leaving the Falklands on December 17 'without one regret,' they proceded south on the meridian of 55° W., seeking for a continuation of Louis Philippe Land to the south-east. They met the pack on Christmas Day, and three days later sighted Joinville Land in the South Shetlands. Extended exploration was made and various islands discovered, while the ships were nearly wrecked on Darwin Islet. On New Year's Day, 1843, Mount Haddington was discovered; on January 5, Cockburn Island, in 64° 12′ S., of which formal possession was taken. Ross's 'Voyage' contains Hooker's special report, five pages long, of its rare vegetation.

Landing on the 'very singular crater-shaped, conical Island,' he writes, 'I procured the ghosts of eighteen Cryptogamic plants, but no Phenogamic, all very scarce indeed but one or two Lichens.' Among his finds he mentions:

> *Ulva crispa!* also I see found in Ross Islet, according to your list of Parry's plants, [and here are pencilled in the words] apparently exactly that of Europe, &c., so that unless the Red Snow of Forster should prove the real plant of Antarctic regions, this is the only plant common to both extremities of this globe, and it would be interesting to ascertain which intermediate positions it inhabited. It is probably found in Europe generally.

This voyage was like to have had an untoward interruption, if not termination, for the ships were nearly frozen in between the islands, and only escaped after six days' struggle with the ice. Another fortnight was spent in trying to pierce the main pack, when again they were nearly frozen in; but once clear of the pack, on February 4, they made for Weddell's track in long. 40° W., where earlier in the century he had found clear water as far as 74° S. But now this line was blocked by a dense pack, while the weather was unpropitious. Crossing the Antarctic Circle on March 1, next day they saw the sun unclouded, the first time for six weeks, and on the 5th turned back at 71½° S., long. 14° 15′ W., only to be overtaken by a fierce gale lasting three days, during which they were repeatedly in danger of shipwreck in the ice or of collision. On the 11th they recrossed the Antarctic Circle, as all devoutly hoped for the last time, and bore up for the Cape, which was reached on April 4, 1843.

Officers and men alike were growing weary of the prolonged voyage, and the threatened addition of a fifth year was as unwelcome as it was unusual. The fatigues and monotony of the South outweighed the solid allurements of double pay. Ross, with his keen interest in the magnetic work and his ambitions as an explorer, and Hooker, with new fields of science opening before him and his heart in his work, were, as the latter confesses, the only two who could have both pleasure and gain in a fifth year or even longer voyage. 'It is nothing to me if they keep us out *six*, except the want of seeing my friends, for I am always improving myself, and it will give me a greater claim on the scientific world.' The unscientific officers, though doing their arduous work devotedly, were buoyed up by no scientific enthusiasms, and with no chance of withdrawing honourably from the task, felt it a hardship to be kept in harness so long, having only calculated on a three years' cruise. They were being outstripped by others on active service, and the promotions that came to them in the guise of special reward were already due for length of service, while the 'Terrors' especially were nettled that when the Geographical Society gave Captain Ross their Gold Medal, no word was uttered in recognition of the officers and crews by whose labour and loyalty he had been able to push his explorations so far. And Hooker writes home of a rumour that they had wintered in the lonely Falkland Islands lest at any other port the seamen might desert rather than face another expedition to the ice. All were delighted when they learned at the Cape that they were to make their way slowly homeward by St. Helena, Ascension, and Rio.

The Admiralty rule that all collections, journals, and charts made on the voyage should be handed over to the Department, and Ross's keen desire that his account of the voyage should not be forestalled by any public leakage of news, geographical or scientific, hedged private letters round with difficulties. It was expected that finally both Hooker's Journals and his botanical collections would come back to him. Before leaving England he had written to thank his grandfather, Dawson Turner, for offering to help in getting his Journals ready for the press when he returned, and added, 'My Journal will be, I hope, very full if not very good, and I shall send home extracts to all my friends in the shape of letters to my father and grandfather. These Journals on my return are to be given up to the Admiralty, who will, I hope, send them to my Father, since Capt. Ross has promised that he will use his endeavours that the Botanical collections shall be sent to him.' Meantime Hooker had urged his parents to keep his letters strictly within the family circle. Even the sending home of an occasional sketch to illustrate his travels, or of a pretty shell for his sister, allowing brotherly affection to outweigh patriotism,' was strictly speaking a contravention of rules, which, if it reached official ears, might get him into hot water with his commander. The young officers, securing spare specimens for themselves *sub rosa,* were occasionally hard put to it to escape detection.

> The Captain [he writes to his father on November 25, 1842] has a noble collection of Birds in casks,—a most noble one. I do not let him know that I skin any at all, for he is a capital specimen himself of a *Naturalist,* no more do Smith or Oakeley, and you would laugh to see us playing bopeep along the deck as he comes along, for he has an eye like a hawk, and the moment he suspects,—the sooner you give up with a good grace the better. I had a narrow escape the other day with a noble Maccaroni Penguin with gold feathers and crest, by jumping down the main hatch as he came up the after one.

The spare sets of specimens for his father had to pass officially through the hands of the Admiralty and the British Museum; but at the Museum, Robert Brown was 'better than the regulations,' and facilitated Sir William's examination of the plants.

Hence, accordingly, the urgent tone of the following passages from a letter to Sir William (December 5, 1842), though lightened by a reference to Ross's epistolary anxieties which, as will be seen later, very nearly chanced on the explanation.

> There is another subject which annoys me exceedingly, and is the only one in the course of the Expedition which does: it is the following passage in a letter from my mother dated August 1:
>
> '. . . Your drawings (you need not tell Captain Ross, unless he would like to hear it) are known far and wide.'
>
> I thought in my letters I explained my wishes on that subject fully to you all, so much so that I feared to trouble you too often by positive desire that they should be known but to few, and as to 'unless Captain Ross would like to hear it,' I surely have said often enough, or at least given it fully to be understood, that I had no business whatever to send

them home at all, and that did it come to his ears I should not so soon hear the end of it. Nothing but affection for you all prompted me to make them, it was a pleasure to me to do so, although my conscience told me that I was not acting properly to an Expedition whose orders I have often told you are 'all journals, charts, drawings, &c.' to be given up. That it will now come to Captain Ross's ears there can be no doubt, I have difficulty enough in weathering him who know him well, I must however blame myself for sending them at all. If you have made Davis's drawing of the ships in the Pack also to be known 'far and wide' you will run every chance of doing him a serious injury who is dependent on the service. Again, a midshipman from the *Philomel*, a youngster of the name of Fox, comes up to me on a cricket ground where I was enjoying a little exercise with the Philomels after the *General Halkett* sailed and tells me he has heard my letters read in Dublin by his Aunt and Mrs. Butler, some relations of some one of the name of 'Innes.' Who these Foxes, Butlers, and Innes are I do not know nor care, but my letters were never written to be made so public or to leave the house further than Yarmouth or Hampstead, nor do I choose to be the gossip of half the friends' friends who may like to see them. My own wishes with regard to them have been expressed often enough, and surely I am old enough to know my own mind on such matters; they were written for *my near relations alone*, and contain such messages to others as are requisite for them to know; my repugnance to any such notoriety is so strong that if these wishes cannot be complied with I must give up writing anything but simple statements. You may remember that I was always very averse to any society but that of persons whose pursuits were similar to mine, and more particularly to that of four-fifths of our Glasgow and other friends with whom my parents, brother and sisters were on terms of intimacy; this may be owing to a peculiar temperament of mine or more probably to a fault; still I cannot help it, and care to be known by few but Botanists and men of Science. With them my own industry must introduce me, and what other real friends I have I can write to. Do not be angry with me for writing the above; as a duty to myself it was in my opinion necessary for me to state that I fear my letters and drawings are given far more publicity to than I warranted, and I cannot help speaking firmly, perhaps too strongly, on the subject. You are doubtless surrounded by many and very kind friends at Kew, and no one can be more grateful to God than I am; you are calculated to shine in their society and have an open heart to receive their friendship, it is however totally different with me—a few friends are all my narrow mind has room for, and I often think they are kept better on that very account. My ambition to rise in one branch of science will soon cause them to think themselves neglected if I should make their acquaintance and not keep it up. I should have mentioned this subject in my mother's letter but shall not; we are men and may talk to one another without feeling that annoyance which women often will, and I am sure you know my feelings well on the subject, though my dear Mother's love may have prompted her to make me the subject of all conversation everywhere. Do remember then that I do extremely dislike having my letters shown to those I do not know, and that with regard to the drawings it is not fair to me to make them known far and wide, inasmuch as I have defrauded the Expedition of them.

However, all's well that ends well. The publicity, such as it was, arose from a command visit to Buckingham Palace. Sir William was bidden bring his news of the Expedition to Prince Albert, who listened with extreme attention, repeating the main points accurately to a visitor who came later, and taking to the Queen Fitch's drawing from Davis's sketch of the ships in the pack. This put a very different complexion on the affair. The unfeigned interest of the Queen and the Prince Consort in the doings of the Expedition made up for seeming neglect elsewhere, and could not be objected to by Captain Ross, himself a correspondent of the Prince by royal command. Sir William's explanation cleared the air, and had answer (April 20 and March 7, 1843):

You have now quite explained the mystery about my drawings which hung over yours and my Mother's Falkland Island letters. Of course the honour is quite too flattering to allow me to be angry, even had I cause. I often speak testily when I do not mean it, as you know; and hope I said nothing in my letter that gave offence, but I must say I was then annoyed to hear that 'my drawings and letters were known far and wide.' We did take possession of the land (landing on the little island) in the name of Her M.G.M.Q. Victoria, and so we did last January, and on another little island. I wish His R.H. much joy of Her Majesty's acquisitions, nothing but Her wish will get me near them again, for I suppose if the Queen tells you, go you must *nolens volens*. Their Majesties' interest and attention is most flattering to a poor Asst. Surgeon, beyond everything flattering.

Capt. Ross wrote Prince Albert a long letter from the Falklands which caused him many hours' deep study and the purser many candles. . . . If he should show any more interest in the Expedition he may like to hear the particulars of the cruise, all of which I leave to your judgement, only premising that I do not at all like my letters to be sent about *whole*. Use your discretion about any parts you like, but you must see that I may say many things intended only for the four walls of West Park. Had I my own way I would forward occasional notices of the cruise to the 'Athenæum,' but I feel sure Capt. Ross would not like it, nor do I wish to be the mouthpiece for both ships, trumpeting our own fame.

It seemed likely that Ross's calculated economy of news might defeat its own ends.

Capt. Ross told me the other day that 'the "Athenæum" was never friendly to him and took no notice of our proceedings.' I thought the latter part very true but did not tell him, telling him instead that the papers had no means of getting news about us; he did not, or would not, take the hint. He seems to wish all the news to come home with him, to astonish the world like a thunder clap; but will find himself much mistaken I fear; 'out of sight, out of mind,' and if the knowledge of our proceedings be stifled it will beget indifference, instead of pent-up curiosity, ready to burst out on our firing one gun at Spithead. I do not believe he tells Sabine too much, or his own father.

Indeed, his lifelong friend, Archibald Smith,[1] writing on August 3, 1842, tells Hooker that the public have less interest in the expedition than should be if they understood its aims. 'But,' he adds, 'Ross will deserve a peerage if he gets to the pole, and I have got a motto from Virgil ready for him—"Polo dimoverat umbram."' And Dr. Sinclair, returning from New Zealand, found himself greatly in demand. He had seen the half fabulous *Discoverers* with his own eyes.

> People read so much fiction nowadays [he writes from Edinburgh in January 1843], and your labours have had sufficient of it to make a similar impression, that they were glad to hear a living man and not a book express his readiness to swear he saw you going on a-discovering as daily work.

Moreover, when in March 1843 Sir William Hooker obtained the Admiralty's permission to draw up for his 'Journal of Botany' a general account of what Joseph had done, he found that already in Paris they had begun to publish the Botany of D'Urville's last voyage, including some of Joseph's best and newest plants, though without any text so far, while a specimen of the white Chionis, sent home by some member of the Expedition, was bought by a German and described in Germany. Clearly there should have been a Committee, as in France, to issue a preliminary report, reserving full descriptions till the return of the Expedition.

Sir William's article, when it appeared, pleased Captain Ross and the officers generally, excepting Captain Crozier, who was much offended—so sailors love their ships—by the description of the *Terror* as a 'heavy sailer.'

For the sake of contrast with to-day, an impression of Capetown in the forties may be recorded at some length,

[1] Archibald Smith (1813–72) was the only son of 'Smith of Jordan Hill.' He was Senior Wrangler in 1836, and entering Lincoln's Inn, became a distinguished real property lawyer. His most living interest, however, remained in mathematics, both pure and applied, and his working out of the practical formulæ for the correction of observations on board ship and especially for determining the effect of the iron in a ship on the compass, incorporated in an Admiralty Manual of 1862, were of the highest value. In 1865 he was awarded a gold medal by the Royal Society, of which he had been a Fellow since 1856.

where, on the first visit in March 1840, he tells his cousin, Mrs. Fleming:

> We went to Simon's Bay near to Cape Town, where the Naval dockyard and stores are; as we lay there for upwards of a fortnight, many excursions were made to Cape Town, distant twenty-one miles, and as we always went on horseback or in a gig, we had our full proportion of accidents; little damage was however done, except to the horses and vehicles, for though some say that sailors are bad drivers, I am quite of the contrary opinion, for landsmen generally break their heads or limbs and the horse gets off, while you never almost hear of a sailor riding or driving without an accident; that accident never affects him further than his pocket, an instance of sagacity in the members of the Naval profession too often overlooked, while their modesty is so great that they never own to meeting with an adventure of the sort, which would infer that they had the address to rescue themselves when their animals are killed and vehicles smashed.

On the second visit he writes more fully to his mother (April 9, 1843):

> The cliffs of the Mountain are here the grandest for effect I ever saw, at least I always thought so; perhaps from coming off the sea,—they quite frown down on the road though 3000 ft. overhead; the worst of them is that they are essentially sterile, and there is a something in the look of the empty and silent water courses which the verdure and beauty of the slope below will not make up for. I quite felt that I should have heard the murmur of the many distant cataracts, which ought to have poured down each little gully. One of the first houses on the road is called Feldhausen and was of great interest to us, as there Sir John Herschel[1] lived and set up the telescope with which he catalogued the stars of the Southern Hemisphere. It is a very nice white house with a long avenue of dark Fir trees, which give it anything but an inviting appearance; near

[1] Sir John Herschel (1792–1871) continued and expanded the astronomical work of his father, Sir William. From the beginning of 1834 he spent four years at the Cape mapping the southern heavens as he had the northern.

it is a little monument erected on the position of the Telescope. One could not help looking at the place where England's greatest Philosopher lived; the man too who paid us the compliment of calling our Expedition 'the Forlorn Hope of Science,'—perhaps though that was because it was a forlorn hope to expect any good out of such a set as we are,—whether it was intended to flatter, frighten, or stimulate us, we take it as the greatest compliment ever received.

A little further on and Cape Town bursts at once into full view, and a most wretched view it is; the slope of the road is bare of trees, the town lies, not nestled but dabbed on a gradual slope at the foot of the opposite side of Table Mt. to what I described above; the great bay is before it, Lion's Mt. to the right, the high inaccessible (except in one narrow gorge) cliffs at the back, and Devil's Mt. on the left; not a tree anywhere, either on the road, town, or hills. The houses look mean, are square, generally low, arranged in squares, glaringly white-washed, with blue or red tiles. You enter by some dirty hovels and mud walls on a road covered with an impalpable red dust, which covers and paints three or four wretched fir trees, which are bent at an angle of 45° by the S.E. winds; approaching, it does not improve, a short turn of the road almost brings horse and gig up against the castle ramparts, which are of a *lively* gray color, abutting on the road, with a foss all round dug out of red clay earth, and some dirty hamlets scattered without order all round. To avoid this you turn your head to the left and meet a glaring white-washed house with a red roof, which in such weather at once puts one in mind of a red heat and white heat, and further on the sterile cliffs of the mountain. Entering the town is, as I have described, most unpromising, and as to itself I cannot say much more for it. There is a large open space of red clay, surrounded with a low wall and ditch, having walks inside under stunted Oaks and the vile Firs. This gives shade and that is all; grass will not grow; and to make it attractive, to Ladies I suppose who are naturally fond of shopping, there are dirty women sitting on the walk sides selling gingerbread, stale fruit, and lollypops. A little further on is a large building which, with Ludwig's Gardens, is the saving clause of Cape Town. This building contains a fine reading room with every good paper in proper order and at hand; one wing, prettily planted round with rose briars and climbing convolvuluses, contains a Library of 30,000 volumes, all in most excellent order, with the tables covered with magazines. . . .

I found the streets all narrow, ill-paved, hot and dusty, the houses generally mean and irregular, some of the shops good but little shade anywhere: most of the houses have a long narrow terrace just before the door, with a seat for smoking at each end and an ugly fir tree or stunted acacia planted over each settee. Now these terraces cannot be walked over, and as they take up all the room where the pavement should be, there is walking straight on, but in the middle of the street; and then the poor advantage of the shady side is lost, without you hug the wall and double every terrace, crossing and recrossing the zigzag gutter, most ingeniously contrived to go the shortest distance by the longest way. The Natives are of mixed breed. Hottentots are scarcely seen anywhere, Malays are very common, both men and women, generally with a red Bandana handkerchief round the head; they have a separate meeting house and burying place. Next are the Dutch breed, often round built, especially the ladies, and inclined to be swarthy. They roll handsomely along the streets, are plump and often well looking, sometimes very handsome,—the men are as often thin and smoke many cigars. All Dutch born in the colony are called *Africandoes* as the colonial Australians are called *Currency* and the St. Helenas *Yam stocks*. Except the shopkeepers the English are not much seen; they compose the upper classes, generally live out of town, and drive in to shop, etc. The Governor, though viceroy of the Colony, keeps a very poor table and only gives one ball a year; the society is quite divided between the Dutch and English; they do not mingle much, though I suspect much of the former class to be far superior to the latter. Amongst the strangers and occasional visitors none are so conspicuous as the Indians [i.e. officers of the Indian army]; they saunter about slowly with white jackets, straw hats, and whips in their hands, though ten to one they belong to foot regiments; they may be descried at once by having long yellow hatchet faces, curious noses of sorts, yellow whites to their eyes, and are said to have no livers, whence I suppose the bile is deposited elsewhere, in the face, eyes, etc., and even so much as to affect their tempers, for some are hypochondriac and others highly irritable; they are gregarious, and frequently live in boarding houses. . . . Baron Ludwig[1] received me with the greatest kindness and wished me to stay at his house, which I declined, not seeing any occasion to trouble him, and having a great deal of shopping to do, which I wished to effect in the cool of the evening, when he would expect me to sit at home. I breakfasted and lunched there, however. His house is one of the best in Cape Town, with a noble drawing-room, handsomely furnished with two busts of his noble self, one of the late Baroness and one of the poet Schiller. My Father's picture used to hang there before, but was not now, and of course I did not ask for it. He, my Father, has given way to William of Würtemberg, who so graciously showered down the crosses and snuff-box on him of Cape Town, which emblems you may remember in the Crescent. I found 'Peter Schlemihl' in his Library and could not help reading part of it for old acquaintance sake; it was the very copy my Grandfather gave him; tell this to the dear old man and how many associations and thoughts of him it brought up; his own handwriting ascribing it to Chamisso was on the title page. I think I was more pleased to have found that book of my dear Grandfather's than with anything else in Cape Town; I had a great mind to steal it.

It has struck me very forcibly during both my visits to the Cape, that there is in the Colony a most remarkable want of a love for flowers, which I always thought so peculiarly a Dutch taste, but so it is. Look here, the only *Eucalypti* and *Casuarinas* I have anywhere seen, are in Ludwig's garden; but though they are planted by him for

[1] Baron C. F. H. von Ludwig (*ca.* 1784–1847), Ph.D., chemist and botanist, left his native Würtemberg in 1804 for the Cape, where he founded a Botanic Garden, Ludwigsburg, and became Vice-President of the South African Literary and Scientific Institute, and a member of the Cape Association for Exploring Central Africa. He was a correspondent of Sir William Hooker, who, in dedicating to him the 62nd vol. of the Botanical Magazine in 1835, made special mention of the rare and beautiful plants with which he had enriched Europe, and called him the Friend and Patron of Botany. He visited Great Britain in 1836–7.

the purpose, and are the best trees possible to break the violence of the S.E. winds, still on the outside of the town the road is sometimes (where anything is) planted with pudding-headed Pines, which are blown at angles of 45 with the ground, beastly black in color above, and covered with the red fine dust of the sand below.

Except Ludwig's garden I enjoyed nothing in Cape Town, for you would not care to hear how the days were sultry without a breath of wind, the streets full of a fine red dust, so light as to be always floating, or how often I had to go to the same shop to get things changed, etc. It was my intention to go up Table Mountain, but Ludwig has no one who could take me up, and the heat was so scorching that all my enthusiasm fairly oozed out of my finger ends, and except for catering for Kew in cool large rooms, Botany was at a standstill.

CHAPTER VII

THE ANTARCTIC VOYAGE: PERSONAL

The voyage left its mark on the young naturalist. His physique was strengthened: the long spells of isolation, though depriving him of much that he longed for, helped to fix the lines of his thought and character and aims.[1]

> The cruize [he writes to his mother, June 29, 1841] has proved me quite hardy. Except a slight cold and its concomitant discomfort, I have had nothing to complain of, and that has been since my arriving here (Tasmania). During all the time I was in the Southward I did not know an hour's illness of any kind whatever: the cold is healthy in the extreme, and an occasional ducking of sea-water proves rather beneficial. I always accustom myself to taking moderate exercise in hauling the ropes, setting sails, putting the ship about, &c. Thus my chest expands, my arms get hard, and the former *rings* almost when struck.

And when he reached the Cape in 1843 he tells her that, as they felt the weather stifling and hot, 'to dine on board the Flagship the other day I had to borrow garments; not one of my 3½ dozen white trousers will go on: so much for my rude health.'

[1] Mrs. Richardson, Franklin's niece, writing to Hooker on August 3, 1842, remarks that she would never have recommended the Navy to him as a career —and that it might even be unsatisfactory as a means of travel and experience when a cautious reserve is wisest: adding sagely, 'As a piece of mental training I cannot think lightly of that retirement into oneself which is the natural consequence of not entirely liking our associates, and not agreeing with their views or notions. Mrs. Barbauld calls this sort of thing the "Education of circumstances," and notices how it contributes to form the character.'

So far as science went, the lengthening chain of months enlarged his powers and strengthened his professional position. Without counting the inevitable separation from friends, the chief thing he found lacking on the voyage was music, though he could not profess to be a musician any more than an artist. He tells his sister Elizabeth (May 12, 1843):

> On board this ship I want music more than anything, and am always ready to break my leave for the sake of hearing it. I often wish I understood it, and perhaps oftener still (am glad ?) that I do not; since, as matters now are, I cannot perceive those faults that would grate upon the ear of a musician.

He does not care for 'modern ballad music' but likes the older English and Scotch airs, e.g. 'Where the bee sucks,'—good sacred music, such as Handel, 'Israel in Egypt,' and Haydn's 'Creation'; and some operatic music of which he is kept in mind by the naval and military bands, and is delighted that the girls and his mother are practising his favourite songs and glees against his return.

Thus it may be imagined what a double disappointment awaited him at Rio on the homeward voyage.

To his Sisters

Rio de Janeiro: June 20, 1843.

> The Americans have an immense fifty-gun ship as Commodore ship stationed quite close to us, and would you believe it ? the Goths have no band on board but some huge drums and squeaking fifes, which they make a terrible din upon every night, and beat off with Yankee Doodle at 8 P.M. Not only is the noise horrible, but at that time a tolerable band plays on board the Brazilian flagship, whose music is consequently drowned before it reaches us.

A letter of November 28, 1842, to his old friend, Mrs. Boott, gives the fullest account of his artistic tastes and education.

> I often regret that I never saw any pictures that can be called good. A relish for this branch of the Fine Arts has not yet extended to the Colonies, whose children cannot

be expected to exercise taste, when the parents have no models to show them. My own taste on such subjects was never formed; though, like most persons, I knew what pleased me, and was much soothed when I was told (on regretting the circumstance) that Sir Joshua Reynolds never could appreciate any part of a painting till he had seen it several times. Sir Walter Scott, I think, in 'Paul's Letters to his Kinsfolk,' says, when speaking of the Louvre in its palmy days, that the beauties of the finest pictures do not strike him at once. Without comparing myself to either of these great men, I must say that next to the want of Society, the want of music and painting is one of the most irksome which a sea Voyager is bound to endure. When I have been weary of work, even a tinkling musical-box has sounded most charming; but all the boxes have, at last, been either broken or given away, and my sole consolation remains in whistling those tunes which most recall pleasant scenes to my memory,—though this is sorely to the annoyance of my neighbours, who growl, like free-born Britons, at the noise I make.

Letters already quoted point to the smallness of his intimate circle. It embraced his nearest relations, and beyond these but a few who could really be called friends. This inner circle was grievously broken during his absence. First his brother William died suddenly of yellow fever in Jamaica. Then his two sisters, Elizabeth and Mary Harriette, at school in Kensington at Little Campden House, were threatened with consumption and taken away for special treatment at Leamington, afterwards wintering in Jersey. Elizabeth, the first to give anxiety, gradually recovered; Mary Harriette, who fell ill later, faded away all too swiftly. Joseph had expected to hear of his grandfather Hooker's death before long; but the octogenarian, with the vitality which he handed on to his male descendants who passed much of their youth in the open air, lived on and was happily moved from Glasgow to Kew, a heavy journey in those days.

The first bad news caught him cruelly at a moment of joyful expectation. Save for a letter sent to Madeira, which had overtaken him at the Cape, his first budget of news met him at Tasmania, in August 1840, eleven months after he had left home. The black-edged letter beginning, 'My dear and only son,' turned all his delight into mourning. He was devoted to his brother William, 'so warm-hearted a fellow that he would cut his right hand off to help even a stranger.' The brother who had been 'hourly in his thoughts' these many days had been dead since the first day of the year. From the Cape he had written to his mother:

> So poor William has gone to Jamaica; if you but knew how often I think and dream of him you would not be surprized at the sorrow I felt that he should have parted from you, though it is doubtless for the best. Poor Isabella[1] is left behind. . . . I feel sure it will be a delight especially to my sisters to take charge of the child till my return when I shall consider it my own should it be better to leave it behind than take it to a foreign country, or should any other circumstances demand another father for it. [He knew William was out of health, though he did not believe, as some did, that he was threatened with consumption.] I wish very much that I had received that letter before, as I had intended to send my brother a check which I can well spare; it is now too late—and I am sure money must be wanted; he need not look upon it as a gift, at any rate it would be but a poor recompense for all the kindness I have received from the poor fellow's hands. The child I do hope to bring up, and you must tell that to my future *housekeeper Maria*, to whom I send my best love.

It was to this favourite sister that he unbosomed himself; the poignant contrast of exchanging the hardships of 'the most tempestuous latitude in the worst season of the year' for the calm beauty of the Derwent with Hobart set in tall trees under a snow-capped hill, only to find in his envied package of fifteen or sixteen letters the news that should make him the one sorrowful man in the ship: 'now he is gone, and there will be none of my childhood's playmates when I return to talk over bygone times with, for he was at school my only companion.'

[1] He married Isabella Smith, April 22, 1839.

The characteristic note of his early religious training appears in his words:

> Mr. Nelson and Susan have now, I trust, met with him, and little as worldly affairs have to do with the state above, I can never divest myself of the idea, that one, though a small share of the pleasures that attend the good, is the meeting of those whom our God and duty have sanctioned our loving. . . . Do not think I repine at this dispensation, nor at the additional and not less felt one of my Grandpapa's illness. I have far too much to be thankful for both for myself and for those who are left, and if there is one thing that cheers my thoughts of home, it is having a faithful sister of my own age. You perhaps do not know how responsible your situation at home is, and it is my great happiness to think that when sorrow weighs down my parents they can feel full confidence in you. Were I not sure that this is the case, it would make me miserable indeed.

To his father, who had also warned him of his sister's illness, he writes (July 6, 1841):

> For my part I can hardly bear to think upon the probability that I shall return to the house I left so lively and merry, and not hear a single gladsome voice, no music and none of the attractions that used to welcome me home every winter night from college. My affection for those who remain will indeed be greater, but of how much sadder a nature will their welcome be than what my vivid fancy has been accustomed to paint when thoughts of home were my only solace.

As to the prospect of his father leaving Glasgow for Kew:

> I sincerely hope he may for his own sake; for my own I am quite indifferent; except Jas. Mitchell, I have no friends that I care about except Adamson now that Thomson and the Steuarts are gone. I shall, however, always look upon the dirty Town as the only place connected with old associations, and whatever attractions other places may have for me, none can have localities so endeared to me as that Town which is the same as my birthplace. It is true I have no friends there, but equally I have none elsewhere;

wherever he and you all live, should circumstances favour my living at home on my return, there I shall be happy to find you, though now no spot is dearer to me than Invereck; *two* sketches of it hang in my cabin.

The best anodyne, however, was hard work and busy occupation: so that he writes to his father on September 7th:

> Still I have been very happy here, and never before could I have so deeply felt how much the study of our mutual pursuit tends to alleviate our distress.

The uncertainty made him 'afraid to mention names of those so far off and in such precarious health.' But warned of Mary's decline, and eagerly following the successive hopes and fears for so dear a life, he schooled himself to meet the inevitable, and the pathetic accounts of the child's last months found him prepared as much as might be to accept his own irremediable loss with the resignation to the will of an inscrutable Providence that was an integral part of his parents' faith. Still, resignation involved a sharp struggle with feeling, and as he drew near the Falklands after the second voyage to the ice, he wrote to his father (April 5, 1842; the words are quoted from a copy only):

> Much as I long for tidings of you all, I cannot but feel sure that they must be woeful; and to own the truth, one of my reasons for beginning this letter before we cast anchor is that I may be able to communicate to you some of the cheerfulness I now feel, and that my letter shall not be tinged with that sorrow and moroseness which I fear may have characterised some of my former epistles: these were written on the spur of the moment, when to my shame present griefs obliterated the recollection of past mercies, and whilst pining over what had occurred, I had forgotten how much I of all others had to be thankful for, and how little it was my duty to trouble you with such complaints. Whatever the tidings may prove to be, I have too long suffered from hope delayed and been kindly by you all too well prepared, ever to feel again the poignant anguish with which I received the first letters that awaited me at Van Diemen's Land.

The movements of the exploring ships, the irregularity of the post carried by sailing vessels, the occasional vagaries of the Admiralty letter-bags going from one naval station to another, made the receipt of news from home spasmodic. For instance, he tells his sister on May 26, 1842, 'My latest news from home is March 29, 1841, and that is in answer to a nearly two year old one of mine from Hobarton." Such news was often anticipated by the English newspapers found at ports of call; the 'Athenæum' in particular giving news of persons and events in scientific circles. To this he owed his first intimation of 'the first and last piece of good tidings that has greeted me about our own family.' This was the appointment of Sir William to Kew at Lady Day, 1841. He found a copy of the journal for March 23 with the news when he was at Sydney early in August. His father's letter about the appointment, written six days later, reached him at the Bay of Islands on November 23. On the strength of it he persuaded Captain Ross to relax the strict rule of the Expedition and let him send Sir William a box of plants he had collected.

Hope deferred was at length satisfied; a month before hearing the news he had written:

> What to think about Kew I do not know; the ministers have put you off so very often that they may do so longer. Next to my poor little Mary, that subject lies nearest my heart, and most sincerely do I hope you may not be after all disappointed. To live near your friends is now your chief aim and must be essential to your comfort; and to be able to raise Kew to the rank of a tolerably good national establishment would be the most honourable service a Botanist could render his country, besides being the most pleasant one you could set your mind to.

Kew, he had felt strongly from the moment of his father's appointment as Director, *must* eventually become a National establishment. He is amused to find from a newspaper of 1842 that Lord Lincoln, head of the official department that ruled Kew, opposed Sir William's scheme of opening of the gardens to the public on the ground that they were 'the only gardens near town to which Her Majesty could repair for exercise,' seeing that Kew had never been so used since Kew Palace was given up. Futile pretext for obstructing public progress. A liberal policy must prevail, the Upper House being won over by reason of 'our noblemen and statesmen being so fond of trees and their gardens' and finding that Kew disseminates new plants; all the more successfully because it has secured the new palm stove. Already (March 7, 1843) his ideal is to see the gardens on an equal footing with the British Museum, and under a body of Directors chosen one half of Botanists at least.

> My mother tells that Invereck [their cottage on the Clyde] is sold, and I much fear that the great expense your family now puts you to is in some measure your reason for parting with it. Everything seems to have gone wrong from the very day on which I first left Glasgow, and believe me that could I with honour give up the Expedition it would not be long before I should be at your side to take my share of your labors; as it is, even were I uncomfortable in the ship, I could not give it up without it being said I was afraid to go on, and further I hope ere this will reach you, you will be snugly ensconced not ten miles from Aunt Palgrave's.

Now he could expand affectionately over his father's advancement in the sphere of their 'mutual interest.' He discusses plans for the future; caters for his new command by making Colenso and Ronald Gunn[1] promise to send interesting plants to Kew from New Zealand and Tasmania; looks forward eagerly to the day when he will himself share in his father's labours. 'My father always works too hard' he agrees with Dr. Boott, the old friend of the family (November 29, 1842).

Now that his employment means more exercise out of doors, he will grow stronger. 'Walking, in particular, always agreed with him, and good walkers invariably enjoy good health; who ever saw a sick two-penny postman? or Police-runner?'

And to his father he writes (April 20, 1843):

[1] See pp. 107 and 124.

You must not work too hard at your plants and Library; rather get on in the gardens, which is more healthy, and in which I shall not at first be the slightest assistance to you, from downright ignorance; I will get up as much back work as you like in the books and Herbarium.

The double link of affection and common intellectual interest runs through all the letters to his father, and may be noted even in money matters. He has no use for his double pay on the voyage; and his father's valuable publications, the 'Icones Plantarum' and the 'Journal of Botany,' are entirely unremunerative. Let him use the money for these; popularise the Journal by portraits of living botanists. If he will not let Joseph pay for the books sent out to him, at least he must accept something for the keep of his pet dog.

You must not refuse to make use of my bills for all such purposes [e.g. looking after dog 'Skye,' which was not allowed to accompany its master to the Antarctic. The *Erebus*, he tells his sister, only carried some fowls—for colonising purposes and two cats. Therefore 'Love me, love my dog'], the money is no use to me. I have enough to spend and to waste, for one cannot help wasting when port is so seldom seen; as sure as a bill is cashed it all goes, and they are sent home instead to be made use of and not buried in a bank. You may be sure I should not scruple to draw on your liberality were I to be extravagant or foolish, and my outfit cost you a great deal more than it should have done had I been judicious or in any place but Chatham, and you should not therefore scruple to use the bills, especially in any way of forwarding your works. You have too many calls on your purse to attend to many things which strike others; for instance, I would far rather pay for a new plate than see such a rotten lithograph of Richard [1] after the excellent ones of Cunningham and Swartz.

Do not let the Journal die for want of funds so long as I have a bill to send home. I have no work that pleases me so much.

[1] Achille Richard (1794–1859), doctor and botanist, Professor in the Medical School of Paris from 1831. Besides various monographs and studies in medical botany, he wrote *Nouveaux Elements de botanique et de physiologie*, 1819, and with Lesson described the botany of D'Urville's voyage.

He had wished to send a present to his brother, but it was too late; or to other relations, if his mother were still obstinate about making use of what he did not want; and to her he writes (December 6, 1842):

> I wish you would not lay aside the few pounds I sent home for me, for I shall not want it; if I can only get enough to keep me respectably I shall be content to live from hand to mouth, and I would not give a penny for a fortune which is sure to prove a curse to most men and a breeder of idleness; however, it is all very well to talk so when there is no chance of getting one,—but I should much prefer that the bills were used,—indeed had I not thought they would be, I should have put them into a V.D.L. bank, or invested it there in land and sheep; however, it is all the same to me.

The years of service in one of His Majesty's ships gave Hooker, as it gave both Darwin and Huxley,[1] an invaluable acquaintance with the realities of things, and there was 'a masonic bond' between these friends 'in being well salted in early life.' But the voyage did not alter his career as it altered the career of the other two. He was already a naturalist enlisted in the ranks of pure science; a rising botanist when he set out, a botanist of higher repute when he returned. From Sir William's point of view, the only serious danger was that he might desert botany for zoology. Hence, as has been said, his delight to receive early assurance that Joseph cared most for botany and intended to devote himself to it when he came home. Here he could best help on his son, with the added satisfaction of knowing that his collections and library

[1] Thomas Henry Huxley (1825–95) studied at Charing Cross Hospital and entered the Navy as Assistant Surgeon. Through Sir John Richardson at Haslar, who had noticed his scientific ability, he was appointed to the expedition under Captain Owen Stanley in the *Rattlesnake* frigate, which was to survey the east coast of Australia and the islands as far as New Guinea (1846–50). His work on the Oceanic Hydrozoa won him the F.R.S. at the age of twenty-seven, and the Royal Medal the following year. In 1854 he obtained a professorship at the Royal School of Mines, whence sprang the Royal College of Science at S. Kensington, where he was Professor of Biology and afterwards Dean. President of the British Association 1870; of the Royal Society 1883–5; Privy Councillor 1892. As Darwin's most vigorous upholder and expositor, as an educational reformer and a brilliant and forceful essayist and speaker, he was one of the chief factors in breaking the shackles imposed on thought and opinion.

would be inherited by some one who could make a good use of them.

Plans for the future are first outlined in a letter of February 3, 1840, written from St. Helena.

> One of your last questions to me on leaving Chatham was: 'What do you think of doing on your return?' To this, if I remember right, I gave an indirect answer from not knowing the service I was bound for. As I know, from your affection to me, you would like a good reply, now that I can form an opinion, I shall give it honestly. The Naval Service generally is very bad for a Naturalist; the particular branch, however, in which I serve, is very good. Though there is not such a scope for the Botanist as I could desire, there is a splendid opportunity of improving myself as a general Naturalist. I am very fond of the lower orders, though farther than studying them here, and perhaps aiding in their future publication, I never intend to follow them up nor any other branch but botany.
>
> Gaiety of any kind has still less charms than ever for me. Even at sea, I am quite happy drawing Mollusca in the Captain's cabin, and I only wish that I had more books and were drawing plants. If ever on my return I am enabled to follow up botany ashore, I shall live the life of a hermit, as far as society is concerned; like Brown perhaps, without his genius.[1] If I have to serve again on board ship, it will be in a service like this, congenial to my taste and pursuits, and not in the regular King's Service. The sea agrees with me, and I am very happy on it, as long as I can work. I am never sick, nor have been so since leaving Chatham. This hot weather is my only and bitter enemy, and from it I suffer very much, in several ways.
>
> What I said of my life and prospects, my dear father, is, of course, strictly private. I am quite happy where I

[1] To this comparison his father replied: 'I am neither surprized nor sorry that you have no taste for the gaieties of life; but neither do I wish you to turn "hermit." If you are no more of a hermit than Brown, indeed, I shall not complain. That is, whether you know it or not, he is really fond of Society and calculated to shine in it: and to my certain knowledge, never so happy as when he is in it. But he has unfortunately sceptical notions on religion, which often make life itself a burden to him: and which bring him no comfort in the prospect of eternity. I really wish that he were now in this house that he might see what is the death-bed of a Christian' (the elder Hooker).

> am, and see my way clearly before me till we return, after which no foresight can tell what will become of me. I can always fall back on the service as a livelihood. I shall never regret having joined this expedition. We must, along with Captain Ross, fail completely so as never to try again, —or succeed. No future Botanist will probably ever visit the countries whither I am going, and that is a great attraction.

For a time, however, in 1841, his plans were sorely shaken by the barrenness of the first Antarctic cruise and the shortness of the stay in Tasmania, which seemed fatal to his project of writing a Flora of the island. The rest of the cruise threatened to waste two good years of a botanist's time. At this juncture his Tasmanian friends conceived the plan that he should be invalided and left in Tasmania, where he could continue his botanical work. His health had suffered, in sober fact, from brooding over his brother's death and the other bad news from home. His friend Ronald Gunn, a botanist himself and officially private secretary to Sir John Franklin, suggested, in the spirit of Midshipman Easy, that he should work up a cough and hoarseness, symptoms of impending consumption, for the benefit of that keen-eyed disciplinarian, Captain Ross. He pointed out the obvious drawbacks to going so far as to quit the service, and the burden it would be on his father if Hooker could not live on his half pay while publishing his collections; but he was ready and able to help him in fifty ways in taking this short cut to botanical fame.

Happily the plan was dropped on reflection; the considerations *contra* were very strong, and there was the further chance that as he recovered his scientific holiday might be cut short by an order to join some other vessel. Moreover, Sir William's next letter urged him not to leave the service till he was fairly home and could see at least what could be done about publishing the collections, and though this only reached him later, it confirmed his new resolutions to go on with the expedition which he could not honourably leave. His gleanings in less abundant fields were richer in scientific results than the harvest he looked for as a collector.

Such regrets as he felt appear in a letter to George Bentham, the botanist (Falkland Islands, November 27, 1842):

> It does sometimes make me sigh, to hear of and to see the rapid strides which Botany is taking both at home and abroad, and to contrast it with my present narrow sphere of exertion; nor can I forget how young De Candolle asked me at your house 'why I was going to such a barren country as the Antarctic regions.' I am far from regretting that I joined this expedition, and I shall always look back on its progress with infinite pleasure; still, the few plants I have obtained are dearly won, and unless my friends will kindly help me by allowing all the Antarctic plants already in England to be added, the results will be meagre enough in Phaenogamic Botany. Of the *Cryptogamia* I do not despair, but this tribe is sadly neglected and finds small favor in the eyes of most Botanists.

By the end of the voyage the practical issues before him take shape in a letter to his father, written from St. Helena on his way home (May 18, 1843), when his eager desire to travel again—but for a shorter time and in a less barren botanical area—is balanced against the necessity of staying at home to publish results.

> St. Helena Roads.
>
> I have a long yarn to spin you about my future prospects, Capt. Ross having been sounding me. He wants me to remain in the service, to serve *only* for Scientific Expeditions; and has, or is going to write home, about my promotion. He told me that he must write for Lyall's [1] and mine at once; and had delayed it, expecting me to have spoken of the subject to him, which I of course never dreamed of doing, it being out of my place. As he said, it was a piece of injustice to delay writing for Lyall; and that he could not do that without doing so for me also and stating my superior claims, provided I remained in the service: he desired an answer. I told him that I did not intend remaining, provided I could get any good or decent shore employment; but that I had no idea of giving up the Navy till I felt my way on land, which I could not do before arriving in England.

[1] Assistant Surgeon on the *Terror*.

Unlikely as it is, there is a possibility of your not being able to help me five months hence, and how foolish I should be to have thrown away the certainty of promotion for the uncertainty of anything else! I also told him that I had no idea of being applied for, until our arrival in England; but as he was good enough to do so before, I should take advantage of his offer, provided that he would not be offended at my throwing away that offer on my arrival, adding, that I believed and expected I should be worth being employed by you for my living; that nothing but absolute necessity should make me enter the ordinary service; and that it was highly improbable that I should ever feel myself at liberty to enter any Government Expedition, which would employ me more than ten or twelve months. I have no wish to be a drag on the service by remaining in it and not serving; and when I explained this to him, he answered, 'it would be a piece of great injustice in the Navy to employ me in any way but Natl. Hist.,' and said a great many flattering things which I divided by two, and appropriated one half (perhaps the better). He also told me that he would apply for a sum of money to defray the publication of the Natl. Hist., the Botany of which should be recommended to me; and that I ought to be employed still on pay (perhaps half-pay), in the service, till they were done, as very inadequate compensation for my trouble; to this, of course, I had no objections, except on the grounds of passing the boards. On this head I am told the regulations are altered, and that having a diploma from Edinbro', I am not required to pass *anywhere* but before Sir William Burnett; such was not the case when we sailed, but I am told is now; a matter of very great consequence, as I have no notion of working up to pass Edinbro' again, which would cost three to five months' study in classes.

The long and short of the matter being—that Capt. Ross must either apply for my promotion, or write home and state that I would not take it if offered me, I of course (having no competency of my own) took the promotion offer, being at liberty to decline it on my arrival in England, without giving him offence for having put him to trouble for nothing. I took two days to think over the matter before giving him a final answer, and hope you will approve of what I have done. I weighed the question in all its bearings, and my only objection is that I should like to leave the service, as I entered it, for the Expedition, and not for any benefit the service would give me in return. However, as you know, I am not independent, and must not be too proud; if I cannot be a Naturalist with a fortune, I must not be too vain to take honourable compensation for my trouble.

You, to whom I owe everything, and on whom I am entirely dependent out of the service, are the best judge as to whether I should accept the commission and the half-pay of 5*s*. a day; at any rate, until the plants be published. Were an Expedition to go (like Parry's last) for eight or nine months to the North; or the more especially any land one, for about the same time, and offer to take me as Naturalist, it is my present expectation to avail myself of it. It must be something *very good* which would put me off doing so.

You have above a full, true, and particular account of my Navy prospects, and have nothing to add on the subject but the hope that you will not have any reason to find fault with the course I have taken.

This letter is endorsed by Lady Hooker:

I do hope I am thankful for Joseph's good sense and modest appreciation of himself, even more than for his Captain's praise, or than the sweet prospect of his preference of his father's roof and employment at home (July 1, 1843).

These plans met Sir William's full approval. Two years' leave on half-pay must surely be granted him for bringing out his scientific results.

'Were I still in Glasgow,' he writes, 'and Professor of Botany, I might have had the means of securing for you my Chair or of resigning it in your favour ere long. But I am of opinion you would not like the drudgery of lecturing.' But 'Merit is generally sure to secure interest,' and the alternative suggestion is to come to Kew, to help in the Herbarium, and by dint of his publications and botanical studies establish in course of time a claim to succeed to the post of Director.

Such work would be congenial and would bring him into

contact with men of science; moreover, its scope was elastic, and could easily admit the schemes for further travel which he had formed.

> You wish [he writes to Bentham[1] in a letter of November 27, 1842] that I should see a little of Tropical Vegetation after my Antarctic herborizations, and I am much obliged to you for your kind desire, which I doubt not is good; but, please Sir, I would rather go home, and have no notion of jumping from cold to hot, and cracking like a glass tumbler. Have not you Botanists killed collectors a-plenty in the Tropics? And I have payed dear enough for the little I have got in a healthy climate.
>
> On my return to England I shall have plenty to do, working in my father's herbarium, and when I can get enough money I should like to visit the capital continents and especially N. America. If entirely my own master, I would not object to embark once more for a distant climate for the purpose of Botany, and to explore the Islands of the South Seas, especially the Society and Sandwich groups. I might prefer the Himalaya regions; but these ought to be investigated and are in progress, by the officers of the Hon. E. India Company: besides the expense of travelling there is dreadful. The only circumstance which has disappointed me is the not having visited the S. Seas. Poor Western Africa remains still unknown, and the Niger Expedition worse than a total failure.

[1] George Bentham (1800–84) was the youngest son of Sir Samuel Bentham, the naval architect and engineer, and nephew of Jeremy Bentham, the writer on jurisprudence. His facility in learning languages was stimulated by early residence in Russia, Sweden, and France (1814–27), and in later life he was able to read botanical works in fourteen modern languages, as well as Latin. His pursuit of natural history, especially scientific botany, took second place to his work in philosophy, logic and law, until set free from other ties by the death of his father and uncle (1831–2). Then he devoted himself to botany, becoming, with his legal and philosophic mind, one of our greatest systematists. It will be seen later in this volume how in 1854, when certain difficulties made him contemplate retirement from his work, the Hookers and Lindley saved him for botany. He was given the run of Kew, and co-operated in the newly started Colonial Floras, undertaking those of Hongkong and Australia, and later projected and wrote with Joseph Hooker the monumental *Genera Plantarum*. He was President of the Linnean Society 1861–74.

CHAPTER VIII

RETURN TO ENGLAND: AND VISIT TO PARIS

THE ships reached Woolwich on September 7 and were paid off on the 23rd, after a commission of four years and five months Captain Ross had landed at Folkestone and hurried to London. For some days the Hookers had to be content with his news that all was well; Joseph, as a junior officer, could not get away from his ship, and it was not till the evening of the 9th that he reached home on a week's leave 'in high health and spirits.' 'He is not stouter,' writes Sir William to Dawson Turner, 'than when he left us, and very unaltered—more manly—broader in the shoulder. He is badly off for clothes, and we had to assist him from my wardrobe to enable him to go to church yesterday.'

Soon he settled down to a six months' spell of hard work, enjoying everything at home and about Kew, and working at his father's side on his plants, 'when not impeded by frequent calls to London and numerous engagements'; working, as his mother puts it, 'like a dragon, like a grandson of my dear Father's, and always happy when so employed.'

First came the Antarctic Flora. But though Ross had made formal application for a grant towards publication, the official wheels moved with discouraging slowness.

> I have no heart [he exclaims to Bentham, February 10, 1844] to do much at my Antarctic plants, having been five years more or less working at them, and my prospects of publishing in a nice form are waning very fast indeed. I most heartily wish that I had at first published a rough

> short synopsis of all the new species with terse diagnoses and nothing more, it would have been printed in the Journal and no one would have known of it at the Admiralty; while it would secure the priority of discovery. It is not having my name at the tail of a specific one that I care about, but I do want our Expedition and country to have the merit.

Next is the *Species Filicum*, in which he was helping his father, working 'as the man does who blows the Organist's bellows, at the rougher part,' a work among the lesser studied plants profitable to the student, though one of the most difficult and laborious that could be picked out in all Botany.

Then came a task suggested by Darwin; he continues to Bentham:

> I am also working up very slowly a paper on Galapagos Island plants, from Mr. Darwin's and Macrae's collections. I find it a very slow job indeed, as there are very few species of a genus or Nat. Ord. and so dissecting one plant is no help to another. There are more new species than I expected, but then I have begun at the small orders and Cryptogamia; I have done the Ferns, twenty-eight in number, and am now amongst the grasses, which are terrible. Fancy two new Panicums; I cannot make them agree with any others, and yet every one will say I only made them new species to save the trouble of finding out their proper names—then there is a vile Eragrostis Poa identical with an Afghanistan one! but undescribed, and another group of the genus Eutriana whose spikelets vary in a most instructive manner, some abortive, some ♀, some ♂, some ⚥, some with two flowers, some with more, and altogether the most unsatisfactory thing possible to describe.

Finally the long accumulations of his father's Herbarium were clamouring to be set in order, 'probably by arranging together all the loose bundles, thus making a grand total of all the Herbarium, and then going through the whole, taking each Nat. Ord. by itself, taking from it what is wanted for the Herbarium, and putting the rest aside as duplicates. Would not this be a grand work?'

It is already October when he reports to Dr. Harvey, who had earlier shared in some of the sorting, a quasi-final descent upon the 'Augean stables' of the Indian and Australian collections,—'stable occupation,' as he calls it next spring when picking out duplicates for his Paris friends, in continuation of the same familiar jest, for in default of proper accommodation these things were housed above the stable at West Park,[1] where 'Elizabeth's pony makes Jenkins sweetly damp' (i.e. Colonel Jenkins' Assam collection), and their favourite 'little Catty! Catty! Meaw!' sometimes 'kicked dreadful bobbery among the things,' until, pleasant reminder though she was of Harvey's visit, she was convicted of 'eating hens and chickens without salt, wherefore she is to be expelled the domains. Will you have your old darling?'

By March 1844 the official wheels had revolved, and the sum of £1000 was promised for publishing the Botany of the Antarctic voyage. This money was to be spent upon making 500 plates of illustrations, 'which there are ample materials for in the Floras of V. D. Land, N. Zealand, Fuegia, and other Antarctic Latitudes.' For his support whilst he was working at the book, Sir William would have liked him to continue receiving the double pay of £250 a year which had been allowed on the expedition; Joseph himself, who did not even wish to be passed for full surgeon and draw the higher pay attached to a rank in which he never meant to serve, was content to ask for the ordinary pay of assistant surgeon, £118. This was more than granted, with an appointment to one of the Queen's yachts, without duty; the pay was about £136, 10*s.*, without living allowance. Through Lord Minto, however, who was warmly interested as having been First Lord of the Admiralty when the Expedition was sent out, Sir William urged the precedent of the allowance to Robert Brown; there were further precedents in the case of Naval surveyors who received a small allowance for living on shore while they worked out their results. Thus the pay finally allowed was raised to £200 a year.

[1] West Park was Sir William Hooker's house, until in 1852 he was given an official residence in the Gardens.

To find a publisher for the book was a matter, Sir William confesses to Dawson Turner, of very great difficulty. But at last a young publisher in King William Street, named Lovell Reeve, undertook it on condition of receiving all the material of drawings, plates, and text without further payment, and that not one copy should be given away to a person likely to buy it.

Coupled with this news of the book Sir William gave another piece of news scarcely less interesting to Dawson Turner. On the following day, April 2, 1844, Joseph was to be received into the Linnean Society, to which he had been elected during his absence from England. His grandfather had been a member since 1797.

A fortnight later: 'Joseph is very hard at work on his Flora and three or four plates are prepared. But I do not think he is yet aware of the great labour in store for him—eight plates a month and two sheets of letterpress.' No one was more aware of this than Sir William, with his long experience of botanical books and journals; and Dawson Turner, to whom he submitted the proofs for notes and suggestions, knew something of it also.

The work was to appear in three parts: the first, or Antarctic portion, to be dedicated to Ross; the second (Flora of New Zealand) to Prince Albert, and the third (Flora of Van Diemen's Land) either to Sir Robert Peel or to Robert Brown. Sir William asked Dawson Turner to draw up the dedication to Ross. The publication of the first instalment early in June calls forth congratulations from Mr. Lyell of Kinnordy on Joseph's début as an author.

At the same time he furnished Ross with various material for his account of the Antarctic Voyage. On the one hand were short botanical sketches of such places as Ross desired, with the full identifications of plants now possible. Thus 'the *liliaceous* plant' mentioned in his first account of the Auckland and Campbell Islands (he trounces the French botanists for calling it a Veratrum in the account of D'Urville's voyage) is now individualised as *Chrysobactron Rossii*. These islands he found to be 'the richest spots we visited anywhere for new and beautiful plants, and the number of species I collected, on examination far exceeds my most sanguine expectations—330 in all' (September 1844). Sending his notes to Ross in November 1844 he writes:

> These have been drawn up in the rough for some time, but the most important parts, concerning the proportional amount of the different orders, present such curious results, that I was anxious to go over all the figuring again, which is (as you may perhaps remember) *to me* very laborious and slow work. As it is I do not know whether they are too *short*, but the vegetation was so *very* remarkable and so unlike any other flora to compare with it, that I feared making so prosy a thing longer. On the other hand they may be too long, but I did not know how to say less. All I can do is to repeat my hopes that you will use your discretion with it. My Father has looked it over and approved it, but says with me that the Flora is too novel to say less of, and by being so, too unintelligible to most to render much more readable. So I hope I have steered a middle course. Certainly no spot on the globe has so large a proportion of new plants and far less of such beauties.

The last of these botanical sketches asked for by Ross was that of Cockburn Island. This took some time, for (December 15, 1845) he had to compare the species with the Polar ones before venturing to write anything definite upon them.

As the book went through the press he saw proofs of the earlier part, and to his horror found that Ross had reproduced his account of the Fossil Tree which had appeared without his wish or knowledge in the Tasmanian Journal. It had not been written for publication, and with Ronald Gunn's conjectural emendations, was in places unintelligible. The great Robert Brown on seeing this had dubbed it 'a very careless production.' He at once begged Ross (January 30, 1847) to correct the unintelligible words, offering as an alternative to rewrite the whole thing.

On the other hand, he helped Ross materially by lending

him his Journal, writing an account of the cattle-hunting in the Falklands at John Murray the publisher's suggestion —the subject being only scantily referred to in the Journal —and supplying a number of illustrations (see p. 86). These were vignetted for wood-cutting from Hooker's original sketches by Walter Fitch, the Kew draughtsman. Fitch was accuracy itself when drawing plants; but in landscape Hooker found that he 'refined upon Mount Sabine without improving it,' and soberly pencilled above it a more faithful outline of the mountain.

Of the specialists who lent their aid in working out certain sections of the Cryptogams, Dr. Harvey was the most valued helper as well as intimate friend, to whom he could write with entire freedom. One of his other helpers indeed 'describes by steam, and all I can say is, I hope I shall not have so many *remarks* upon yours as his; *remarks* is an uncommon modest word here I assure you.' In fact, Hooker had to do that work all over again. But as to Harvey, no one should touch the many seaweeds until he had a fair chance. 'I send,' writes Hooker (May 21, 1844), 'everything on which I can lay my hands—because you must see whole suites of things to judge of them.' His intention was to keep the Antarctic Algae from Cape Horn, Falkland Islands, Southern Ocean, and Kerguelen's Land 'distinct from the Auckland and Campbell Isld. ones, as the phenogamic Floras of those regions are very distinct.'

> . . . I think the sets of *Macrocystis* will prove that too many species have been made of the genus—but I should like all the forms, made by Bory[1] into species, to be acknowledged under some form or other, as my great anxiety throughout will be by my book to show that the English have done as much for Crypt. Bot. as the French [apropos of Montagne's brochure on the subject], and I wish particularly always to state who was the first discoverer of a species. . . . I am also particularly anxious that the

[1] Jean Baptiste Bory de Saint Vincent (1780–1846), naturalist, soldier, and geographer. He sailed in 1800 with Baudius, the geographer and naturalist, to explore the Australian coasts. Owing to illness he was left at Bourbon, and proceeded to study its natural history.

Geog. district of the species should be mentioned under each. I am sure you can give me vast help in this. My Father thinks they should be published under our joint names, but I expect your kindness will lead you to do so much before I can begin that I scarce see how I shall be entitled to further credit than as a collector; should you not think my name too presuming, I beg you to understand, that I am quite ready to swear to anything you say, to stand Godfather to any names you may insert, and to believe anything except that the French have made better collections than the English.

As to the question of making new species, he remarks:

Generally speaking the plants (Jungermanniae) are *very* distinct from the European ones, though externally, like all creeping Crypts., they look like them. The fact is that all those who now have continued the study of Hepaticae for many years, find that besides the Europ. species having wide ranges, there are plenty more with as wide elsewhere and others that are local too. Taylor has discriminated well, but not compared well with other discriminators.

But:

I am proving all or most of the Lycopod. to be the same.

As to mere changes of nomenclature:

I am not the least frightened at your changes of names. I always liked to call you a *sticked algologist*, but that is only in comparison with myself. The changes being for the better are signs of your improving! The greatest men change their minds oftenest, e.g. Brougham, Stanley, Graham,[1] and your own dear Don,[2] who is a trump in my opinion.

[1] Graham, the Home Secretary of 1845 (see p. 204), was a lesser political luminary than Lord Brougham and Stanley, 'the Rupert of Debate.'

[2] David Don (1800–41), botanist, son of George Don, for some time Curator of the Edinburgh Botanical Garden. Through Robert Brown he was employed at the Apothecaries' Garden, Chelsea, where he became Librarian, and in 1822 succeeded Brown as Librarian at the Linnean Society. In 1836 he was appointed Professor of Botany at King's College, London.

But excessive or ignorant species-making is to be dealt with relentlessly, especially when made at second-hand, as in a given case by Montagne, resting himself upon the supposed infallibility of a certain observer. And he adds:

> My dear friend, I want no enlightenment or refreshment about *Ballia Hombroniana*; I examined them native hundreds of times; it is one of the most common southern Algae, and I often tried if that state was a different species; Brown would not make me believe it a good one.
>
> I shall give Montagne a rap over the knuckles if he does not look out; we are not all fools because he is so double-barrelled knowing; it is childish of him to insist against the testimony we have and which he has no grounds whatever to disprove; it is silly of him to adduce as an argument that an unbotanical man pronounced them distinct.[1]

Against Montagne there was another score to be chalked up. He was bringing out a book on the Algae himself, and Hooker had sent him a copy of his best plate of Alga drawings. With this Montagne was so much delighted that he promptly incorporated it in his book, a most undesirable form of compliment. To Harvey, who was much upset by the incident, Hooker writes:

> With regard to your *cher confrère*, I have had a hearty laugh at your distress. I am wholly to blame for being *so weak* as to send him it; feeling as I did at the time how dangerous a thing I was doing. . . . However, I try to laugh off my disappointment at being chiselled so dirtily out of my pet plate amongst the Algae. Confound his

[1] A little later, the same point is amusingly exemplified in the description of Planchon, the Kew assistant, given to Bentham, September 25, 1846:

'Planchon thrives, i.e. grows leaner and looks yellower and hungrier. He is getting up his geography with a vengeance, and now no two plants can be the same, if gathered two miles apart: he is hammering away at the Compositae splendidly, and after having abused D. C. for making infinitely too many species on other genera he now wants to make more of Senecio! even of the S. American, all *except* the Antarctic of which he says I have made too many. There never was such a compound of contradictions. I benefited enormously by his views and "ça touche's" on genera and orders, but on species he fairly drives me mad. We are capital friends, however, only bicker a bit. He is now trying to get some friend's picture of a water-lily exhibited at the R.A. next year; I tell him he might as well try to get himself into the Book of Beauty.' Cp. p. 344.

impudence to ask for Hepat. etc. in the same letter as he so coolly boasts his guilt and shame. I have promised, however, and shall send them, 'sans lettre' however. I shall drop *cher confrère* quietly, as our friend Berkeley has H.; and place him 'inter eos maxime vitandos.' . . .

One of these Southern Algae, contributed by Darwin, was difficult to identify, and called forth the following to Harvey, November 11, 1844.

Do not bother about Darwin's Alga till I tell you; such a chap as that will, after all, require some of the double-barrelled powers here in London to solve it, and after I get your verdict I shall ask Berkeley. I shall be amused to know how many genera I can get it put in by a good many observers. When you have done with it I will have a crack at it myself, and after I get all verdicts separately, I will acquaint you. I shall let no one know that another has examined it.

Meantime Sir William was keeping a prudent eye on the possibilities of any permanent post that might suit Joseph, whose own views on the subject are shown in a letter to Dr. Harvey (March 10, 1844), when, speaking of Harvey's candidature for the Dublin chair, he says:

For my own part I should have preferred the Curatorship with half the salary, to the Professorship, which would have obliged me to give two courses of Botany, besides having the fear of being obliged to take Medical duties (i.e. Clinical lectures), for which I am neither competent nor inclined. I could not be a good Botanist and Medical man too.

For a moment there seemed a chance of the Curatorship of the Dublin Herbarium, left vacant by the death of Dr. Coulter, till it was resolved that this be attached to the professorship of Botany, which would be given elsewhere. Robert Brown's health was failing, but succession to his important post at the British Museum was out of the question. 'We must never think of Brown's situation for Joseph' (writes Sir William on December 14, 1843), for 'Bennett[1] [his assistant] would in all human probability outlive and succeed him.' In November 1844 came news of a vacant Curatorship of the Botanical Gardens at Sydney, but this would hardly suit his views, even even if the salary were better. In the course of the winter came the proposal to lecture for Professor Graham at Edinburgh, with a fair prospect of succeeding him in the Botanical chair. The story of this is told in the next chapter.

In the meanwhile, Hooker proceeded to fulfil his intention of seeing the chief Continental botanists, and comparing their Gardens and collections with those of Kew. He hoped also to effect exchanges of specimens and living plants.

Midwinter certainly was not the ideal season for such a visit, but Schomburgk,[2] another distinguished traveller, was going to Germany, and promised to act as his 'chaperon' there; moreover, any permanent appointment at home might interfere for a long time with further travel, which in itself was one passport to good society in such a place as Edinburgh. And at this moment it would involve no delay in his book; the next two monthly parts were ready for press. He planned an extensive journey, including a visit to 'a man of the name of Alexander Braun, who has written on the development of leaves and branches in a spiral direction, and who has *developed* the laws of their *development* and future directions on the plant. Mr. Brown thinks Braun a very first rate man, though a little known one, and considers him as well worth my seeing as any man abroad.' (To D. Turner, January 26, 1845. Cf. p. 425*n*).

But Sir William warned him that all the time at his disposal would be taken up with seeing what was to be seen at

[1] John Joseph Bennett (1801–76), botanist, was Robert Brown's assistant in charge of the Banksian Herbarium and Library on its transfer to the British Museum in 1827, succeeding him as keeper in 1858. He was secretary to the Linnean Society, 1840–60; F.R.S. 1841; and published various botanical papers.

[2] Probably Sir Robert Schomburgk (1804–65), discoverer of the Victoria Regia lily, who was knighted at the end of 1844 on his return from his three years' travel delimiting the frontiers of British Guiana. His brother Richard, who had accompanied him as botanist, had returned to Germany in 1842. After the political troubles of 1848, he fled to Australia, where he cultivated the vine with great success, and in 1866 became director of the botanical gardens at Adelaide. He survived till 1890.

Paris and Berlin, and he gave up the idea of a longer journey. Finally, time growing short, he contented himself with Brussels and the Dutch towns instead of Berlin.

He reached Paris on January 30, travelling by way of Southampton and Havre.

> This route takes me through Rouen, which I should hope to be able to see a little of, though the object of my journey is so entirely to see *men* more than *things*, that I cannot afford to delay much.

His promised fellow-travellers did not make their appearance; but he scraped acquaintance with other travellers, including one Reimers from St. Domingo, whose brother he had met at Rio, and a Frenchman from Rio, who could not speak a word of English; 'a very shrewd fellow and liked everything English but Sundays, which were quite insupportable, there being no innocent amusements in which he could take part on that day.' Leaving at 2.30, they reached Havre at 1.30 A.M., when

> we were immediately roused out of our beds, no one, according to Customs Laws, being allowed to remain on board after arrival. . . . Havre is very dirty, the houses very narrow and tall; those along the quays are composed of sundry bits of all the (rotten ?) vessels that ever were stranded; the air of the whole place was that of Greenock, though not quite so noisome.

The Customs next morning had troubles of their own.

> My things were overhauled in a house and turned out for me to repack in the street. . . . They charged for Brown's *Rafflesia* books, against my earnest remonstrances,—I showed them the names of the illustrious Bobby himself, of Humboldt, Ehrenberg, &c., &c., written in one or other, but they were inexorable; it was the plates they charged for, and if I had told them that I deserved a premium for importing the works of Bauer, they would not, I expect, have regarded it.
>
> [On the diligence to Rouen.] The stages are about three leagues long on an average, and a new driver to every one.

The same guard goes throughout dressed in a magnificent silver-lace uniform, covered with a blue blouse. Altogether he was an ill-conditioned dog, and fitted his garments like a hog in armour. The drag is curious, being a sort of compressor, worked by this guard who sits in this Phaeton with me and others, turning a thing like a coffee-mill handle, which produces a pressure on the axle of one wheel, aiding the diligence in turning and taking the pressure off the horses in descent.

By dark they reached Rouen; thence by rail to Paris; '100 miles for 16 francs, 14 stoppages, 4 hours in passage, 3 tunnels, one 3 miles long.'

Thanks to Baron Delessert, a wealthy amateur, to whose collection alone Sir William's took second place, he was able to move from his first hotel, where 'last night I had some of my *Erebus* friends in bed,' for clean rooms at the Hotel de Londres in the Rue des Petits Augustins, 'but and ben with Baron Humboldt.' One or two impressions of Paris in 1845 may be quoted from a letter to his mother (February 2).

My way led through the Champs Elysées, which are very dirty indeed, and I soon got terribly splashed with mud. I do not think these town avenues at all in good keeping; they are half rural and that is all; the broad flagged pavements and macadamised roads, covered with carts and coaches, do not suit the noble trees at all, so that I could not in any way compare the Champs Elysées with the avenue at Bushey Park or at Inverary—the trees look much more to advantage in our parks, where we have not rows of shops at their backs and restaurateurs under their shade. [Apart from the individual beauties of such buildings as the Louvre] there is here nothing so good as Regent Street, though a little bit of the rue Rivoli and the rue Royale are better than any equal portion taken out of that London thoroughfare. [Going to the rue St. Honoré to call upon Lord Howden] the street is very narrow, so that two can scarcely walk abreast upon the pavement, and the stoppages of carriages and carts are ten times worse and more numerous than (in the) Strand at Temple Bar.

His first meeting with the famous Humboldt is thus described:

On putting up here I sent in my card with Mr. Brown's books to Baron Humboldt; he was not at home, but sent his flunkey (Scoticè Footman) to my bedroom at 8 o'clock yesterday morning to say his master wished to see me at 9. Ten minutes after his Lord had grown impatient and sent to say he was all ready, so I went in and saw to my horror a *punchy little German*, instead of a Humboldt. There was no mistaking his head, however, which is exceedingly like all the portraits, though now powdered with white. I expected to see a fine fellow 6 feet without his boots, who would make as few steps to get up Chimborazo as thoughts to solve a problem. I cannot now at all fancy his trotting along the Cordillera as I once supposed he would have *stalked*. However, he received me most kindly and made a great many enquiries about all at Kew and in England, particularly about Mr. Brown and my father.

In a letter of the same date to his sister Maria he draws a keenly etched picture of several distinguished botanists then in Paris, a companion picture to his careful comparison of the Jardin des Plantes, the libraries, collections, and glass houses with the establishment at Kew.

I have seen a great many men here, but they are so swallowed up, in general, with self-conceit that the only way to make oneself agreeable is to hold your own tongue and allow them to rattle away; each begins by telling you *literally* of the magnitude of their works, whilst of those of their neighbours they seem to know very little indeed. To this there are exceptions, of course. There are truly a large concourse of Botanists here, but they do not appear to me such sterling men as we have by any means. There are six Botanists at the Jardin des Plantes, three heads and three subs of the heads. Only one loves Botany for its own sake, who is M. Mirbel,[1] who was out when I

[1] Charles Francois Brisseau de Mirbel (1776–1854), artist and botanist, deserted science for ten years in favour of civil administration, but returned in 1827 to a professorship at the Paris Museum of Natural History. He was one of the pioneers in microscopic anatomy and vegetable physiology. Of the friends Sir William had made among the French botanists when he visited Paris in 1814, Mirbel and Bory were the only survivors.

called. M. de Jussieu, son of the mighty Jussieu,[1] does not really love Botany, but wears his father's shoes though they pinch him. Being clever, all that he does is good, but that is not much; he is extremely kind and amiable, but close, and buys no books. He took me for five hours round the garden in the kindest manner, but never once opened his lips to ask about Botany in English gardens or plants; he is the teacher of Botany. M. Brongniart, a clever youngish man (he looks twenty-eight and is forty-eight), is the second head, and his department is to name the garden plants; he is considered hardly a Botanist at all, but is fond of fossils though there he has done nothing lately. Mirbel is the third head, who cultivates the plants, and a pretty mess he makes of it, I assure you, for worse grown things I never saw; in their best houses they look like our smoke stoves exactly.

Now the great aim of every French man of Science is to become a *member of the Institute*, of whom there are but very few, and only added to by the death of one of the original members; all having one aim and that being ambition, they quarrel like cat and dog, and excepting Brongniart and Jussieu there *is not one* who has not many enemies, as it is said these two would have did they study Botany and were they not members already, very much because they were their fathers' sons.

To his Father

February 13, 1845.

I have been very busy since I wrote last, chiefly in the Herbarium of the Jardin des Plantes, which grows in magnitude under my eyes ['though it must be confessed. he adds four days later, 'that the want of space and proportion of paper are enormous']; its riches are very great, and the persons connected with it are all so extremely kind to me that I can hardly thank them enough; they have given me 300 species of New Zealand plants, chiefly from the Middle Island, and where they have duplicates of

[1] Adrien de Jussieu (1797–1853) succeeded his father as Professor of Botany at the Paris Botanical Garden in 1826. In addition to several important botanical memoirs, he wrote a very successful *Cours Elementaire de Botanique*, while many botanists of all nations were trained by him. His father, Antoine, Sir William's friend, wrote the *Genera Plantarum*, the principles of which were adopted and enlarged by De Candolle.

other things are quite willing to send the first set to your Herbarium.

I spent a whole day with Decaisne [the third aide] over his drawings, &c.; they are most beautiful, masterly, and truly botanical; he is too a most amiable and excellent fellow, is modest and well informed, by far the best Botanist here on all points. He sent to Normandy on purpose for Seaweed to show me his marvellous discovery of the animalcules in the organs of Fuci; I suppose it is the most curious of recent discoveries and opens the widest field for discovery. I am quite astonished with what he has shown me. He has arranged the Fuci of the Herbarium most beautifully. . . . His whole pay is £62 per annum, and yet he takes my book; but every one here considers him a model of generosity.

The question of buying Lenormand's collection of Algae when so small a proportion were new, prompts the reluctant advice to his father to 'give up purchasing for the present wholly. We have far more plants than we know how to keep in order, far more expenses, which are annually increasing, than we have the means to cover,' not to mention the growing expense of books, for 'plants without books are useless.' His fortune was not, as the Paris botanists fondly imagined, equal to that of Delessert, his only rival in purchasing in Europe, and 'I do feel quite sure that you cannot on your own means support a Herbarium which is, as you wish, to *keep pace with the progress of Botany.*'

The following passages from a long letter to Harvey towards the end of his stay in Paris deserve quotation as illustrating not only the kindness of his hosts and their respect for his father, but his own readiness to readjust his personal preconceptions.

February 25th, or thereabouts.

I ought to have written to you before, from this great mother of Babylons, but have been too busy enjoying myself selfishly, to think much of my neighbours. This is indeed a wonderful place, and the natives are most uncommon polite, not only in word but in deed, for they pour upon me such loads of pamphlets and little presents as obliges me to make up a parcel for England, to go without me, to the land of my Fathers. . . . (All thanks to my father's name, for I have done nothing to please the French ; but his name carries me everywhere.)

My great allies here are Montagne and Decaisne, both of whom are extremely kind to me, and very remarkable persons in their way ; they have both fairly gammoned me into liking them, by force of good words and good offices, and the latter particularly I find to be an exceedingly good fellow, of whom I had formed a very wrong notion. My Hotel being close by Montagne, I see him every day for an hour ; he is a clever, active, little old man, who took up Cryptog. Bot. when nearly 50 years old, and has continued it ever since ; his knowledge of species is very great, and his collections kept in beautiful order ; of structural Botany he knows nothing, and is much too old to learn at 61 (as he calls himself). I have had sad work with the Antarctic Algae ; you never saw such specimens. Montagne very fairly says that he does not hope that his work is at all to be depended upon !

You know well how apt I am to form uncharitable opinions of people ; I hope I may prove as ready to make the amende honorable as I know them better, for now I must confess Decaisne to be the most remarkable Botanist for his age I have ever seen. In structural, anatomical, and physiological Botany, better judges than I say he is deep, nay profound, and his descriptive knowledge is very great, as is that of the Nat. Ords., and that of both live and dead plants specifically. His drawings are also very talented, and every one likes him but Montagne. The latter I have always found a most excellent and warm friend, truly anxious and willing to go to any trouble to serve me, never tired of showing me his beautifully kept and named specimens and atrociously vile drawings ; he is always pleasant and agreeable, but has the character of a tricky temper, with £100 a year as retired army surgeon, in which capacity he served with Napoleon in Egypt ; he keeps both house, library, and collection up, and subscribes to sundry concerts, the delight of his old age, for he is passionately fond of music ; he is also very generous and kind, a warm friend and generous.

Happily Hooker was able to maintain friendship with both these men, though they were of opposite temperaments and at personal variance with one another.

> The fact is that poor Montagne does make awful mistakes from neglecting structural Botany, and is very obstinate too; Decaisne, on the contrary, owns a fault on the spot, and is both frank and generous; his indifference to Montagne certainly does not mend matters. The latter is infinitely the most careful observer, though the more ignorant, his faults arise from giving over value to trivial characters and from misunderstanding the relation and structures of plants; the faults of the other are owing to carelessness. Montagne works slowly, steadily, carefully, and by a fixed method, examining a plant piece by piece, never making any great discovery, and but few remarks characterised by originality. Decaisne works like a horse, till his strength is exhausted and he is fairly ill, for he works himself to death; takes wide general views of things, appreciates an organic change, and comprehends it in all its bearings at once, but instead of thinking upon his discovery, jumps at a conclusion right or wrong.

Thus, returning to the question of the animalcules in the antheridia, which Decaisne showed him in the specimens of seaweed specially brought up from Normandy, he adds:

> They were all perfectly simple and easy to be seen. The vegetable origin of these, which have hitherto been considered animalcules, is very positive, though it may still be doubted whether they are a sex of the plant, which the diœcious, monœcious, or hermaphrodite nature of the several species would argue, as also their analogy to the so-called sexes of mosses—on the other hand, they may have more analogy to the motive spores of *Vaucheria* and of *Protococcus*; be that as it may, Decaisne not only believes them sexes, but forthwith cuts old Fucus up into three genera, depending on the monœcious, diœcious, or hermaph. state of the species!! You will no doubt agree with me that this is heinous and needs no proof of absurdity to any reasoning mind, and how so talented a man as Decaisne can behave so is a puzzle to me, for I know no Botanist but Brown so skilled

as he is in *all* that concerns Botany. I think I have reasoned him out of this or shall have before long, for he is both modest and open to reason.

His drawings of the genera of Algae are wonderfully numerous and beautiful; I often thought how numerous your exclamations of *comè bella* would have been, had you seen them.

The Botanists here have not ceased being kind to me, and such a three weeks of being lionised I never at all expected. I am quite aware that this is owing to my bearing your name, but so far out of sight as you are, it was very unexpected. Were it not that the style of living—(or rather killing oneself) here is very prejudicial, I should wish you to come here one spring, but I am sure you would be made ill, as I have been, and only recovered by dint of sticking to *Seine water* and letting *vin ordinaire* alone. This was a fortnight ago, and my poisoner was M. Gay, who eternally complained of the badness of his dinner, and made Webb[1] and me eat and especially drink more than we liked by dint of a similar pressing to what you underwent in Ireland. The poor man evidently thought us great guests, and that we were too proud for his table perhaps. . . .

(*February* 27.) . . . Humboldt I saw very often, sometimes three times a day, for he was never tired of coming to ask me questions about my voyage; he certainly is still a most wonderful man, with a sagacity and memory and capability for generalising that are quite marvellous. I gave him my book, which delighted him much; he read through the first three numbers, and I suppose noted down thirty or forty things which he asked me particulars about. I left him at the third number, and as he paid me two visits whilst I was out on the morning I left, he has doubtless not *digested* it all. I bade him three goodbyes the day before

[1] Philip Barker Webb (1793–1854) of Milford House, Surrey, early came into a fortune which enabled him to travel and pursue his studies in geology and botany. His observations on the Troad and his *Iter Hispaniense* were followed by his work on Madeira and the Canaries, where he spent 1828–30 with Berthelot, a young Frenchman who had already been eight years studying the islands. In 1833 they established themselves in Paris, where their great work, *Histoire naturelle des Iles Canaries* took fourteen years to produce (1836–50). The years 1848–50 he spent botanising in Italy, as a sequel to which he left his large collections and herbarium to the museum at Florence, then under his friend Parlatore.

> and the next day; he, as I said before, came twice for me in my absence. He talked in the warmest manner of Mr. Brown, Murchison,[1] and yourself, also of Darwin and Herschell. . . .

His plan was now to visit the botanists at Brussels, and to bring back the plants that Blume and Siebold [2] had promised his father by taking Leyden and The Hague on his way home (with a digression, if possible, to Haarlem to hear the organ, and to Amsterdam to see Linnæus' Lapland dress), and he adds later, 'I have seen such fine things lately from Blume and especially from Siebold that my regret is not so great at missing sight of Germany as it was a week ago.'

But one or two difficulties loom ahead on this Netherland visit, though the kindly French botanists gave him no less than twenty-six letters of introduction. Siebold and Blume, to whom he wishes one of the four remaining copies of the 'Genera Filicum' to be given as a return for gifts of plants, 'are on dreadful terms; I must manage between them.' More personal to himself is the result of an outspoken review in the 'London Journal of Botany.'

> Hombron is in very bad odour; I want to see him, but Decaisne and Jussieu say he is boiling with rage at us, and that I must not go or there will be a row. I find that that critique was well received here by those whose opinions are best worth having. At the Jardin the critique is considered quite fair as his work is a disgrace to France indeed, and that it is well to scold bad books as that gives a character to the Journal, and the latter is very well thought of here, especially the review part.

[1] Sir Roderick Impey Murchison (1792–1871) took up the study of geology after his marriage and retirement from the army. His chief studies lay among the ancient rocks of Wales and the Highlands of Scandinavia and Russia, where he assisted in the geological survey. His fame was secured by the establishment of the Silurian system. As President of the Geological Society twice, and of the Geographical for fifteen years, and director of the Geological Survey from 1855, he possessed large influence, enhanced by his wealth and social position.

[2] Philip Franz Siebold (1796–1866) spent six years from 1823 in Japan as doctor to a Dutch embassy, and became an authority on Japanese language, literature, and natural history. Then till 1859 he lived in Holland; revisited Japan 1859–62, and thereafter settled in his native city, Wurzburg. Besides introducing many Japanese plants into Europe, he introduced the tea plant into Java.

However, Hooker's natural tact brought him safely through.

The formalities of travel on the Continent in the forties were exasperating, his passport having to be signed by the Belgian and English Ambassadors in Paris and twice countersigned by the Prefect of Police. Ten days were filled with fruitless errands, and to crown matters, diligence and train failed to make connection at Valenciennes.

Brussels, where he stayed a second day to make acquaintance with Quételet,[1] at a meeting of the Brussels Academy, is summed up as 'a very interesting city, but not strong in Botanists,' though in the Garden 'the collection of Palms was excellent; . . . of other things they have no great store.'

At Ghent, where he did not fail to see the Rubens pictures, he went over Van Houtte's nursery gardens, 'most extraordinary, both for the number of species of Botanical plants and of Camellias and other such.' After arranging for exchanges of plants, he was invited to dinner by Van Houtte, who was as hospitable as he was liberal. One point especially in his botanical interests struck his visitor: 'he takes the Magazine and is going to have the Journal and the Flora Antarctica.'

Meantime the discomforts and difficulties of travel in such an Arctic winter are worth recording. March 4 saw delay of trains, the missing of diligence connexions, and consequent midnight journeys. 'I began to think,' he writes, 'that I should never get to Holland at all.' March 5 was worse than ever;

> the roads and rivers were so bad that several passengers were frightened and went round by some place South. Such a cruise I never had by land . the cold was intense, the thermometer at 7° with a keen wind. We crossed three rivers, one all frozen and covered with Hummocks and piles of ice, the second, the Maes, 1½ miles broad, loaded with huge masses of *Pack* and *Berg* ice, rushing down to the sea; the navigation

[1] Lambert Adolphe Jacques Quételet (1796–1874), a Belgian statistician and astronomer, Director of the Brussels Observatory 1828, and Professor of Astronomy 1836, and from 1834 Perpetual Secretary of the Belgian Royal Academy. Apart from mathematical treatises, his most important work was the book *Sur l'homme et le developpement de ses facultes* (1835), and later he turned his mathematical mind to the study of anthropometry.

was very bad and performed in boats, which were shot down from a bank on to the stream and pulled up and down the river, working many diagonals, at times fixed in the Pack and at others free again. In about 1½ hours we were across in safety, but wet and cold enough. As, however, all the little Cabarets have hot coffee, the cold did not much matter. The third river was half fixed Ice with great holes of water, and the boats were dragged or pushed or rowed according to circumstances. We arrived late at Rotterdam.

On the way home, a week later, all this had to be traversed again, it being impracticable to pick up the mail boat in the Rotterdam direction.

I went the first thing next morning (March 6) to Miquel, an intelligent and agreeable man, full of Botany, and who will prove an acquisition to us. I spent the day with him. . . .

Leyden, March 7.—Blume received me most warmly, and has shown me such wonders in the Museum and at his house as are almost incredible; he has all the Japan things. Blume promises me much, but he says I must take them myself, as he has no aid and no time to make selections.

. . . You have no idea of the richness of this place, such beautiful drawings, as good as Fitch's or very nearly; they beat the Paris ones, as Decaisne acknowledges. The B rd collection is superb, specimens, stuffing and attitudes. Here is a Penguin perfect, such a specimen I never saw alive; it is a truly wonderful place.

The Jardin des Plantes and this place are truly two epochs in my life. I must work very hard when I get home. I do not fear the lectures, but I am backward in British Botany.

Next day, the 8th, he writes:

Of all the Botanists I have seen, except Decaisne, Miquel is the one I like best and think the most promising; he has an excellent and rare knowledge of structure and of exotic genera and species, and his respect for you is very great. . . . Next to yourself and Mr. Brown I think I am asked more for Darwin than anyone; his book[1] has made him so many

[1] *The Voyage of the 'Beagle.'*

friends where he is not personally known. Reinwardt is in raptures with it.

Once back in England, he was busily engaged throughout the spring in sorting out plants as return gifts to his French hosts, in preparing for his Edinburgh lectures, in working at his Flora Antarctica and at the Niger Flora, based on the specimens brought back by the Expedition of 1841 under Captain A. D. Trotter. All these things, and especially the progress of the Flora, and detestation of mere species-mongering, are reflected in frequent letters to Harvey—a correspondence continued all through his stay at Edinburgh, for Harvey, who had recently stayed at Kew and worked there before being elected to the Dublin chair, was busily working out the Antarctic Algae, both Hooker's and D'Urville's from Paris, and was moreover a friend to whom he could scribble with the careless freedom of intimacy, now chaffing his friend, now poking fun at his own efforts as a lecturer, when lecturing turned out to be a less terrible ordeal than he had expected ; for, as his mother said, 'Joseph is not a sanguine or hopeful person : but he becomes attached to his work : thus we trust he will take interest in lecturing and *warm towards it,* as he proceeds.'

The book suffered many vicissitudes ; Harvey took up lithography and drew his own plates ; occasionally carefully drawn plates were spoiled by the engraver or colourist, and a monthly part was delayed ; so that the disheartened author exclaims, 'Never will I undertake such a work again. The Icones is the only model for what a Botanical work should be. I wish they would have let me publish in that form, and yet I sighed for glory too' (April 29, 1845). Then for a time Hooker, lacking the necessary books of reference at Edinburgh, resolved to end the publication with Part X. But the work was approved by those whose approval was worth having. His Edinburgh lectures over, he took it up again, and in October, being rejected for the Edinburgh chair, he was left free to complete it on the original scale, taking care that Smith's, Davis', Lyall's, Crozier's and Ross's names should be attached to five of the fine Algae that required figuring.

> Such scrubs as that *Pol*[*ysiphonia*] [he declares to Harvey] are rather *infra dig.* for an 'officer and a gentleman.' Cannot you spare some of those dandy *Delesseria*, or some showy things that will require a whole red plate? I do hate too much of this sort of thing, but I think they ought to come in. (April 14, 1845.)

Harvey carried out his wishes, for not only is there a *Polysiphonia Davisii*, but two *Delesserias* are named *D. Davisii* and *D. Lyallii*.

Meantime details are scrutinised; carelessness about species ruthlessly exposed. D'Urville's collection assigns a certain Alga to Lord Auckland's Island, where it was inconceivable that Hooker and Lyall should have overlooked it. He reminds Harvey how he proved in Paris that specimens were wrongly ticketed, and as for the so-called species itself (*Rhodomenia ornata*), which Brown enters as *Ballia Hombroniana*, 'I am convinced,' he writes, 'of its being no species at all, and long to restore the name *callitricha*, but "am not game"!'

Similarly, in an undated letter of 1845:

> I am now hammer and tongs at my Lichens, which are an Augean stable. The British species are humbugged by the introduction of varieties; if ever I publish an Ed. of Eng. Bot. I shall not hesitate to cut down Usnea and Ramalina to one species, all the intermediate forms of every-day occurrence.

CHAPTER IX

EDINBURGH

On October 17, 1844, appears the first reference to the Edinburgh Professorship of Botany,[1] which takes definite shape by Christmas Eve. Dr. Graham's health was very precarious; he was likely to resign his Chair soon, and as a first step, perhaps, require a substitute to deliver a course of lectures in the following spring. This substitute, if he did well, would be a strong candidate for the Chair with the backing of the retiring Professor. The Professor of Botany generally united two appointments in his single person, the College professorship, in the gift of the Town Council, and the less lucrative but more important Regius professorship attached to the Curatorship of the Botanical Gardens. This latter, being a Crown appointment, was in the gift of Sir James Graham, then Home Secretary, with whom Sir William's official friends would naturally have considerable influence. Acceptable as the prospect of £100 for the course of lectures would be to the young botanist, to interrupt his more serious work on the Flora without aiming at the permanent post would be against his best interests. 'It is indeed not easy

[1] *J. D. H. to W. H. Harvey.* October 17, 1844.

'I am not much nearer my fortune now than when you were here, and am getting very anxious to be doing something that will pay me—*on dit* that poor Dr. Graham of Edinbro' is on his last legs, and my friends want me, should he go off the hooks (which I from my heart say heaven forefend), to stand for the chair of Botany there (don't laugh). I suppose you like my impudence. I should not be sanguine, as the opposition would be very strong, and if Forbes stands he will be by far the most eligible: I have no great notion of lecturing, but I must pick up a livelihood somehow. How I shall quaque at my first lecture. You must not say anything about this, at present, visionary subject.'

for a Botanist to obtain a situation altogether agreeable to him, and that will afford him means of support.' Sir William might have said this with equal truth of any branch of science, and not at that time only.

At the same time Hooker fully realised the importance of completing his *magnum opus*. The arrangements for its publication in parts, month after month, rendered it impossible to carry out the scheme anywhere but at Kew. 'The value of my library and Herbarium,' writes Sir William, 'was never more fully evinced than in his preparation for his work. The British Museum, though invaluable in some respects, does not afford him a tythe of the information that my collections do.' With his usual generosity, Sir William hoped to make over the Herbarium to his son once he was established in Edinburgh, when it could be kept either at the Garden or in the College.

As it soon appeared, there was no question of payment for this course of lectures. Professor Graham had just suffered severe money losses, and was fatally ill. Indeed his increasing weakness prevented him from helping at all in the lecturing as he first hoped; and although he offered rooms at his own house, the good prospect of the succession to the professorship was regarded by the Hookers as sufficient material reward. To undertake the temporary course was both to make a trial of lecturing and to do his old friend a service, 'and I think,' writes his father, 'that alone will go a great way with Joseph.'

After Professor Graham's death, however, when his affairs had been wound up, Mrs. Graham wrote begging him to accept £100 for his great services. Hooker writes to Dawson Turner (April 25, 1846):

> She says it was only a portion of what her husband would have done, and entreats me to accept it if only to gratify her and all the rest of it, in such a strain as you can well understand without my repeating. I believe that no one could be more grateful for real services on my part than Mrs. Graham is for supposed ones. But if she would not add these testimonies of the sincerity of her regard, I should be much better pleased. To have felt as I did, that I had the

confidence of all the family under circumstances very trying to both parties, was reward in full for me. However, after due pondering on the affair and casting up the pros and cons, I determined to write and accept it, gratefully, for to accept it as if I really did not want money, would have been implying a falsehood on my part, and appearing proud to her. After all her feelings ought more to be regarded than mine, much tried as she has been, poor thing, and it will be a gratification to her to suppose that she has repaid me in part at any rate.

The matter was set in train; Robert Brown gave him a strong recommendation, and Professor Graham privately invited his help for the forthcoming course of lectures, with promise of support for the succession to the chair. The invitation was forwarded to him, for he was then in Paris, on February 3. It seemed the first and sure step to the professorship. 'The "Golden Durham" of Botany,' exclaims Lady Hooker to her father, 'the object for twenty years of his father's aspirations, is now, without Joseph's seeking, apparently put within his reach.' It would be very hard work to lecture for three months in addition to writing at the Antarctic Flora, but 'he loves labour,' she adds, 'and can turn off much work, and really takes such a pleasure in strenuous exertion, as a descendant of yours ought to do; to say nothing of his dear father and of my beloved mother's share in his parentage.' The Admiralty letter granting a month's leave of absence for travel abroad enjoined him 'not to enter the service of any foreign Power: this will not apply, 'tis to be hoped, to the service of Professor of Botany in Edinburgh!'

At the advice of his father and Robert Brown, and especially of his grandfather, he accepted the proposal, albeit lecturing was not to his taste, though he might 'like it better upon trial.' He was by no means inclined to become a botanical or any other professor, and but for Dawson Turner's advice would have declined the Edinburgh chair if it came his way. There was more in this reluctance than mere dislike: and he took his grandfather into his confidence before resolving to proceed and overcome it as best he might.

To Dawson Turner

January 16, 1845.

As to lecturing in London, there is at present no opening for it, nor should I like it except it was surely profitable. You do not know, nor do I like to tell my Parents, how wholly unfitted I am to be a Lecturer, constitutionally in particular. I am really nervous to a degree, and though I joined debating societies on purpose and studied speeches and stood up too to deliver them, I never could get two sentences on—I have earnestly endeavoured to conquer this, but without avail. I have consulted medical men, who tell me I have irritability in the action of the heart, which some have pronounced a slight disease of that organ; and this I know well, that I could never even stand up before my fellow scholars to say my lesson at school or college without violent palpitation. You know me too well to think me a coward, or, still less, to accuse me of affectation, but this I do certainly think, that I am naturally unfitted for any situation calling for a public exhibition of myself. My case is not as if I never had to *parse* or *construe* before a body of fellow mortals, for surely if this feeling was ever to be overcome, it would have been in eight years of college-life and with my efforts at debating, where I have always had to sit down in shame and confusion, however carefully I had conned my speech. This, and this alone, has led me always to hope that I should pick up some situation where hard work and good manners were all that should be required of me, though in leaving the public path I should not so soon rise into notoriety.

Of course I should forego all this dislike, or, as I believe, physical incapacity for lecturing, were anything so tempting as Edinbro' offered, and even then one's own students would form a more private body than the miscellaneous assembly of a London institution. Do not think that I am frightening myself with any such bugbear as a *Heartdisease*, for I assure you I give no thought to the matter, though I cannot help feeling, from the frequency and pain of my palpitations, that I have a nervous affection there. I have no idea of its *calling me away early*, though I shall probably not live to your age in the ordinary course of things, but even if I did, I should not al er my opinion or be alarmed, knowing by experience that I could, though ill-prepared, face my end with more calmness than I should a miscellaneous assembly of students. . . .

My dear Grandfather,—Your kindness has tempted me to lay my heart open in a way I have done to no other person. What I say here is not the result of a month's or a year's opinion, but of the experience of the greater part of my lifetime—I would not for the world that my Father or Mother knew that I had ever been to a Medical man about myself, which I have done both before my voyage and after my return, and received a very similar verdict which, though it contained nothing to alarm me, was sufficient to prove that I need not expect ever to attain a freedom in public delivery.

Pray do not hint on this subject in your letter here, it would only vex and do no good. I think my father rather inclines to keep me here, and though I do not want to be a burthen to him, I hope I am not altogether useless. My aim is not, however, to live always in this house, if I could only get some situation elsewhere. That some opening will come I cannot doubt, in the meantime my income is not much under £300 a year as long as this work lasts.

Hotel de Londres,
Rue des petits Augustins, Paris:
February 5, 1845.

My dear Grandfather,—I cannot let this post go without a letter, however short, to tell you that I have accepted the office of Lecturer for Graham, unconditionally for itself and its consequences. Though it is an expensive procedure, I would prefer commencing as assistant without the *onus* of being the Professor; as being more advantageous towards so young a lecturer and one so unfitted for lecturing as I shall at first be. I shall hope to get over my nervousness in time. There appears no doubt of my future success, when a candidate for the chair, in the meantime I only do a kind office for my poor friend, without emolument and indeed with great expense to some one or other, for he says that he has nothing whatever to give to the assistant. I hope he will not ask me to live in his house, which I should most decidedly refuse to do.

However little suited to my taste and my habits a Scotch Professorship is, and however much I shall regret giving up my book (the aim of the last twelve years of my life), all *that* shall not interfere with my determination, in whatever situation in life God may place me, therein to excel. I shall not only use every exertion to be Graham's best assistant, but also to raise the Botanical chair to Botanical excellence, and to have it a useful appendage to the College; and no longer a burthen to students' pockets, without Museum or any advantages for making men Botanists; I should also like to raise the standard of that lowest of all classes of students, the medical; but that shall be a secondary object.

I do feel a deep regret in having to desert my book, which I have lived so long for. Money, time, and labour, all my preliminary education, all my holidays from the first day I entered college, were devoted to laying myself out for making a voyage and publishing the results. Except that this chair allows me to continue a Botanist, I would just as soon turn to the law or to business as anything else that took me off the travail of so many years. I shall, however, hope for better times, and though the Government will take (and properly take) my pay and perhaps grant away, I shall live one day to finish my book. If I do get the chair, I shall commence laying up money to enable me to house my father's plants, whenever they may come to me, for I am determined no one but myself shall have them.

Here is Humboldt often speaking of you; he wants me to write the distribution of Plants for his grand work 'Cosmos'; pray say nothing of this to anyone. I can but live and hope, but Humboldt is so old that it may never appear.

Of the impending lectures he writes to Harvey (April 2, 1845):

Graham tells me he has not a single lecture written out! and that I must dwell much on physiology, chemistry, and morphology, in which my Father's lectures are particularly poor. This is no joke to me; what with Cryptog., Paris duplicates, and these lectures my hands are full indeed. Graham's lectures are always considered useless by and to his students, and so I am in a regular fix, nor have I cheek enough for an audience. I would rather go to the S. Pole again by far than to Edinbro', but it is no use growling. . . . [And later] I am in a stew already, but must trust to providence and my middling good fates.

Harvey, who on the 9th had written, 'My letters come as quickly as events in the life of Solomon Grundy,'' replied on the 10th with good advice :

I pity you the mess you are in about Edinburgh, knowing well what a fuss I should be in, in your case—but I expect you will wriggle out of it bravely. Be provided with written lectures for the parts you are not glib in, and skeletons for the rest—plenty of pictures—and talk much about these. Hand about specimens, and 'twill all get on right well. Here we had Allman [1] last year taking half a dozen lectures to describe the cellular and vascular tissue alone! and by the time he got to the end of the structure and physiology the course was expended, and he had to sum up arrangement &c., in a few words. Very convenient for him, but query, what for his Class ?

Hooker's response on April 14 asks :

Who is Solomon Grundy ?—but I am very behindhand in polite literature ; how do you find time to read what a gentleman should know ? I have given up all hopes and intentions of being accomplished,

and proceeds to set forth the difficulties of the situation, which left him sometimes, as he told his father in June, 'in a pretty fix between my own mind, my master, and my men.'

Graham has not *one* lecture written out and he has given me a syllabus of the course ; you never saw such a thing ; he goes through with no order, introduces his subjects higgledy piggledy every day, and does not give one really instructive lesson throughout the course. I have no idea what I am to do, I heartily wish he would leave it all in my hands or write me lectures ; I sincerely say that no human being could lecture for him as he desires, certainly no student

[1] George James Allman (1812–98), was Professor of Botany at Dublin 1844–54, and of Natural History at Edinburgh 1854–70, and President of the Linnean Society 1874–83. His special branch of science was marine zoology.

could follow him through such a medley of subjects, introduced wholly without method and order, and with no relation to one another, he follows neither a book nor his subject. He says he finds the students will not follow a regular course. I am in a deplorable state of uncertainty : nor can I write out a lecture to include, as each and all his seventy do, a little of all branches of Botanical Science, including the original production of species ! in some.

He also presses me, disagreeably hard, to take up my quarters with him, which I have fifty reasons for not doing and not wishing to do. I never more heartily wished a man well in my life.

To this Harvey replied on April 17 :

I pity your lot about Graham—to me it seems absolutely impossible to follow the course of such a Sun—and therefore I would cut out a new line for myself—were I you—digest my subject into seventy discourses (if there be that fearful number) and write out at least the heads—with a grand oratorical first lecture—in which you should talk of matters and things in general—and, like a friend of mine on a similar occasion, mention 'Oscillatoria *trembling* on the borders of animal and vegetable life,' or like the old gentleman formerly, looking two ways at once.

The Professor of polite literature sends you the following:

Solomon Grundy was born on Monday,
Was christened on Tuesday
Was married on Wednesday
Took sick upon Thursday
Died upon Friday
And waked on Saturday
And buried on Sunday—
And this was the life of Solomon Grundy.

By the beginning of May he was settled in lodgings in Edinburgh at 20 Abercromby Place. For personal reasons, and wishing above all to be quite independent in his movements, he declined Professor Graham's urgent invitation to be his guest, though painfully conscious that he might be accounted churlish in thus refusing the only form of return which, as has been said, was possible on the part of his old friend.

As regards the lectures, the arrangement was that he should deliver Professor Graham's own course. As has already appeared, he early felt some doubt of their complete sufficiency, and even while still in France he contemplated using some of his father's Glasgow lectures, as well as writing others of his own. But the event outran expectation; Graham's syllabus was unsuitable. Some even of the most recent discourses were on budding and grafting, composed at a period when the appointment of a Professor of Horticulture was threatened. Thus he was compelled at the shortest notice to write new ones of his own in the scanty hours left by a multiplicity of occupations. He was slowly at work—with little progress for want of time and special books—on the Flora Antarctica; was following a course of lectures on Organic Chemistry; straightening out Professor Graham's affairs, preparing the campaign for the election to the chair of Botany. If the professors, for the most part, seemed to take little trouble to seek him out, Edinburgh society overwhelmed him with attentions. Some account of these things is taken from letters of the day, beginning with the first lecture on May 5, and Graham's extraordinary effort in presenting him to the students.

> My dear Father,—The weather being fine there was a tolerable attendance at the class this morning of about 120 people, who came with itching ears to see a reed shaken by the wind. I plucked up courage enough to get through without any outward or visible signs of my own want of confidence in the treat I had prepared for them.
>
> It was my own composition, and I read it so fast that no one could follow me and find out the mistakes.

And on the following day he continues the story to Harvey:

> I am lecturing away like a house on fire. I was not in the funk I expected, though I had every reason to be in a far greater one.
>
> On my arrival here I found Graham very bad in bed, he had not been out of his room for weeks and did not expect ever to be again. The day before my 1st he took the determination of going down to introduce me to the students,

though no better and wholly unfit for the task. We all opposed it most strongly but unavailingly. A Fly was hired and Mrs. G. went too and sat in the back room. On the road we passed Principal Faith going down to hear me go off, and him Dr. Graham enlisted too. At the door we fell foul of Arnott, and he and his brother also were impressed. We all went into the class-room together, myself like a candidate amongst his constituents. Graham first introduced me, he could hardly stand but did not faint; the Principal did the same, myself looking like a fool and muttering angry words to myself. After which I read them a screed on the influence of vegetation on creation, wholly opposed to Graham's teaching and doctrines, for he holds that plants and animals are in all functions precisely the same, and I that they are diametrically opposite. Altogether the being shown up as I was, and having Brown's far too flattering testimonial of my attainments and *moral character* read by the Principal, was hateful to me.

The class is small apparently; the room holds 160, but has never yet been full. I do not expect there will be much over 100 altogether. All hands are very friendly to me, and I suppose that I stand a good chance of being booked for exactly half my life in Edinburgh, for I shall never stay here more than half of each year if I can help it.

Forbes [1] does not think of the chair; he told me so the other day voluntarily, but that he would like that of Nat. Hist. Jameson's [2]—who has long been in most precarious health.

[1] Edward Forbes (1815–54) was a brilliant worker in botany, geology, marine zoology, and palæontology, who travelled widely in Europe as well as in Syria and Algeria, and was naturalist on board the *Beacon* in 1841. After holding the chair of Botany at King's College, London, from 1842, he was appointed Palæontologist to the Geological Survey in 1844, leaving this for the chair of Natural History at Edinburgh in 1854. In 1853 he became President of the Geological Society at the unprecedentedly early age of thirty-eight. His important paper 'On the Connection between the Distribution of the existing Fauna and Flora of the British Isles and the Geological Changes which have affected their Area' (1846) dealt with a subject in which both Darwin and Hooker were then at work. Forbes was not only a witty writer and the genial founder of the Red Lion Club, but a personality equally beloved and admired.

[2] Robert Jameson (1774–1854) was appointed Regius Professor of Natural History and Keeper of the University Museum at Edinburgh in 1804. His main work was in mineralogy, but he also wrote on geography, ornithology, and travel. With Sir David Brewster he was the joint founder in 1819 of the *Edinburgh Philosophical Journal*, and for the last twenty-five years of his life sole editor.

May 9, 1845.

My dear Harvey,—999,999 congratulations on Van Voorst's happy appreciation of your algological properties:[1] 10,000 I reserve for myself alone, some day: when I have as much reason to be as thankful as I sometimes tried to be for mercies vouchsafed in the old *Erebus*. I have positively nothing to say but to congratulate you. For my own part you may also extend to me a little gratulation on my beginning to feel the truest and most heartfelt pleasure in having come here, and in having come with no selfish object in view; and in having overcome my modesty, i.e. metamorphosed it into modest assurance. . . . I never felt so happy in being able to be useful, for Graham is as nearly helpless as possible, and though surrounded by friends, there are none who can help him in his class, garden business, examinations, and many other little things.

To the Same

May 30, 1845.

As to lecturing, that now comes perfectly easy and natural to me, and I can spout an hour of gas, without notes even, by dint of desperate cramming: the fact is I found that human nature, i.e. my nature, could not stand the drudgery of writing out an hour's reading from day to day, so I took to the extempore preaching, and find that it answers to the students even better than to myself: they do seem here to delight in *generalities* however false, if attractively delivered [i.e. without being read], and by dint of never losing an opportunity of comparing the vital phenomena of animals with those of vegetables (right or wrong) I can rivet their attention *au merveille*. I often think how I should blush to see what I speak in print. I often think how you would laugh to see and hear me gull the multitude, for they are like all other crowds.

. . . I have picked up acquaintance here with a funny old fish who devotes himself to fossil Botany and has splendid specimens marvellously cut for the microscope, Nicoll, the great fossil cutter, who has a splendid cabinet of specimens of wood etc. I am really anxious to form a fossil Herb., it suits my generalities about the floras of byegone ages, so pray do

[1] I.e. in undertaking publication of his book.

not lose sight of any you can beg, buy, borrow or steal for me.

I am always up at 6 and go to the garden at 7. At 9½ I go up to Graham's and breakfast and then down to the garden again, where his Herb. is. I work at it the rest of the day or when able go to Gregory's[1] lectures on Organic Chemistry from 3–4; then return and dress for dinner and call to see how Graham is. (I am rather heavily ironed with Society here, and have not paid for one dinner since my arrival—even with a headache.) I generally get home about 11 and cram for lectures like a dragon till 1 or 2—you see I must dine out for two reasons, first because the good people must know me before they elect me (do not say the safest plan would be to stay at home !), and secondly because I hear a great deal of excellent music in this town which is irresistible. Balfour[2] is exerting himself to the utmost with the townspeople and I should not wonder to see him carry the chair: I assure you I shall be quite content to go back without the Professorship if I could only see these unfortunate Grahams safe through their sea of troubles.

No wonder that by the end of June he says:

I get very tired of it towards the end of the week. Wednesday is my favourite day, as three lectures or the half is over; Thursday I get weary in, but the knowledge of Friday being the last lifts me through that hour.

[1] William Gregory (1803–58) was the fifth in lineal descent of his family to hold a professorship at Edinburgh, the first of mathematics, three of medicine, William himself of chemistry. He was a pupil of Liebig, whose works he edited in English, as well as publishing successful handbooks of his own on Organic and Inorganic Chemistry.

[2] John Hutton Balfour (1808–84) gradually gave up a successful medical practice in Edinburgh in favour of botany, to which he had been devoted since his student days under Graham, helping in 1836 to found the Edinburgh Botanical Society, whose library and herbarium were eventually acquired by the Crown as the basis of the collections at the Edinburgh Botanic Gardens. In 1842 he succeeded Sir W. Hooker at Glasgow, and three years later was elected to the Edinburgh chair on the death of Graham, defeating J. D. Hooker. This chair he held till 1879, writing successful text-books, developing the Gardens and the museum, and proving himself an inspiring teacher. He not only extended the field work already established, but was the pioneer in Edinburgh of practical laboratory work with the microscope. But though stimulating his pupils to consider the wider problems of botany, his religious views led to his opposing the Darwinian movement.

A letter of June 27 to his cousin, Francis Turner Palgrave,[1] whose inherited interest in art and art-criticism had displayed itself very early, deserves passing reference as showing Hooker's sustained interest in pictures as well as music. The letter is too long to quote save for a few personal passages. Palgrave, the younger by seven years, had won a scholarship at Balliol in 1842. Now 'the reappearance of some quondam Scotchmen, who return hitherward with good Scotch seriously damaged through long continued unsuccessful attempts to speak English,' reminds him that Francis is to be congratulated on the beginning of the summer vacation; but it was Francis who had the credit of 'breaking the ice that has frozen up the current (ever sluggish) of correspondence that runs (creeps) between us.'

> I heartily wish that you would come down to this place before I go. You would I am sure enjoy it extremely, for it is a most liveable place, with plenty to see and admire in the neighbourhood. The only exhibition that I have seen was one of Scotch artists, open, or rather which shut on the day of my arrival; it was very bad as far as Scotch performances were concerned; some Stanfields, Turners, Landseers, and young Phillip's 'Borrow' were far the best things in the room.

Next he speaks of ten of the prize cartoons for the decoration of the Houses of Parliament, which had been shown two years before in Westminster Hall. These were now exhibited in Edinburgh in connection with a proposed book of lithographs. He criticises them as if Francis remembered all about them, which very likely is not the case; noting the relation of the best among them to the Hampton Court cartoons, of which no one in Edinburgh knew anything; and quoting the story of the best picture if the least original, Caractacus led through Rome, namely, that the artist studied a lion's head to pourtray the British Captive's from.

Of Old Masters he could show his cousin the collection at Dalkeith, where 'the place is very badly kept, but the scenery

[1] See family pedigree, p. 18.

is exquisitely beautiful.' And so of the recent adornments of Edinburgh:

> Certainly these modern Athenians have not improved their Athens lately; the much-vaunted 'Scott Monument' is, to my mind, *vile*, bad in composition, situation, and in all other particulars, saving the handycraft. It is very like the top of the steeple of a Belgian Hotel de Ville, taken down and placed on the side of a road. Here it is thought perfection, and Scott is conceived to be unspeakably honored, both in the design and execution.
>
> I do wish you would come here and let me talk you into my likings and dislikes. Have you seen Cennini's book on old Fresco paintings? I think you would care to look at it, as you were once addicted to frescoing stables and outhouses; there are also some few graceful little outlines in it. I often think that a nice book of lithographed outlines of good pictures would sell well. I am sure that you and I, who could not afford better, would buy such things. . . .

To return to the Botanical Professorship—canvassing for which he found 'detestable work.'

As has been explained, the Crown appointed to the less valuable Regius Professorship and the Botanic Garden, the Town to the valuable College Professorship. The Town Council felt aggrieved that, without consulting them, Sir James Graham, the Home Secretary, had decided on Hooker as the Crown nominee; and indeed gratuitously aggrieved, as there was a large majority for him at their first meeting, the Edinburgh candidate, Balfour, having refused to stand if the two appointments were separated.

The Provost cannily tried to better the situation by proposing a bargain. The Natural History chair was under the same dual control, the Crown appointing to the Museum, the Town to the chair. Let the Crown take over the whole of the former and relinquish the latter entirely to the Town, who would on this occasion bestow it on the Crown nominee, Hooker. The Crown, however, could hardly look on such a proposal with favour, having spent full £20,000 on the Garden and more on the Professorship, while the Town had done nothing for either.

At this juncture Balfour revoked his refusal to take the Chair without the Garden. The Town Council were put on their mettle to show the Crown that they had a power, and as they truly said, they wanted a lecturer rather than a botanist pure and simple, however overwhelming his testimonials might be.

Tactically, had Hooker wished to push his claims, this move would have left him in a strong, if rather absurd position. Suppose the two chairs separated; it was the Regius Professor with his £150 a year whose ticket must be accepted by all the faculties for the University degree, and the College professor would be 'dished.' But for all reasons, including Government goodwill, it was preferable to conciliate the Town Council, and far preferable indeed, were it only possible, to have the Garden alone with £300 a year than a Professorship at twice the salary and College troubles and Town Council odium.

One councillor, unaware of the great difference in attractiveness between the two posts, proposed that the Edinburgh man should stay in Edinburgh, while Hooker received the Glasgow chair, thus keeping both in Scotland. Hooker undeceived him; this consummation was only possible by electing him to Edinburgh.

Finally the election became wholly a matter of politics, even with the Provost, and local interests prevailed.

CHAPTER X

THE GEOLOGICAL SURVEY

EDINBURGH failing, Sir J. Graham offered Hooker the Glasgow chair.

Sir William felt it his unwilling duty to point out such advantages as attached to this offer; he was unfeignedly glad when Joseph's own decision kept him at Kew. Father and son were equally attached and equally generous one to the other; this time it is Joseph who, from a chance word dropped about finances, is suspected of 'having paid something to my account' for his share in Fitch's artistic services. Sir William protests; after all he is paying Fitch no more than before; no wonder Joseph has little or nothing in the bank if he makes such a use of his money!

His hopes that some opening might be found for Joseph at Kew itself were revived when in November Bentham told him that having just made his will, he had appointed Joseph one of his executors and had left his fine Herbarium to the Royal Gardens, if proper accommodation were provided for it. The Kew establishment even now was being enlarged, and here was the prospect of further material for the projected Museum. If the Commissioners were not likely to require more than one Director, at least an assistant would be wanted, and, so far as qualifications go, he confidently asserts, 'if his life be spared, there are few men that will rank higher as a Botanist than Joseph.'

Through the winter Joseph Hooker continued at work on the Niger Flora as well as the Antarctic Flora, remarking of the former to Harvey (December 30, 1845):

> I am doing my utmost to the Niger Flora and hope to succeed, but it is a terrible task from the badness of the specimens, the worseness of the published descriptions, and the necessity of comparing everything with both American and Asiatic species ; you will be surprised at the quantity of species in common these countries possess.

But in February a post was found for him. Sir Henry de la Bèche,[1] head of the Geological Survey, was in search of a botanist to work out the British Flora, extant and fossil, in relation to Geology, and consulted Sir William. After brief consideration, the latter proposed the name of his son, who was instantly accepted. The salary was £150 with travelling allowances for the local research to be carried out from time to time ; the work, much of which could be done at home, would not prevent him from continuing the Antarctic Flora with its contingent allowances from the Admiralty, while not only would fossil research widen his botanical outlook, but with such an intimate local knowledge as he could acquire of Great Britain and Ireland, he would be able to carry on his father's book on the British Flora. Nor did his father forget that the Survey was under the same Department, the Woods and Forests, as Kew, and the official connexion might well help to bring him as assistant to Kew when the projected extensions were carried and the Museum established, possibly within a year.

The work was agreeable, moreover, it threw him very much into a new world and class of society in London, such as the Lyells, Owen,[2] and Horner, as well as brought him into touch

[1] Sir Henry Thomas De La Bèche (1796–1835), the geologist whose enterprise in making the new ordnance survey the basis of a geological map of each county led to the establishment of the Geological Survey in 1832, under his directorship. To him also were due the Jermyn Street Museum of Geology and the School of Mines (1851).

[2] Sir Richard Owen (1804–92), the famous anatomist. He was assistant to Clift at the Hunterian Museum of the Royal College of Surgeons from 1827, succeeding him as conservator in 1842 till 1856, when he was appointed superintendent of the natural history departments of the British Museum, retiring in 1883. Unrivalled though he was in the amount and general value of his work in comparative anatomy and palæontology, it was different when he came to speculative theory. His doctrine of the Archetype was founded unstably on Oken's transcendentalism, and his proposed division of the mammalia into four sub-classes, according to the difference of their brains, was unsatisfactory, while very little of the classification in his great work, *The Anatomy*

with Robert Hunt, Keeper of Mining Records; Lyon Playfair, the chemist (afterwards Lord Playfair); John Phillips, Professor of Geology at Trinity College, Dublin; and Edward Forbes, the naturalist, his colleagues in the Geological Survey.

The new appointment and its relation to his outstanding work are discussed in the following:

To Sir James Ross

The object [of the Geological Survey] is to have the connection between the plants and the geological formation they occupy investigated, and the Fossil plants arranged as they are collected. The first object will require my visiting the ground they are surveying once or twice a year, probably with Sir H. De la Bèche and Prof. Forbes (who are the Geologist and Palæontologist to the Survey), and the arrangement of my observations for publication, as well as the directing what vegetables should be gathered for analysis. The duties will leave me more than enough of time to carry on my Flora as fast as the plates can possibly appear, but I do not know what the Admiralty will say to my taking the duty. My work has in many ways cost me already nearly £100, and I believe I have never made 6*d.* by it and never shall. If the new duty were to interfere with my Flora, or were my salary so good as to make me independent of the Admiralty, I should not think about drawing any further Admiralty pay, but as that is not the case and as I have never made a farthing by my Botany work, I think of making a push for the continuance of my pay when I enter upon my new duties. I should feel very much obliged for your opinion of how their Lordships are likely to regard my views. As the new appointment is a most honorable one, and one worth *to me* twice the income it offers, I have made up my mind to accept it at all hazards, even if it should entail the leaving the Service. Had I gained the Edinburgh Chair I would have gone on with my Flora on my own resources and have given up the Admiralty pay without waiting to be asked, as a point of honor. And

and Physiology of the Vertebrates, 1866–8, was accepted by other zoologists. His bitterness against any possible scientific rival and his disingenuous attitude towards Darwin and his work ended by leaving him isolated in the scientific world.

were my expected pay sufficient to justify me in carrying on so expensive a publication on my own resources, I should equally be now ready to act in the same manner. Nor need I conceal *from yourself* that the Flora Antarctica portion shall be carried on as hitherto whether my request is granted or not, though I should not think it very generous of their Lordships to expect me to continue the work without some reward, even did it cost me nothing.

[He explains that publishing at the extreme limit of eight plates a month, the work would last another four years, and adduces precedents for Naval pay being continued till it was finished.]

I am quite sorry to trouble you about this, but should not wish to act without your sanction, and feel it a duty at any rate to lay before you my prospects. My hope is that, before my present work is over, other national voyages may have brought home stores worthy of publication, and that as long as I can be usefully employed and busily too on works of that sort, I may also draw pay for it, but no longer.

With respectful compliments to Lady Ross,
Believe me ever,
Yours most respectfully and truly,
Jos. D. Hooker.

The two sets of work fitted in well together: 'Happily my duties at the Geol. Survey,' he tells Ross in an undated letter, probably 1846, 'are (like the pay) very light; they employ me first of all to draw up a catalogue of the known British fossil plants previous to my arranging those of the Geolog. Survey Museum, and corresponding for more. My work *never* went on so fast, having appeared unremittingly for five months and will for two more; but then the struggle must cease for one month, to get up the Cryptogamia plates, which are very heavy work.'

Kew at this time was two hours' distant from London by omnibus, for the railway had not yet reached it, and riding presented itself as a speedier alternative, especially as his delicate sister Elizabeth could also use the horse for the exercise prescribed by the doctors. In the winter he found it convenient to take rooms for some time at 3 Great Ryder Street, near the temporary quarters of the Survey, and Jermyn Street, where the Museum of Practical Geology was being built for its accommodation.

Of his occupations at this time he writes to Dawson Turner (April 31, 1846):

> At present I am worked rather hard, having to go into town every day to study fossil Botany, until the proposed Museum is built in Piccadilly. The apartments now filling up are thus only temporary, and are granted by the Dean of Westminster in the shape of servants' rooms over his stable. Though small, they are neat and quite suitable, looking into Dean's Yard and entering by a respectable little doorway on the courtyard. The Dean is very civil and busy in his improvements of the badly dilapidated yard; he is giving us a fine lamp opposite our door and otherwise takes a great interest in all that is going on.
>
> The great difference between my father's and all other Government employments evidently consists in his not being supplied with tools, as I am in my humble capacity, and as Brown and all other public officers whose real income is thus *apparently* not so good as my father's; but it is apparently only, for if they had to purchase their books and plants they would all be ruined.

In May and June his work took him into South Wales, to examine the coal-beds for fossil plants *in situ*; in August and September to the Bristol coalfield. In South Wales, where 'De la Bèche appears very pleased with what I have done,' his headquarters were near Swansea, with his grandfather's old friends the Dillwyns,[1] whom he delighted by discovering the Lesser Wintergreen (*Pyrola minor*), which had not been found in the neighbourhood before. Their son, Lewis Dillwyn,

[1] Lewis Weston Dillwyn (1778–1855), botanist, conchologist, and potter, was born at Ipswich, within touch of the Turner-Hooker circle. It was not till 1803 that he moved to Swansea to take charge of the pottery bought by his father. He had already begun his *Natural History of British Confervae*, and collaborated with Dawson Turner in the *Botanist's Guide through England and Wales*, 1805. At Swansea he wrote on the local flora and fauna and the history of the city, as well as sharing in civic affairs. He was M.P. from 1832–7.

who 'worked the old *family pottery* in Swansea,' had married De la Bèche's daughter. He was Hooker's special companion, being a good ornithologist and fond of Natural History in general. Another good companion was Mr. Dillwyn's son-in-law, Moggridge, whose hobby was British Botany. An additional attraction of the house, which appeals to Dawson Turner, is the collection of pictures, and specially Cuyp's Burgomaster of Haarlem.

Lecturing was still a trial to him, but wishing to make some return for the great kindness with which he had been received in Swansea, he offered to give a lecture on the Antarctic Voyage. This was duly delivered with great success at the Royal Institution of South Wales on June 17. The advertisement of the lecture makes the interesting announcement that in addition to members of the Institution and affiliated societies, who were admitted free, 'Thirty free admissions to the back seats will be distributed by the Council to persons of the working class not connected with the above Societies.'

He writes to his grandfather, June 21, 1846 :

> You will be surprised to hear of my lecturing here, but I not only could not get off the task, but *hating it* as I do, I felt a real pleasure in gratifying my many friends in Swansea. The lecture has added seven new subscribers to the Swansea Institution, and I have had thanks and innumerable requests for another, which however I cannot comply with. You can have no idea how easily these people are pleased with my compliance with their wishes in lecturing, nor how good-naturedly attentive they were to the lecture itself.
>
> I have been travelling about a great deal in South Wales, visiting the Collieries, collecting fossil plants, and gaining information on all subjects connected with the ancient Botany of our globe. The subject is a deeply interesting one, and though it decidedly interferes with the progress of my studies in recent Botany, it will, I hope, in the long run, turn to good account. The work is very hard in this hot weather, especially when the coal-dust and other annoyances attendant on my investigations in these dirty districts are almost insupportable. Still I like the work and my master, and hope to get on with this accessory to my pursuits.

His visit to the Forest of Dean, in company with the brother of his old friend Thomas Thomson, whom he picked up at Bath, invalided from India, precluded a pleasant 'touch at recent botany in W. Ireland' with Harvey and Ward, who had been making various 'finds'; and he writes to the former (August 7, 1846):

> I do long intensely to go to the field with you and especially to take the water. Well done, Ward, but I won't knock under, having youth on my side and better eyes. I look forward to no greater pleasure in British Botany than to see the Delesserias growing in Ireland as they did at Cape Horn, and under such perfectly similar conditions. I want to see how the Antarctic seaweeds are replaced on the British coast; and no one can do it to my satisfaction but myself. (Pretty well that for a Tyro.)

However, a future visit to Dublin seemed possible if an Irish collector should have to be appointed in connexion with the Geological Survey scheme to form a complete British Herbarium with special reference to the distribution of species.

> I have persuaded Sir H. that no results can be obtained as to dependence of plants on soil, till a good many complete floras of counties with different formations are formed; he and I draw well [together], by reason of his profound ignorance of Botany. He has an idea that the difference of the vegetations of the sandstone and limestone is something more marked than between Lat. 0 and Lat. 90 or the top of Ben Nevis and low water at Roundstone.

To Mr. Bentham he writes (September 13–25) of his researches in fossil botany, the interest of which grew

> as the impossibility of relating all but the Ferns of the coal strata to any existing Nat. Ord. becomes more evident. Hitherto the collections formed are not large, as such are only to be obtained to any extent by employing men about the pits, but I have been grounding myself underground in the elements of the study by noting the conditions of their preservation and their association, so as to know what of the various broken pieces belong to the same genus

or species, for the majority of the genera of some of the tribes of coal plants are merely names applied to individual plants, sometimes of the same plant; thus Calamites are all stems, Lepidodendron all branches, Lepidostrobus all cones.

[After this] I took to recent Botany, crossing and re-crossing from the village to the heart of the forest, to observe what difference in the native vegetation may occur in progressing from New to Old red sandstone, then Mt. Limestone, and lastly the sandstone of the coal; all these rocks lie here in parallel stripes as it were. The scenery was most beautiful, and from some of the hills I caught sights of the Sugar Loaf, Garway, Graig[1] and the long back of the Black Mountains.

One enjoys so much the sight of familiar objects in the new aspect they wear when viewed from other points than those we have been accustomed to. Another year I hope to take your part of the country, though I do not expect there are many rare plants there, still as my Master wants the Botanical features of each soil, I will condescend to accommodate him when my other interests suit my duties. This will appear possibly a curious way of doing duty, but Forbes and I try to drum into Sir H. the dogma, that all scientific work is duty, whether he may be able to appreciate the immediate bearing of its results on the Geol. Survey or no.

But this British Botany had to give way to the Fossil Botany at the Geological Survey; it was impossible to deal with both. However, the latter had the greater attraction in the novelty and interest of the field, and the need for perfecting a knowledge of anatomy and physiology. Still there was plenty to be done in British Botany, and later he fulfilled Bentham's word that the work ought to be done, despite the opposition which might be expected from those who already occupied the field.

The winter and early spring of 1846–7 are filled up in part with arranging the autumn's collection of fossils and preparing

[1] These hills, familiar points in the landscape around Pontrilas, called up many recollections of the Benthams.

three essays on the Coal plants, which involved both the drawing of woodcuts and personal superintendence of slicing and polishing fossils. These essays were printed in the 'Memoirs' of the Geological Survey for 1848; two dealt with the structure of *Stigmaria* and *Lepidostrobus*; the third drew a general comparison between the plants of the Coal and of the present day. Here microscopic examination of these sections of 'coal-balls' was made fruitful by his great knowledge of living forms; he was able to demonstrate the actual structure of the fossils, and as Professor W. W. Watts remarks in his Anniversary Address to the Geological Society, 1912, 'these memoirs differ from all others on the subject published at the time—or, indeed, long afterwards—in receiving unstinted praise alike from geologists and from botanists.'

Except for a return in the eighties to the 'enigmatic' *Pachytheca*, on which he first published in 1853, Hooker's short but brilliant work on fossil botany ended with his explanation of *Trigonocarpon*, a fossil fruit of the Coal measures (in 1854–5). India and Kew absorbed his energies, though his early interest was not quenched. True that for many years the rashness of geological identifications led him to dub Fossil Botany 'the most unreliable of sciences'; 'but,' adds Professor Watts, 'when, in recent years, the study of Carboniferous, Jurassic, and Cretaceous plants yielded such new and startling results to investigators in this country, France, Germany, and the United States, all his old enthusiasm returned.'

The other part of his winter occupations in 1846–7 included completion of the Antarctic Flora and the Niger Flora, which had grown too bulky for printing more than the opening part in the 'Journal of Botany.' 'I have had,' he complains, 'to write something rather "Flowery, Bowery" for a Botanist, to please the "Emancipators," but it is not very much, happily.' The Galapagos Florula was to appear in the Linnean Society Transactions, and to be followed with notes on the botanical distribution of the flora. Another task was the naming of all his own and R. Gunn's Tasmanian Compositae and Coniferae, with publication of diagnoses of the many new species in the Journal, for the prospects of bringing out the Tasmanian

Flora were for the moment visionary. Indeed, it did not appear until 1859.

During the autumn of 1846 Sir William made another effort to secure his son's future. The Woods and Forests Department being unwilling to take over the cost of housing and increasing the Herbarium, the notable addition brought to Kew by the elder Hooker, on the ground that his plants could not be marked, as were his books in the Library, to keep them distinct from later additions, Sir William offered to present the Herbarium to the nation, on condition that Joseph should be appointed his assistant and successor at £800 a year. Lord Morpeth was friendly, but would not guarantee the succession with the salary proposed. Future arrangements were uncertain.

Kew was still too much a mere object of aristocratic patronage. Joseph Hooker was too proud to press his claims on any but scientific grounds. He was revolted by the suggestion that he should make friends with the Mammon of Society, by helping his father to pay the required attentions to aristocratic sight-seers. It was all very well to meet old friends or officials or scientific persons, high or low; but when his father would introduce him to these others, he knew himself to be in a false position, to which he could not submit, officiously thrust forward and wasting his valuable time to boot. His father was used to making use of patronage in the days when patronage was the road to progress; but even so, Hooker writes bitterly to his grandfather (July 25, 1847):

> My Mother and Sister will tell you that of the hundreds of aristocrats who detain my father at the Garden for hours *waiting their arrival,* and then drag him through every house and acre, there are not half a dozen whom he could ask to back even an application for himself or for me, or who have shown him the smallest politeness in return.

Meantime Hooker himself was growing more and more eager for another Botanical journey, this time to the mountains of the tropics, either the Andes or the Himalayas. His father would have been content for him to stay in England, filling up the time till some satisfactory post offered with his botanical publications and a big travel book in two volumes, 'Journals of a Naturalist on the *Erebus* and *Terror*,' which the John Murray of that day, meeting Sir William at dinner, declared his readiness to publish as a companion to Darwin's famous 'Voyage of the *Beagle*.' But a year botanising abroad was worth five of study at home. The Admiralty were planning a scientific voyage to Borneo, and might appoint him as naturalist; again, 'If I could only get the W. and F. to pay expenses and Admiralty to give leave, I would go to India and collect fruits, woods, and seeds, &c. &c. The E.I.C. superintendent of W. and F., Dr. Gibson,[1] in the Indian Peninsula offers me a cruise with him to province of Cannar (S. of Goa) at a very cheap rate, and I have a huge yearning that way; his is only a four or five months' trip or tour of inspection. I wish I had a private fortune.' Again, in July, he writes to his grandfather:

> I shall be ready to make any sacrifice to get to the tropics for a year, so convinced am I that it will give me the lift I want, in acquiring a knowledge of exotic Botany. My friend, Falconer, goes out on December 20, to the charge of Calcutta Bot. Gard., and I hope to be ready to share his cabin. I shall then spend some months at Calcutta and the neighbourhood (Gurney,[2] &c., &c.), get up to the Himalaya betimes, and return the following winter *via* Bombay.

He had strong hope of joining the Tibet mission, which was to go from Ladak to Yarkand and Kashgar over wholly unexplored country north of the Himalaya, and in September 1847 was in active correspondence about this. The work already in hand would not suffer, for as he wrote to Ross:

Kew: September 7, 1847.

MY DEAR CAPT. ROSS,—I have delayed answering your letter till I should know something more definite regard-

[1] Alexander Gibson (1800–67), went to India in the medical service of the Company, and became superintendent of the Botanical Garden of Dapuri in 1838, and Conservator of Forests in Bombay 1847–60.

[2] Gurney Turner, his cousin, in the medical service of the E.I.C.

ing my plans. The Woods and Forests seem very desirous of sending me out, and as I do not see any other prospect of my doing better, and being extremely anxious to undertake any exploratory expedition, I need hardly say that I do hope they will employ me.

The last ½ sheet of the Flora Ant. is in the press, and it contains a vast amount more matter than I had ever contemplated bringing in; it has cost me *out of pocket* upwards of £100, and Lord Auckland has not yet had his copy, which will cost me £8 10*s*. I feel it to be now quite time that I were looking out for a livelihood, and as my future hopes and prospects all will be with the Woods and Forests I feel that in justice to myself I ought not to throw away the present opportunity of improving myself, and the science to which I am attached, and of establishing a claim upon them in the proper quarter.

Neither the Flora of New Zealand nor of Van Diemen's Land will suffer by the delay, as Mr. Gunn and Colenso are still employed in making collections in all parts of these islands and are paid by my Father and self for doing so, from our private pockets. Under any circumstances I did not think of beginning the publication of either Flora before some months, when their latest collections shall have arrived.

Failing anything else, he was even ready to go out and report on the nature of the Island of Ascension, a barren rock, in connexion with the Admiralty plan of improving the vegetation there. Unexpected encouragement of the Indian plan came from De la Bèche, who desired to retain him on the Survey staff, while taking the fossils he might collect for the Geological Museum, and letting the plants go to Kew.

The first point then was to secure a Government grant for the Indian expedition, and the support of the East India Company. The latter was easier to win than the former, finance at the moment being unpropitious. The Admiralty, moreover, to whom Hooker owed allegiance, thought India out of their proper sphere, and suggested that if he wanted botanical travel he should join the official expedition to the Malay Islands, planned for 1848, though this would not be a very well paid post. Difficulties, however, evaporated in personal discussion, when at the beginning of October Hooker met Lord Auckland, then First Lord of the Admiralty, during a visit in the Isle of Man to his brother the Bishop. Then it was arranged that if he went to India first, he should go on to join the frigate *Mœander* at Borneo during the healthier season and prepare a botanical report on the British possessions there, keeping his half pay till he arrived and then being put on full pay, with botanical allowances of £300 during his term of service.[1] This paved the way for an appeal to the Treasury for a grant of £400 a year for two years on behalf of the Gardens to cover their botanist's expenses in collecting.

The Eastern Himalayas were practically unknown. Lord Auckland and Dr. Falconer [2] alike proposed that he should explore the Sikkim valley up to the snows on the Tibetan frontier. It was under our protectorate, and Hooker, on his official mission, would be accredited to the British Resident.

Reinforced by a striking letter from the veteran Humboldt pointing out to Hooker what could be done by him in the Himalaya for science, Lord Morpeth, of the Woods and Forests, prevailed on the Treasury at the eleventh hour to give the grant. On October 20 came an official intimation that the Chancellor of the Exchequer had given his hearty consent to the Indian Mission, and the Admiralty proposed that a free passage should be granted as far as Alexandria at least in the *Sidon,* which was to sail on November 9, conveying Lord Dalhousie, the new Governor-General, to India. This proposal was made subject to Lord Dalhousie's consent. Sir William immediately called upon him, when so far from raising objections, he insisted that Joseph should continue the whole journey with him to India, thus overcoming the various difficulties raised by the East India Company in regard to the journey from Aden to Calcutta. Indeed, he enjoyed Hooker's

[1] When the Borneo expedition was abandoned, the £300 was allotted to a third year in India.

[2] Hugh Falconer (1808–65), Palæontologist and Botanist; M.A. Aberdeen 1826, M.D. Edinburgh 1829. Assistant Surgeon on the East India Co.'s establishment 1830, and Superintendent of the Saharanpur Botanical Gardens 1832. Superintended the manufacture of the first Indian tea 1834; Professor of Botany at Calcutta Medical College; Vice-President of the Royal Society.

society so much that on reaching Suez he took him into his suite.

With this another early ambition was realised. It has already been told how Cook's Voyages, with the picture of Kerguelen's Land, was one of his earliest recollections in reading. The other was Turner's 'Travels in Tibet.' Here his imagination was gripped by the description of Lama worship and the great mountain Chumalari. There he notes, 'It is singular that K. Land should have been the first strange country I ever visited, and that in the first King's ship which has touched there since Cook's voyage,' and that later 'I have been nearly the first European who has approached Chumalari since Turner's embassy' (in 1783).

The disappointment at Edinburgh, despite the fatigue and momentary sense of failure, had never gone very deep. The years of steady work since returning from the Antarctic, though not bringing him an important appointment, had done more by preparing him for the new venture, which had unexpectedly created the long-desired link between his scientific work and official Kew. Now his second great scientific ambition was fulfilled, following but a few months after a more intimate felicity. No wonder that during these last days in England his father could write, 'I think I never saw him so cheerful and happy.' For in the beginning of July he became engaged to Frances Henslow, eldest daughter of the Cambridge Professor of Botany, so widely beloved for his personal qualities, who is still remembered outside the circle of specialists as the man who first made nature study a living pursuit among the school children of his village, and the man who greatly helped to turn Charles Darwin to a scientific career. Frances was a close friend of his sister Elizabeth; and now matters came to a head during the 'week's holiday and idleness,' as he called it, at Oxford during the meeting of the British Association, to the great joy of Elizabeth.

It is characteristic of the strict family régime of the Hookers that in his announcement of this happy event to Dawson Turner 'no flowers' were permissible—no approach even to 'flowers.' Joseph opines that, 'as an affectionate grandfather (and man of business), you may be glad to hear the reasons for my preference '—and to the man of business rather than to the affectionate grandfather sets forth their mutual suitability, her industry, energy, education, good principles and scientific sympathies, her literary helpfulness, for 'she is much cleverer than I am.' But enough of 'reasons'; there was another and more personal side to all this, and if he should not speak of it, the sister friend might perhaps speak more warmly, so 'for the rest I must refer you to my sister Elizabeth.'

The high-stepping Johnsonese chosen by Sir William for discussing 'Joseph's attachment and his prospects' with Dawson Turner is irresistible. 'I believe,' he writes, 'Miss Henslow to be an amiable and well-educated person of most respectable, though not high connections, and from all that I have seen of her, well suited to Joseph's habits and pursuits. He himself seems well pleased with his choice.' Formal propriety could go no further in concealing a warm heart.

The work already mentioned on the Antarctic and Niger Floras and travel on Survey business alternately occupy the rest of 1846 and most of the next year. March saw him in Ireland. From South Wales, his mother notes, he returns brown and well, carrying out his grandfather's dictum that six hours' sleep is enough for any healthy man. In August he was away again; 'busy and happy he seems.' For most of the first three months of 1847 his father was ill; 'Joseph,' writes Lady Hooker, 'is most helpful to me with his father; always glad to assist, calm and quiet. He knows too what is fit to be done and is very handy.' He would not, however, take the opportunity of his father's temporary absence from work to 'put himself forward at the Garden,' as his mother inwardly wished, with a view to the future.

On April 17 he went to Cambridge for a fortnight to see a collection of coral plants from Australia; then after a few days with Berkeley[1] the mycologist at Oundle, proceeded on

[1] The Rev. Miles Joseph Berkeley (1803–89) as a botanist devoted himself to the Cryptogams. He wrote the volume on Fungi in Smith's *English Flora*, 1836, *Outlines of British Fungology*, 1860, and a *Handbook of British Mosses*, 1863, besides an *Introduction to Cryptogamic Botany*, 1860. The collections of fungi made by Darwin and other travellers came to him for description. His

Survey work to Wolverhampton, Manchester, Leeds, Barnsley, and Birmingham, stealing a few days off his Survey duty to spend at pure Botany at Warrington with Wilson the botanist, who had been working at the mosses in his Flora Antarctica. 'We are now pulling my Tasmanian specimens of *Dawsonia* to pieces, and can hardly make out whether it be a new species or variety' (May 20).

On April 21 of this year he was elected to the Royal Society, as Wallich[1] described it, 'by a vast majority, . . . a majority much greater than any among the eight candidates that were successful. He had ninety-five votes, nor was any one candidate's certificate so amply and gloriously filled up as his!'

Of this scientific success he writes with his usual diffidence in his own powers to his grandfather, to whom he owed so much scientific encouragment.

St. John's College, Cambridge: April 26, 1847.

MY DEAR GRANDFATHER,—I thank you very much for the kind congratulations you have sent me on my election to the R.S. You I can thank with more ease than any one, for you are one of the very few who can see to the full how entirely I am indebted to those who have gone before and stood by me, for what superiority in position over my contemporaries their good offices have obtained for me. My advantages in Ross's voyage; the procuring of the after grant; the launching of my book into the world in the form it boasts and the continuation of that work in a creditable state up to the present day; my testimonials for Edinburgh; my appointment to the Government Survey (small though it be)—are all advantages for which I am indebted to the position my father has gained for himself and which has enabled him to lay my little merits before

special knowledge was of great value to the Commission on potato disease, 1845, On his retirement in 1879 he presented his herbarium of fungi and his books to Kew. He was elected F.L.S. 1836, F.R.S. 1879, receiving the Royal Medal in 1863.

[1] Nathaniel Wallich (1786–1854) was a Danish surgeon at Serampore who, when the place fell into English hands in 1813, entered the service of the E.I.C., and in 1815 was made superintendent of the Calcutta Botanical Garden, a post he held till 1850. He returned finally to England in 1847, having done immense work as a botanical explorer, and brought back vast collections, the final distribution of which was completed by Hooker.

the world under the most advantageous conditions. I know myself to be deficient in education and I can feel my abilities to be only second-rate, and so can only feel truly thankful that I have light enough to see to whom I owe the appreciation of my works by the public.

I have done a good deal here both with marsh and fossil plants. From one of your letters to my Father I think you possibly mistake the nature of my studies as connected with the Survey. I am no Geologist: my work is fossil botany; as legitimately a branch of *Botany* as is Muscology; fossil plants, though imperfect, are still *pure* plants; and, though dead as species, they form and show links between existing forms, upon which they throw a marvellous light.

Here also must be noted the beginnings of the close friendship with Charles Darwin which was to be lifelong. They had already been in close touch over botanical matters; Hooker had been working out Darwin's plants from the Galapagos Islands; now on October 10 he has gone to stay with Darwin in Kent for three days, and on January 14, 1847, again he goes for a visit of a week or ten days.

CHAPTER XI

THE VOYAGE TO INDIA

The *Sidon* left England on November 11, 1847, calling at Lisbon, Gibraltar, and Malta on the way to Alexandria. As a matter of course, the voyagers made the fourteen mile excursion from Lisbon to Cintra. Most of the party, mounted on jackasses, visited the Convent of Our Lady of the Rock; Hooker climbed a rocky hill hard by, and believed he had the best of it, for outstretched before him were typical groves of fruit and timber trees, and many miles of vast, grassy undulating plains of Portugal, conspicuous upon them the lines of Torres Vedras and many another place of note in the Peninsular War, 'for which see Napier (a book I never could and never shall get through).'

The botanist sees at once in 'the multitude of Lichens, which coated the granite rocks as completely (though not with such fine species) as in the Antarctic plains,' a proof of the prevalent dampness of the atmosphere. The traveller, marvelling that a nation of discoverers should have fallen so low, reflects that it was gold alone that stirred them from indolence, and exclaims sadly:

> What is to become of them it is hard to say. The land is rich and productive; the climate delicious; and they are neither warlike nor romantic people, such as the Spaniards, whose temperament keeps them in hot water. I have now seen them in Madeira, the Cape Verdes, Brazil, and at home: and they are the same all over the world. I hope never to see them again.

As a world-voyager himself his one regret for taking a new way back into Lisbon was not

> to have looked once more at Belem Church, where Columbus dreamed that an Angel directed him to the discovery of the New World, if I remember aright; and where especially Vasco da Gama and his successors offered up, some their prayers, and others their thanksgivings (to St. Nicholas, by the way) on the occasion of their several voyages to the Eastern Indies, or return therefrom.

Still the quarter of Lisbon by which they returned was magnificent by night, albeit the high and handsome squares were perhaps whited sepulchres. Night also offered another advantage: 'After the heat of the day is over the many *smells* are in a great measure dissipated; the dogs gone to kennel; and little else but drunken seamen to disturb one's reveries.'

The fortified rock of Malta provokes agreeable comparison with St. Helena and Gibraltar: for here the heat that is fervid on the black soil of St. Helena and scorches at Gibraltar is tempered by the yellow stone, which neither attracts like the one nor reflects like the other the powerful rays of the sun. There is a thumbnail sketch of the town with its magnificent entrance to the harbour, its 'church and convent bell-towers innumerable, ringing all day long, many with good voices, some with bad,' its rocks bare of any green save the Caper plant, and its picturesque streets, which

> form a sort of square telescope, with busy people along the bottom, handsome yellow carved stone balconies projecting on either side, bright blue sky above, and the sea like a perfect jewel at the further end.

Apropos of the carved stone work everywhere (of which he bought some for the Geological Museum):

> Stone cutting and carving is indeed the besetting employment of the Maltese; and the facility afforded by the limestone has the same effect on this, their hereditary disposition, that a soft deal bench has on a schoolboy.

At Citta Vecchia, he tells Miss Henslow,

everything is attributed to St. Paul, and your father would have laughed had he had presented to him for sale (as I had) some fossil sharks' teeth, 3 inches long, as the teeth of the Apostle himself!

At this distance of time it is curious to recapture the impression made on an old naval man by the 'terrible-looking' steamers among the white-sailed ships of all nations, the noble line-of-battle ships, and the smart frigates; and the same epithet is repeated soon after when it is recorded that the passage to Alexandria was long, 'owing to contrary winds and a head-sea, which though slight, were sufficient to retard the *Sidon*, which despite her size and terribly grand look, is a very poor steamer or sailer, after all.'

The Alexandria of 1847 was a 'ruinous city of dirty white houses straggling round a broad bay' with 'outskirts horrible to a degree,' consisting of clusters of huts, or rather mud hovels not four feet high, grouped in squares about ten feet each way, with a hole for the door and another to serve as a window. Pompey's Pillar and the slave market were the two extremes of interest for the sightseer.

But he found the Pillar, 'like all such attempts at effect, a failure, as the mind does not perceive at once the gigantic labour which the erection of such a single stone must have cost.' The sight of it added nothing to the impression gathered from books. The slave market was

a small court about 30 feet square, surrounded with cells of about 12 feet, devoted to the slaves of each nation. These were dark and dirty, full of vermin, in spite of the smoke of a fire in the middle of the earthen floor, which all but suffocated the poor inmates. I saw only the Abyssinians, two or three squalid wretches, in the most abject state of dirt, disease, and suffering, from the smoke which inflamed their poor eyes. They said nothing, but crouched behind the door and up in the corner on my entering.

The most agreeable episode connected with quitting the *Sidon* at Alexandria was Lord Dalhousie's expression of the friendship he had formed on the voyage for Hooker. 'On our arrival,' writes the latter, 'he took me on one side and invited me to belong to his suite for the future, in the most kind and handsome manner.' Hooker accordingly travelled freely in the Governor-General's launch to Cairo, accompanied him to Mehemet Ali's reception, and from Suez sailed not on the ordinary packet-boat but on the East India Company's frigate sent to convey the Governor. This smoothed away many of the minor difficulties of travel, especially the refusal of the India Board in London to give him a passage, because the Company's naval officers disliked the ships being employed as passage-boats.

The journey to Cairo was effected by water. A 'pretty little steamer of the size and shape of a Woolwich boat,' belonging to the transit company, took the party eighty miles to the Nile, along the Mamudieh Canal, Mehemet Ali's vast work carried out by the forced labour of the *corvée*, which drew all the unhappy fellahin from the fields unpaid and unfed, and was followed by a disastrous famine. 'It reminded me'—he draws a homely comparison for his father—'of the canal through the Bog of Allan, if you can suppose that wholly bare of any vegetation except around the very scattered Egyptian or Turks' houses.' From this point Mehemet's own steamer, the size of a Greenwich boat, took them another twenty hours' journey to Cairo.

Cairo he found 'a most interesting place for everything but its botany,' standing as it did 'half in the desert and half on the alluvial deposit, so that you enter it amongst gardens, avenues, and richly cultivated fields, and step from the gates on the other side into utter sterility.'

As for the Rhoda Gardens, originally laid out by Ali Pasha with the oriental desire of getting shade and refreshing masses of green, he frankly confesses to disappointment in them,

> from not previously appreciating the many obstacles Egypt presents to the formation of a real garden of Exotics. It must be near the Nile for water ; and then it must be flooded at one season, and burnt up the next ; a state of things to which few plants will subject themselves, and whence it is that on the fertile banks of the Nile there is little or no native

> vegetation beyond annuals, and the majority of these are planted.

Still it was 'really and truly the Dropmore of Egypt,' 'a noble project' struggling against adverse conditions.

> Everywhere you turn you are greeted by some English or well-known exotic, struggling to accommodate itself to Egyptian bondage, or rebelliously resenting all poor Mr. Traill's kind attentions, and doing the worst a slave can do, dying on the spot, and breaking his master's heart. (To W. J. H., December 24, 1847.)

Far more interesting was a trip into the Desert to the Fossil Forest.

> Though few plants were to be had, I was anxious to make a few observations on the temperature of the soil and dryness of the desert, so that I might know how near the starving and burning point vegetation would exist, as supplementary to our many observations in the Ant. Expedition of how much cold they could bear.

Completing these a few days later by other experiments at the halfway house to Suez, he found that

> even in the winter time the sun's rays give a heat of 100° to the soil, so that the poor plants have to undergo in winter a change of 56° every day. Here the only water they get is by the dew forming during the night. Unhappy plants! if their feelings are like ours, who like to drink best when most heated.

The waste of rolled pebbles and fragments with here and there huge trunks, heaped together in the greatest confusion, all chalcedony and coarse agate, reddish brown against the white of the desert sand, inspires a long disquisition on its geological origin and a smile at Mehemet Ali, for

> At this place the Pasha had sunk a pit for coal: sapiently concluding that so much fossil wood above ground indicated no less below: he, however, did not get through the limestone rock, which is subjacent to the formation to which I presume the fossil wood to belong.

As to the city of Cairo itself,

> the charms of these Eastern houses are all in the abstract and idea; to live in they are truly odious. [Seeking a Turkish bath], we wound through many nasty lanes and streets of shops, which are called Bazaars, but which I should rather y-clep *Vennels*, if you remember those Glasgow holes. After all, a Cairo Bazaar is very like a Greenock street, without the windows.

The visit to Mehemet Ali in the cortège of Lord Dalhousie smacked of the 'Arabian Nights.' He writes to his sister Elizabeth:

> The road was long, through narrow streets and very crowded ones; we were preceded by two attendants, running, with long whips, which they laid about them right and left, to clear the way, utterly regardless of man or beast, who scurry out of the way or cower under their Bernouse cloaks to fend off the blows. I saw an unfortunate Egyptian, whose cart stuck across the street, get a terrible whipping, to which he offered not the least resistance. We were rather late, and arrived just after the Governor [Lord Dalhousie], as the guns were pealing forth a Royal Salute. Passing under the gates, through a most splendid new and half finished alabaster Mosque (see Panorama of Cairo) [i.e. that shown in Leicester Square], we arrived at the Quadrangle, where the Governor was getting out of a splendid six-horse coach, like the Lord Mayor's, with Egyptian Lancers as outriders: the band played a sort of 'God save the Queen' to him, and I know not what to the second carriage, with Fane and Courtenay;[1] but I got the Bohemian Polka for my share of reception outside. The gateway was crowded with tame-looking, fiercely armed Egyptian officers, with gorgeous sashes, diamond-hilted scymitars, and the like. Behind were plainly dressed attendants on a dais, each with a gold badge at his breast (the Turkish crescent and star), who passed us on through gorgeously furnished apart-

[1] Members of Lord Dalhousie's suite. Francis Fane, who succeeded to the Earldom of Westmorland in 1851, was his A.D.C., and F. F. Courtenay, his private secretary since 1843 at the Board of Trade, was retained in that capacity.

ments, sofa'd all round the walls, and covered with rich Turkey carpets, to the private audience-chamber. This was splendid, surrounded by looking-glasses; the walls above pale satin, with worked crimson and gold flowers, the windows some 15 feet high, with transparent blinds, worked also with most superb groups of flowers, exquisitely imitated. All round were sofas and cushions of satin, worked with carnations, fuchsias, and roses. Mehemet, an old cunning-looking man, in a plain, olive-green braided coat, sat in the right hand corner near the window, and received us standing. He conversed with Lord D. by means of a Dragoman interpreter, we all being ranged round and forming a gorgeous cortège. Behind were several other gentlemen, who came, but took no part, including his son and son-in-law, and many plainly attired domestics. In a few minutes each, Lady Dalhousie included, was furnished with a pipe 6 feet long, having amber mouthpieces full of brilliants, the mouthpieces as thick as my arm almost, and 8 inches long. The bowl was placed in a silver dish on the ground; and we all whiffed away. The servants then brought coffee in little egg cups, set in gold filagree holders, blazing with diamonds. . . . The same attendants removed pipes and coffee cups; and we all retired much pleased with all we saw.

The troubles of the old Overland route, even under the gilding of the Viceregal ægis, merit description. The journey to Suez took nineteen hours. Hooker himself barely escaped the hideous discomfort of doing this on a dromedary's back. By some mistake, the disembarkation of the ordinary Indian passengers at Alexandria had not been telegraphed through by signal till after he had set off to visit the Pyramids. It would have been highly inconvenient for both parties to travel together across the desert; accordingly the Governor-General prepared for early departure, and when Hooker returned he found all the luggage had gone on; and he was in consternation, having only two hours to pack, get his fossils sent home, and go to the Consul's, whence they were to start. 'We were prohibited taking anything but a tiny carpet bag; so I hired a fleet Dromedary for my baggage (my very heavy things had gone on to the Palace on arriving and went on with Lord Dalhousie).'

All's well that ends well, however; thanks to Lady Dalhousie, who also had a baggage dromedary, and the members of the Suite, who bullied the Transit officers into providing an extra two-wheeled car, the baggage was safely taken. 'I never was so glad in all my life,' he exclaims, 'as when I got my things all stowed away, though at the expense of relinquishing my scanty collection, and all but a few sheets of small-sized paper, for the Desert and Aden.'

Night had fallen, for it was 8 o'clock, and 'our departure by cresset and torch light was very pretty, surrounded as we were by Orientals in all costumes.' As for the vehicles, the Dalhousies 'mounted a beautiful barouche, as good as ever the Park saw, with six Arab horses and two outriders, and dashed off at full speed, the cressets and torches scampering on before, through the narrow streets, whipping everybody and everything in the way.' The four-horse vans in which the rest followed were exactly like short omnibuses, to hold four each, but had only two wheels with broad tires; 'a cad stands on the step behind; an Egyptian drives at a furious gallop, with a red fez and long whip.' In the first were Dr. Bell, an old Indian, bundled up in all imaginable clothes, European and Oriental, to keep off the cold, and Hooker, with a plaid for the night, and slung round his neck his two precious barometers to save them from the breakage declared to be inevitable in the terrible jolting of the Overland route. The road was worst at the beginning; in many places it became really good, where the flats of pebbles were broad and long; but the Arab tribes who were heavily bribed to keep it in some sort of order, cared little for the Pasha. So long as they were paid, they removed the large stones from the track; as soon as the money stopped, they would replace all the big pieces, and so render the track impassable.

The smooth-seeming, uninterrupted slope of eight miles from the highest level down to the Red Sea was indeed a howling wilderness, and the Desert of Sinai opposite looked no better. Amid the pebbles and rounded lumps of rock as big as one's head the Colocynth was the only plant visible, and that sparingly, so like the soil it straggled over, that the great yellow apples alone betrayed its position. At Suez,

> as the position of the transit of the Children of Israel, one could not help looking about, and trying to grasp one natural feature that should afterwards vividly recall the spot ; but there was none ; looking N., an arm of the sea wound up to where a canal in the more glorious days of Egypt connected the Nile and the Red Sea.

A score of years were still to elapse before de Lesseps renewed that glory of the ancient empire, and incidentally swept away these wearinesses of the old Overland route.

The Governor-General's party were comfortably installed in the hotel long before the ordinary passengers began to arrive, 130 in all, in detachments of six or eight vans every four hours through the night. Next day they embarked on the *Moozuffer*. This was 'a noble ship,' as large as the *Sidon*, but although the captain gave up everything to Lord and Lady Dalhousie, the Indian Government had made no proper accommodation for the large party :

> the rest of us have to *pig* it out in the ship's armoury, a dirty place, next to the engine, intolerably hot and smothered with coal-dust. We lie on mattresses on the deck, and it is all we can do to turn out tidy for meals in the cabin.

In consequence, as he writes later from Madras,

> I have lost nearly all my collections (particularly that made at Aden) from the salt water in our wretched dormitory on board this ship. Not only were much of my collections destroyed, but my spare paper ; so that at Point de Galle I could not collect a single thing.

Aden itself was 'the ugliest, blackest, most desolate and most dislocated piece of land of its size that ever I set eyes upon, and I have seen a good many ugly places.' Unsatisfactory also was the Indian Ocean, 'the most uninteresting sea I ever crossed ; without birds or any fish but flying fish to relieve the monotony of the cruise.'

Lord Dalhousie's friendship, which was built up on the voyage, and in India showed itself in unstinted support to Hooker and to any friend he recommended, was a personal appreciation of the man rather than of the scientific investigator. Hooker, who was no less attached to him, as a man, during the too few years that he still had to live, wrote very frankly of his lack of scientific interests.

> I find Lord Dalhousie an extremely agreeable and intelligent man in everything but Natural History and Science, of which he has a lamentably low opinion, I fear. He is a perfect specimen of the miserable system of education pursued at Oxford, and as ignorant of the origin and working of our most common manufacturing products and arts as he is well informed on all matters of finance, policy, &c. I very carefully drop a little knowledge into him now and then; but I cannot awaken an interest or any sympathy in my pursuits: he is much pleased at my being busy, and especially with my carrying on my Meteorological register three times a day. Lady Dalhousie shares her husband's apathy, but is otherwise a kind-hearted creature. In the Desert I brought them the Gum Arabic *Acacia*, which I thought must interest the late president of the Board of Trade; but he chucked it out of the carriage window: and the Rose of Jericho, with an interest about it of a totally different character, met no better fate.

The thought arises that 'he has so much Scotch caution that he does not like to broach a subject he cannot talk upon'; however this might be, the efforts to interest him in the vegetable products of the East seemed to bear fruit, and later:

> The Governor-General hints to me that he would like reports on the Tea districts of India; so that I shall hope to be made useful by him and to have an opportunity of returning all his kindness. I need not say that I shall lay myself out to attend to his wishes in India. Assam, however, did not enter into my calculations.

And at Point de Galle he took care to present to the Governor-General his friend Gardner, Sir William's protégé, the representative of Botany in Ceylon. Science was likely to benefit by official acquaintance with men of science.

Madras revealed the splendours of Oriental pageantry in the official reception of the Governor-General. The military display, the brilliant colour of the crowds who poured out of the city, amply compensated for the waste of half a day on board ship while arrangements were being completed ashore.

This was India itself; authentic information was to be gleaned, practical arrangements made for forthcoming travel. An old acquaintance turned up in Major Garsten, bluff and burly, whom he remembered as a threadpaper of a lad in Edinburgh. He heard tall stories of the Mysore summer; when wineglasses snap off at the stem, untouched, and tables of teak split across the grain. Through Gideon Thomson, the brother of his Glasgow friend, he had hopes of securing a good plant collector. Five servants were needed for his travels, besides collectors; and Madras servants were reputed better and more faithful than Bengalis. More lessons in Hindustani were required; 'my progress in the lingo,' he laments, 'is very slow. I have no head for languages, especially such a cacophonous one as this.' He spent most of his time in the Horticultural Society Gardens, and seeing Mr. Elliott's collections of birds and animals. But even so, when he began travelling in Bengal he found the plants, presumably common Bengal species, new to him, 'and without books I cannot give even the generic names, so ignorant do I find myself.'

In Calcutta, where he arrived on January 12, he first stayed with an old friend of his father, Sir Lawrence Peel;[1] afterwards at Government House, for

> neither the Governor-General nor Lady Dalhousie will allow me to take up my quarters anywhere but with them. [And a little later]: Both show great friendship to me. He is a very fine fellow, who always means what he says; and

[1] Sir Lawrence Peel (1799–1884) was a cousin of the statesman, Sir Robert. He was knighted in 1842 when promoted from Advocate-General to Chief Justice of the Supreme Court at Calcutta. Returning to England in 1855, he became Indian Assessor to the Judicial Committee of the Privy Council. The love of his beautiful place at Calcutta was recorded in the name of his house at Ventnor, Garden Reach.

I really believe he would be not only *mortified*, but hurt, if I resided elsewhere than under his roof while I am at Calcutta. On one occasion he turned out of his own chamber to give it to me, because I returned from Sir Lawrence Peel's house a day earlier than was expected.

The only drawback to their great kindness was that, though he had entire freedom to follow his own pursuits, Government House was five miles away from his work at the Botanic Gardens, 'and to walk there in this part of Bengal is quite out of the question.'

Sir Lawrence Peel's house on Garden Reach was the Chatsworth of India, with its unrivalled gardens just across the river from the Botanic Garden, classical ground to the naturalist, where Hooker spent most of his time with McLelland, the indefatigable *locum tenens* for Hugh Falconer, then on his way to succeed Griffith,[1] a botanist distinguished alike as draughtsman and collector.

As we see more of one another he opens out; and I think it not difficult to understand him. He is a persevering Scotchman, without much ability, or powers of perception; blinded by Griffith's extraordinary ability, and impressed with the belief that it is better to fail in following Griffith's views and course, than to succeed in any other more suited to his own powers. He has, he considers, a pious duty to perform, imposed on him at Griffith's dying hour, to publish his MSS. and drawings. This he has been doing with great zeal and perseverance, on a wretched salary of £500 a year at

[1] William Griffith (1810–45), a pupil of Lindley, entered the medical service of the East India Company, and in 1835 was employed to report on the suitability of Assam for tea planting. His botanical travels took him through Assam and Burmah and the Khasia mountains: as surgeon he accompanied an embassy to Bhutan, and the army which invaded Afghanistan in 1838 and the following years. Appointed to Malacca, he was recalled to Calcutta (1842–4) to take charge of the Botanical Gardens and lecture to the medical students during Wallich's absence: on his return to Malacca he fell ill and died.

In making his collections he aimed not at species hunting but at giving a general account of the Indian flora on a geographical basis: in his botanical studies he was more of a morphologist than a systematist, and as an accurate and penetrating investigator of plant life, and especially of the problems of reproduction, he was expected by competent judges to have taken the highest place as a botanist had he not been cut off at the age of thirty-five.

the Gardens, out of which he will be turned in a day or two, to return to Europe (his service time having expired) or take military duties, which are disagreeable to a man of his age and long civil servitude. The expenses of the publications are defrayed by the E.I.C. taking 250 copies; the proceeds of the sale of the remainder he generously puts by, as a fund for the orphan boy: this is very noble; and every one says so.'

Of the actual MSS. and drawings on which he was at work Hooker, who lent his help, writes more enthusiastically: 'I am perfectly amazed at Griffith's powers. His exertions were all but superhuman and he was a far better artist than I had supposed.' The misfortune was that they were being given to the world as they stood, the drawings beautifully lithographed, but with many flaws in the descriptions and unelucidated by proper notes which the pious editor could have added.

A full description of the Garden goes to Sir William. McLelland had improved it by clearance of jungle, road cutting, and rearrangement; but without system and judgment, sacrificing noble trees and a thousand fine features without satisfactory result. He failed in his endeavour to turn the Garden into a botanical class book. Though scientifically brilliant, Griffith before him had not the eye of a landscape gardener nor the education of a horticulturist, and the whole establishment had been suffered to get out of order for the last dozen years. 'The Library is in dreadful confusion, just as Wallich left it, and the Herbarium worse.' Still, 'Falconer has, after all, a much easier job than you had at Kew.'

Later he tells how he had written to his friend Falconer giving his notion of what the Garden should be, and wondering how he took it, as it amounted to the annihilation of all Griffith and McLelland had done.

This included the laying out of a good river front, the re-plotting of the systematically arranged garden, with provision of shade and shelter from the fierce sun for plants and visitors alike, above all in the thirty acres outside the house, consisting of dried up grass and red gravel paths all askew, where to go out of the house is going out of the frying-pan on to the gridiron; 'I used to hop along like a bear on hot bricks till I reached the remains of the mahogany grove, some 200 yards off or more." He winds up to his father with some fun on the blending of the popular and the scientific.

> Lastly, there is room (and to spare) around the garden for a good arboretum and pleasure ground. McLelland encourages *Music, Dancers, fish bones,* and orange peel, so that the place looks at times more like Alger's booth at Greenwich Fair, the Cremorne Gardens, or Baron Nathan's Elysium at Gravesend, than a place for profit and instruction. I am sure, if good Lord Morpeth saw what I have, it would be a profitable sight. I declared to McLelland, he ought either to confine this to a pleasure ground or lead the first hops and hob and nob on gin and water himself with chocolate-colored damsels in boots and large ankles, that ogle himself and myself on our scientific vocations. As it is, he is often asked to join, and bring Mrs. McLelland to the picnic and Polka. Whatever you do, *never* let the Pleasure ground open into the garden.

The rest of his time was divided between trying to finish off the Niger Flora in time to be sent home by the February mail, together with instructions as to the remaining illustrations to be drawn for the Niger Flora,[1] and preparations for his first botanical expedition.

[1] These instructions are characteristic of his outlook on Distribution. Certain orders had been assigned to Planchon, the Kew assistant, to prepare for publication. Hooker writes: 'Please see that he alludes to species in too bad a state to describe, at the end of the genus: or if the genus be unknown, of the order to which they belong; this is essential for Botanical Geography, and he won't do it if not told.'

CHAPTER XII

JOURNEY TO THE KYMORE HILLS

TRAVEL to the Himalaya was still impossible for a couple of months; the interval was employed in a botanical excursion to the little explored hills of south-west Bengal, which culminate in the Vindhya range further west, and to the valley of the Soane, a southern tributary of the Ganges some 300 miles from Calcutta. This is reached by the Grand Trunk Road to Benares, seventy miles further on, which passes on its way from Calcutta the Burdwan coal-field and the sacred mountain Parasnath.

Another coal-field was reported higher up the Soane river; Mr. Williams, of the Geological Survey, was proceeding from Burdwan to investigate it. Hooker arranged to join his travelling camp on January 28, after a sixty hours' journey in a wearisome palkee, and from the upper Soane valley traverse the Kymore or Bind Hills to Mirzapore, above Benares (March 8–15), then to take boat down the Ganges to Bhagulpore (April 5–8), and finally by palkee from Caragola Ghat, some thirty miles further down the river, to Darjiling, some 140 miles, which was reached on April 16.

On the way he collected *everything* that the dry season produced in an elevated district which surprised him by its signs of constant dryness. Even when sailing down the Ganges he experienced a gale and blinding dust-storm, swept up from the boundless alluvial plains of the river valley.

> So dry is the wind that drops of water vanish like magic. What Cryptogamiae could stand the transition from parching

> like this to the three months' flood of Midsummer, when the country for miles will be under water?

The specimens he so arranged as to present a good illustrative Flora of the whole Road, gaining finally 'a knowledge of the look of whole botanical regions which, however poor in species, are highly instructive in other points.' From before daylight every day, he was hard at work; but the fatiguing lack of a collector had its compensations.

> My specimens are well dried; this is no difficulty, with a little trouble, at this season: three changings drying the majority: the difficulty is to prevent their drying too fast, yet, would you believe it? Wallich's and Griffith's plant driers were in the habit of pressing *once* in paper, and then spreading all out *in the sun*: no wonder their specimens are so contortuplicate.

Detailed letters home were deferred, but he kept a full journal, corresponded with the Governor-General and Mr. Colvile,[1] President of the Asiatic Society, to which his meteorological observations were communicated. Of these the most remarkable was on the night of February 14,

> when, on going out at 9 P.M., I saw the finest Aurora, on the whole, that I ever witnessed, either N. or S. This is a phenomenon supposed to be so rare in or near the Tropics, that it kept me up till past midnight observing and describing.

This account met with a good deal of incredulity; the sceptics ascribed it to forest fires, the appearance of which would be very different to an observer so long accustomed to the Aurora. Grievously as he grudged the time, he wrote an immediate

[1] Sir James William Colvile (1810–80), an Indian lawyer and sociologist, who, like Sir L. Peel, was knighted on being raised to the Bench in 1848—he was Chief Justice of Bengal from 1855—and on his return to England was appointed Indian Assessor to the Judicial Committee of the Privy Council. He was distinguished for his knowledge of Indian systems of law and of scientific and economic questions affecting India, and was President of the Asiatic Society of Bengal.

account of the Aurora to Wheatstone.[1] Referring to this the following year he tells his father:

> I thought I had said enough of the Aurora, and was only afraid of troubling you with too much unbotanical matter for the Journal; besides I did not consider that phenomenon to be so *very* wonderful as to cause surprise — much less argument. The sceptics may content themselves with 'tant pis pour le fait'; it required no witchcraft to pronounce upon the display which I beheld; and, in such a country as India, where every Englishman eats a heavy dinner at 8 and goes to bed at 10, it is not astonishing that these spectacles have been hitherto unobserved. I suppose I should be snubbed for averring that I have seen others since, and in the daytime.

Meantime he is able to assure his parents 'I am in perfect health and enjoying myself exceedingly.' He spared them the anxiety of knowing what he told Darwin (p. 246) that he still felt the results of his rheumatic fever at Madeira nearly nine years before.

His examination of the Burdwan coal fossils threw no material light on the question of their age, a question which, he tells Darwin, is no less perplexing there than at home. Others boldly assigned most of them to the Lower Oolitic epoch of England, from the prevalence of certain species, also found in Sind and Australia. In his cautious judgment the evidence was insufficient; the form of the fronds alone, especially in fossil fragments, supplied frail characters for specific identification; considering that 'the botanical evidences which geologists too often accept as proofs of specific identities are such as no botanist would attach any importance to in the investigation of existing plants.' Recent ferns were so widely distributed that inspection generally gave no clue to their place of origin, and considering the wide difference in latitude and longitude of Yorkshire, India, and Australia, the natural conclusion is that they could not have supported a similar

[1] Sir Charles Wheatstone (1802–75), the famous discoverer and inventor in the fields of acoustics, optics, and electricity, to whom we owe the practical foundations of telegraphy.

vegetation at the same epoch. And he cites the Cycads especially (Himalayan Journals, i. 8) in support of the statement that,

> finding similar fossil plants at places widely different in latitude, and hence in climate, is, in the present state of our knowledge, rather an argument against than for their having existed contemporaneously.

Later (p. 44) he insists on the point again, contrasting his own difficulty in identifying the impressions of living leaves in the lime-deposits of a spring, with the fact that geologists, unskilled in botany, see no difficulty in referring equally imperfect remains of extinct vegetables to existing genera.

The ascent of Parasnath, the sacred mountain of the Jains, was of vivid interest :

> We went thither on two elephants with a blanket cart and some provisions ; but the jungle was so dense; the elephants having to break away the branches of the trees with their trunks, that we did not arrive till 2 P.M. I got many plants on the route, the elephant getting several inaccessible species for me. You will hardly believe that a well-wooded mountain of (reputed) 7000 feet (but I expect only 5000) could rise out of India all but within the Tropics, and present neither Palm, Tree-Fern, Lycopodium, Scitamineous, Aroid, Piperaceous plant or Orchid-epiphyte of any consequence. No moss or Hepatica below 4000 feet, on trunk or rock, no foliaceous Lichen below that, and scarcely above, and not one fleshy Fungus. Such, however, is the parching effect of the N.W. dry winds, that the soil throughout is crumbly and the Cryptogs. at top, consisting of a few crust-Lichens and mosses (no Hepaticae seen), are withered and brown and covered with a Selaginella equally dead.
>
> There are six tops to Paras-nath, rising from a curved ridge, all very steep and rocky, and each crowned with a platform and little white Temple, of the size of your Temple of Victory. There is, besides, a large temple, a little below the ridge on the N. face, sunk in a hollow, very picturesque, square with a large dome and four spires at the angles. All are neatly covered with white lime. In the little apical ones I was surprised to find a slab of stone with the feet of Boodh engraved in relief, whilst the larger had many marble slabs,

> each with multitudes of little cross-legged Boodhs. My mind was at once carried back to Adam's Peak in Ceylon, and the high places of the N. of India, where Boodhism and not Hinduism prevails, where the less impure form of Heathen worship has taken refuge. Idol worship as it is, it was gratifying to find it taking possession of this lovely spot, to the exclusion of the abominations of Brahminism, which shock the eye as much as the senses.

The three weeks' leisurely sail down the river had a double advantage. He could stop where he would to see things of interest, such as the manufacture of rose-water at Ghazipore or the opium works at Patna; and he had time, most grievously needed, to write up his notes, journal, and correspondence, though the boat, externally very like a floating haystack or thatched cottage, internally became too much of a Noah's Ark for his liking, what with rats that mounted the table and stared him in the face, cockroaches of indomitable courage which 'take the crumbs off the side of my plate with the familiarity of Robin Redbreast, withdrawing but not retreating on my remonstrating,' and insects in swarms from mosquitos to the flying bug, which is no better than a winged skunk *in petto*, not to mention centipedes and monstrous spiders, a hand's spread across, darting about as if they had seven-leagued boots on each of their eight legs.

For the Kew Museum he was indefatigable in collecting vegetable products used in the arts, notably a pair of smelting bellows made entirely of leaves, and all the gums and drugs procurable, with the Hindu name transliterated, whenever possible, in English and Persian characters. With 250 of these already in hand he exclaims:

> The number of things still to be got at every market is infinite: and I shall go on amassing; but I have been only two months here now, and cannot bargain properly—it also takes a great deal of time.

This was not merely the passion for collecting; it had a very practical bearing, and the view of Hooker's work is incomplete without remembering that the practical applications of his science were as interesting to him as pure research. And even the forthcoming expedition to the rich botanical fields of Sikkim included the hope of discovering a trade route to Tibet if the result of war with China were to be the opening up of direct relations with the Forbidden Land, still under Chinese suzerainty.

At the same time the personal friendship with Lord Dalhousie enabled him to send in a memorial regarding the excessive cost of postage and travel, the destruction of timber, and the need of drawing up a good Indian Materia Medica.

Further extracts from letters to his aunt Ellen (Mrs. Jacobson), and to his sister Elizabeth, give some lively impressions of Oriental travel.

To Mrs. Jacobson

I often think of my cousins, little Willie and Mary, when perched on the top of my elephant ; or when I am struck by the peculiarities of this far foreign land. Many things are interesting, through their novelty : others are of a deeply melancholy nature, too much so to be pleasing. The elephant is always an agreeable animal ; he is so docile and gentle, when properly tamed ; and though to ride on a pad on his back is somewhat akin to being tossed in a blanket, one soon becomes accustomed to the motion. Every morning, after he has breakfasted heartily on a stone and a half, or two stone, of boiled rice, relished with large boughs of Fig-trees, the elephant is led to my tent to be mounted. A little active Mohammedan driver sits on his broad neck, and directs his movements by poking his own toes behind either ear, according to the way he desires to turn the beast. He carries a goad, a short spear of iron, which he sticks into the poor elephant's head, if lazy, or inflicts a pat with it which would lay Willy's skull open. When the order is issued to '*butt*,' elephant drops on his knees ; and I climb up, by getting on a hoof and holding by the tail, or with ropes. Or I accomplish the ascent by stepping upon a tusk and gripping at the broad ear. At the word of command, he rises, and walks off, at the rate of 6–8 miles an hour, his broad hoofs crushing the soft soil as he boldly tramps along. If the road be stony, he picks his way with great care, placing the hind hoof in the exact place from which he has lifted the fore one : he is a tender-footed beast, and cannot travel far or fast upon rocky ground. As the heat of the day increases, he drinks at every stream; drawing up the water in his trunk and then putting his long proboscis down his throat, he deposits the fluid in a bag near the stomach, which it takes ten minutes to fill. When this natural water bottle is replenished, the elephant walks on,—every quarter of an hour or thereabouts, poking his trunk down his throat, drawing it out and squirting its contents all over his body to cool himself, for the hot sun beats strongly on his black carcase. Of course I come in for an ample share of his shower-bath, which, as it sprinkles my spectacles, is not desirable. So much for the elephant's fashion of cooling himself by day, and he is not a whit less clever at expedients for retaining his warmth during the 'chill dewy night': he scrapes up, with this view, all the dust he can collect with his foot and trunk, and aided by the curious crozier-like coil at the end of the latter, he dexterously jerks the earth all over himself, so preventing the evaporation from his skin which would make him too cold at night. When crossing rivers, he pulls some carts across and pushes others through the deep sand with his broad forehead. After one morning's work my poor beast had a lump on his brow, as large as a child's head, raw and bloody at top; but all of us had to work so hard that we could not excuse him, and it was touching to see the docile creature lay his expansive brow obliquely to the back of the waggon, first by one temple then by the other, stoop and try with his soft trunk to move the load and avoid the sore place—till, finding all was useless, he gallantly planted the sore bump, and with a short cry of pain, he thrust on, and persevered till all the waggons were fairly over, though aware that every time he lifted his head and set it to the work again, the same suffering must be endured. So, when he has to remove a thorny tree from the path, if he cannot find a smooth part of the trunk, he boldly grasps it, thorns and all, tears it up and lays it on one side. If I drop anything, hat or book, he picks it up with his trunk and adroitly tosses it over his head into my lap. The other day I went to a fair, in the heart of a remote district, and dismounting, went through the whole show. It was just like Glasgow or Greenwich Fair, except that, as in all Eastern and some Western countries, the trades were drawn together in lines. There were children, with trumpets and squeaks, merry-go-rounds and rocking-chairs. The little girls were decking themselves with trinkets and patches of gold leaf for the forehead and pieces of bone thrust through the ear,—the greedy boys were twitching their mothers to the lollipop sellers, and the bigger ones eyeing the ponies on the outskirts. Old men were chaffering for graven gods, and the sick folks were waiting round the doctors' stalls. I was looking on, followed by an immense trail of people whom my presence had diverted from their traffic, when I suddenly heard a fearful yell, which proceeded from the direction of the spot where I had left my elephant, and casting my eyes thither, I saw all about him in an uproar. The men were swearing and flourishing their sticks, the women and children were in full flight, the driver on his neck was banging him with the goad till his skull rang again, or digging it into his forehead till the blood sprang. As to Elephas himself, he would not stir from the place, but kept laying about him with his trunk, bellowing through mouth and nose, retreating or advancing a step or two with fearful violence and continually darting his proboscis at some object,—what I knew not, in the crowd. You may guess my terror: I felt sure he was enraged and wreaking his violence on some of the poor creatures from whom proceeded the dismal shrieks which I heard! I rushed through the throng, overturning some of the stalls in my hurry to reach the place,—when I found—what do you think, Willy! now, guess, Mary!—why my elephant was clearing out a sweetmeat booth: he was eating barley sugar by the pound, and comfits by the peck. I had another anecdote for my cousins about a crocodile which I saw caught, just as he had devoured a poor woman's child who was standing by and looking at the odious brute; but my time is up and I must break off.

Meantime, his scientific and personal standing in India was greatly enhanced by the publication of Ross's account of the Antarctic Voyage.

You have no idea how many people in this country have been reading Ross's work: I am better received in India for having accompanied that voyage, than ever I was on that account in England. Every individual with whom I have stayed, on my way up and down the Ganges, has read it! and knows me through it! . . . On this table in this house [of Dr. Grant of Bhagulpore] lies the *N. British*

> *Review*, containing an article on Ross's Voyage, written, I suspect, by Sir D. Brewster. There is the most flaming flattery in it of my share in the book—especially the chapter on *Cattle Hunting*. Pray tell my mother of this: (I suspect I must be a sort of humbug after all). My Journal shall be copied and sent, as soon as I can get settled: for I know you want it. You may easily suppose that, surrounded with plants to dry, information of every kind to secure, &c., a *Griffin*, like myself, has his hands sufficiently full of occupation. I try hard to understand everything as I go on,—but I am sorry to find the attempt is hopeless!

But people were not to be persuaded that an Indian hill storm which he describes to his sister Elizabeth could be inferior to an Antarctic storm.

> Though more tremendous looking from the thunder and lightning, it was not so strong as many S. Polar squalls I have felt. People won't believe that here, and so I say nothing about it.

A double letter to Darwin (February 20, and March 4 and 16, 1848) which opens with the words, 'Though our correspondence has not ebbed so low for full four years, you have been so constantly in my thoughts that it appears far from strange to be writing to you,' and ends with 'love to the children,' is too long to quote in full. It answers many questions on which Darwin had asked him to obtain information; e.g. on the habits of the Cheetah and the way in which it is used in hunting and its curious refusal to hunt more than one season; on the extension of different species, where he finds an apparently undefined rule; the Soane, for instance, in the case of the antelopes and the gaur, in providing a line of demarcation, like the Obi in Siberia, which Humboldt, when Hooker visited him in 1845, adduced as 'dividing two Botanical regions, and (being) one of the strong arguments *against* the migration of plants, as large rivers do not in other cases prevent what is considered migration.' So of elephants, dogs, cattle, squirrels, swallows, saurians; the desiccation by destruction of forests; local geology: in short, 'I am perfectly bewildered by the facts hourly thrown before me, whose importance I can scarce appreciate from my ignorance of Indian natural history;

all I can do now is to attempt to collect those relating to the larger or more common animals.'

However,

> As in other parts of the world, so here, almost all the animals of the plains will descend; this is a common observation, but it never struck me before coming to India, that in this respect height is not analogous to Latitude; for most of the animals and man himself accommodate themselves rather to an increase of temperature than a diminution. Thus the Englishman, horse, dog, sheep, &c. &c., all thrive in India; but the monkey, man, Bhil, and all the other common tropical animals, are incapable of supporting colder climate, dependent on latitude.
>
> I will give you but one botanical fact, and that is regarding the vegetation of heights. You have often asked if Mts., especially isolated ones, in the tropical and S. lat., had closely allied representations of Asiatic or N. temperate forms; now I have been up but one eminence, and that of no more than some 5000 feet, and there I found a *Barberry* in abundance, and one not unlike our English and (I may say) one smaller Cape Horn species; but one more fact of a different value—do you remember the allusion to *Vallisneria* in your grandfather's 'Bot. Gard.'? I have found what I take to be a second or new species of the genus in the waters of the Soane, with the same wonderful habits: without books, however, and a limited memory, I rather talk at random about new species.
>
> With regard to my health, it is exactly the same: I am still troubled at times with those bothering pains on the left side and palpitations, aching in the axilla and occasionally down the arm. The motions of the heart are on these occasions very irregular, but I have no ringing in the ears, shortness of breath, or any symptoms that alarmed me. Hot or cold days make no difference, and indeed I had a long cessation of all pains for three weeks after my arrival, that I thought the hot weather had cured me. Whatever it is I am *none the worse* of being here, otherwise I never had better health, am thinking of getting fat, and hardly know what a headache is. Please do not show this part of my letter, as this refers to a subject of which my friends know nothing.

CHAPTER XIII

TO DARJILING: THE FIRST HIMALAYAN JOURNEY

It was a weary journey by palki from the Ganges to Darjiling. Whole days were wasted in trying to secure bearers. Frequently none were ready though arranged for with the Post, and those who had already come a stage were obdurate to 'praying, promising, and protesting, bribing and bullying.' Once Hooker had to walk while the men carried the empty palki till they met certain return bearers of a previous party. 'People may say what they like,' he exclaims feelingly to Miss Henslow (April 9, 1848), 'about the "mild Hindoo" and all that sort of thing; they have their good points, but being led by kindness or generous treatment is not amongst them; they *never thank* you and, overpay as much as you like, they growl. Highlanders cannot be worse.'

At Darjiling began a new phase of life in India, and with it a deep and lifelong friendship with a very remarkable character. Brian Hodgson, administrator and scholar, had won equal fame as Resident at the court of Nepal and as a student of Oriental lore. Known to English science as the best Indian zoologist and the donor of the Hodgson natural history collection at the British Museum, he was yet 'far better known as an Oriental linguist, Ethnologist, and Geographer.' Dismissed from his responsible post against the wish alike of the Nepalese and the Government officials by the petulance of Lord Ellenborough,[1]

[1] The Earl of Ellenborough (1790–1871) was Governor-General of India 1842–44, in succession to Lord Auckland, after twice being President of the Board of Control. By the irony of fate, his purpose being 'to restore peace to Asia,' he spent his time waging wars of punishment against China and Afghanistan and of annexation against Scinde. His unpopularity with all classes except the army was due to his vast self-sufficiency and disregard for others' feelings and interests.

'in one of that nobleman's absurd fits of determination to undo everything, good or bad, which Lord Auckland had done,' he had retired in bad health to this lonely eyrie on the edge of the mountain world he knew so well, in close touch with the Asiatic travellers from the Buddhist cities of Tibet.

In Hooker he found a kindred spirit, a personality that inspired confidence, and he placed himself under Hooker's medical care as well as admitting him to his intimacy. From June 1848 Hodgson's house was his home. It stood a good 800 feet above Hooker's first residence, Mr. Barnes' house, 'and like Olympian Jove, I am daily surrounded with the clouds,' for the rains had 'fairly set in, and it sometimes pours for eight, ten, and, I am assured it will, for fifty or sixty hours consecutively.'[1] He enjoyed its retirement, the opportunities for uninterrupted scientific work, the personal charm of his host, and the mine of information on all things Indian ever at his disposal.

> We are working together every evening [he tells his mother on June 23] at Himalayan and Thibetan Geography and Nat. Hist., and though I say it myself, it is true that I ought in a month or two to have a better knowledge of these aspects of India than any man, having every advantage that an excellent library and tutor can afford. We are now arranging a sketch by which to divide the range into natural sections [i.e. divided into districts by the watersheds from the Monster peaks], each of which will bear some illustrations from personal experience and books, and this ground plan will do for others to work upon. . . . I am determined I will not leave off working till I have gained a thorough knowledge of the subject. [And again]: Hodgson 'is a capital helper,' and this stay with him 'the very best chance for me that could have occurred.'

[1] 'Hodgson's house is on a hill and amongst many other hills all heavily timbered, with plants through the wood and lots of new plants close to the door. It is a one-storied house with a broad verandah all round, facing North and the Snowy Mountains. I have two good rooms besides the run of the dining-room and parlour. There are lots of servants to go and come as I please to call or send, cats innumerable, and more "Bishop Barnabees" than at Kew, and exactly like them. (To his sister Elizabeth, August 9, 1848.)

Experience of such friendship inspires him to write home of 'Hodgson, who shows me all the attachment and affection of a brother, and whom I shall always regard as one of my dearest friends on earth,' and later, hoping that his friend would leave Darjiling, which did not agree with him, and go to England in the autumn of 1849, exclaims, 'I am so anxious you should all know him.' He allows that Hodgson was too proud and haughty, but never towards himself. He had lived too long with the power of a prince in his hand not to acquire something of a prince's outlook. The sensitiveness of ill-health, added to absorption in keen intellectual interests, helped to render him impatient of the chatter of a small station, and thus he was not disposed to suffer pettinesses gladly.

> He is said to quarrel with every one, and in truth is as proud a man as I ever met, but we have always got on comfortably, and as we live like brothers our quarrelling would be absurd. We have a tiff now and then, but very rarely. [And July 19]: He and I live like hermits, and hardly ever see anybody but Mr. and Mrs. Campbell and the Müller brothers.[1]

But he opened out at once to a kindred spirit, forestalling every wish before it could be uttered, and what is more, seeing to it that every promised arrangement should be carried out, to Hooker's great relief, during the privations of his journey in Sikkim. Like a prince he gave; with a prince's pride he shrank from any appearance of a return for friendship's favours. In this mood indeed at first he even declined to let Hooker name after him the finest of the new rhododendrons discovered in Sikkim.

If the friendship with Lord Dalhousie provided the key that opened official barriers and made Hooker's journeyings possible, the friendship with Hodgson more than anything else made them a practical success.

[1] These bachelor brothers were here for their health; one being the head of the opium factory at Patna, and both interested in science. They gave Hooker every help in their power, and in particular reduced all his meteorological observations for him.

The other friendship here cemented was with Dr. Campbell, the Political Agent to Sikkim.

> He is well versed in all Tibetan and Frontier affairs; he has given me much information on these subjects, and on the vegetations of the countries beyond the snow, which he has learned from the Thibetans who came hither through the snowy passes (April 28).

Warm, hearty, and helpful as he was, he was not the grand seigneur or professed scholar like Hodgson, nor did he equally possess that fine imagination which would outrun an ordinary welcome, or ensure perfect diplomatic goodwill in the Sikkimese representatives with whom he had to deal. Moreover in his official dealings he had had many small rubs from the Calcutta Government, so that he was at first shy of pushing Hooker's wishes as he for his own part would willingly have done. Thus friendship with him took longer to establish, but was drawn close long before their joint experience of travel and captivity in Sikkim.

The charm of his home at Darjiling was completed by Mrs. Campbell and 'her beautiful children; for the little creatures have taken a vast fancy for "Hooker doctor," who gives them sweetmeats, and who rides "the naughty pony."' To them Hooker was devoted, and to Josephine, born while he and Dr. Campbell were still prisoners in Sikkim, he stood godfather. This friendship also was lifelong, and is prettily illustrated in a letter to Sir William Hooker dated July 19, 1848:

> I wrote and told him this morning that I would ask you to confirm the name of a Rhododendron on his wife, a little compliment that has touched him to the quick; he is very much attached to his wife, and I really never saw a man so heartily appreciate a trifling favor. Now pray don't forget to attach the name to one of the species sent if the one I have given it to be not new. With regard to all the names, pray alter them as you please or name the plants yourself altogether. I have no ambition that way now, and would indeed rather see your initial at their tails than my own, but, I beseech you, don't forget this MacCallum Mor*ae* [for Mrs. Campbell].

The supreme objective of the Himalayan journey was to reach the snows. Between these and the deep, humid valleys of the lower Sikkim lay a whole botanical world, with a range equal to that from the tropics to the pole. There also lay the secret of the Himalayan geography. It was still generally believed that the vast line of snow peaks on the northern horizon formed a continuous ridge, the axis of the chain and the water-parting between India and the Tibetan plateau, instead of being but bastions at the southern end of cross ridges projecting from the true dividing range. From one of the icy passes in this region traversed by the traders from Lhassa there would be the possibility perhaps of entering, at least of surveying, the forbidden land and determining in this quarter the elevation of the great central plateau.

Travel itself would not be easy. The rude paths, alternately plunging into deep valleys and scaling precipitous mountain spurs 5000 feet or more, only to descend again, were constantly liable to destruction by torrential rains and mile-long landslides; rushing streams had to be forded or crossed on frail bridges of swinging bamboo. Forests where a way had to be pushed or hacked through dense vegetation pestilent with leeches and noxious insects, would be exchanged for bare rocky defiles at breathless altitudes where only a few poverty-stricken herdsmen lived and where the Indian carriers suffered from the fierce winds and freezing nights. But the greatest difficulty arose from the political situation. No place could be better than Darjiling for acquiring information from native travellers, but as regards permission for a European to travel, he writes on April 28:

> I fear that even Lord Dalhousie's influence will not enable me to accomplish my wish of visiting the snows. I have written to him, however, on the subject.

The much involved situation is set forth in a letter to Lady Hooker, June 10, 1848:

> My prospects of visiting the snow are somewhat faint. The Sikkim Rajah, whose territories were once the prey of the Nepalese, was replaced on his throne by us, who thus

kept the warlike Ghurkas from over-running Bhotan; unluckily we did not demand even a nominal tribute from the Rajah, who at once fell under the influence of China, whose policy it is to rule the Councils and hearts, but not the people, of these three Border powers; and by teaching them a wholesome dread of the English, they exclude the latter from these several States and prevent our interfering with the Chinese Trade from the East into Thibet. Darjeeling is a narrow slip of land, running north into the heart of Sikkim, about halfway to the snow. It was bought from the Rajah to be a Sanatorium for sick Europeans (as Simla, Mussoorie, Nainee-Tal, Almorah, &c. &c.). We paid 3000 rupees for the freehold, stipulating also that merchants should have a right to trade to Sikkim, but made no agreement of the sort for travellers, surveyors, or any other class of people, whom the saucy Rajah excludes from his kingdom. Had we acted with any vigour in our policy, we might still have retained our power over the Rajah; but I look upon the conduct of the local Government of Calcutta and the Political Resident here as weak to a degree and prejudicial to the interest of the country. The Rajah, who has not a soldier to his name, refused to allow the Surveyor-General (a man whose Indian power and appointments would astonish an Englishman) to visit a mountain twenty miles from hence, and not only the Surveyor-General but the Government who applied for him, only granting it when Col. Waugh,[1] disgusted with both the Rajah and Government, went (as I did a few days ago) without the permission of either. I have explained all this to Lord Dalhousie and asked him to send me to the snow, whether the Rajah likes it or not; offering to be the means of making any overtures to that Prince, which may render my mission less unacceptable than the appearance of any Feringhi must be. Dr. Campbell, the Political Resident, recommended that the Rajah should be asked, knowing as well as I and Lord D. do that, though the Rajah dares not refuse, he does dare to withhold an

[1] Sir Andrew Scott Waugh (1810–78), knighted 1861, reached India in 1829 as a lieutenant in the Bengal Engineers, and in 1832 joined the great trigonometrical survey, in which he distinguished himself so much that the surveyor-general, Everest, when he retired in 1843, obtained his nomination as successor to that important office, though still only a subaltern. Waugh gave the name of his old chief to the Himalayan peak Devidanga, which proved to be the highest in the world.

answer, and thus place our Government in the quandary of putting up with an insult or sending me with an armed force. Such is the Rajah's dread of the English, that he declined receiving an Ambassador, laden with English presents; and when the hot-headed Colonel Lloyd (who bargained for Darjeeling) hunted him like a hare to strike the bargain in person, he would only meet him with a river between. In pushing my own way there is nothing to apprehend but the lack of provisions; the Rajah is too weak even to put a traveller in confinement as China does, and too much afraid of England; but he can withhold supplies and frighten your servants. Hence all my wanderings have been hitherto only so far distances as I could carry provender for myself and the men, and through the least inhabited parts of the country. Towards the snow the country is more populous, the convents, nunneries, and villages are numerous (though small), and the people (Bhoteas) are a disagreeable and morose race, immigrants from the East into Sikkim. What Lord Dalhousie may do I know not. Elliot,[1] the Secretary to Government, proposes the using 'douce violence' with the Rajah, and insisting that he shall behave like a friendly power, but this view cannot be supported in Council. My own conviction is that, if the Rajah allowed me to visit the snowy Passes, China would punish him, not ostensibly but indirectly, and the only profitable part of his revenue is derived from Darjeeling (which did not yield him 200 rupees when we bought it), and a property called Chumbi in Thibet, which he rents from China, and which is a fruitful place yielding turnips, radishes, and Pine-wood! To proceed with Oriental crooked policy, Sir Herbert Maddock, Governor of Bengal during Lord Hardinge's[2] absence, in a fit of spleen assumed that the rent which the Rajah received for Darjeeling, 3000

[1] Sir Henry Miers Elliot (1808–53) entered the E.I.C. service in 1826, and became Secretary to the Governor-General in Council for Foreign Affairs in 1847. With Lord Dalhousie, after the Sikh War, he negotiated the treaty with the Sikh chiefs for the settlement of the Punjaub and Gujerat, receiving the K.C.B. (1849). His valuable historical work dealt especially with India in Mohammedan times.

[2] Sir Henry Hardinge (1785–1856) was the Peninsular veteran and later Secretary at War, so highly esteemed by Wellington, and was Governor-General of India between Ellenborough and Dalhousie (1844–8). At the conclusion of the First Sikh War, he was created Viscount Hardinge of Lahore.

rupees, was too little. He attacked Dr. Campbell, the Political Resident, for allowing the poor Prince to be so shabbily treated by England, voted the 3000 to be doubled, without any sufficient reason, and did this without even stipulating that the Rajah shall behave more civilly to Europeans. Campbell, who ought to have flung the reprimand back in the Governor's teeth and complained of the unjust treatment to the Board, took it all quietly, doubled the Rajah's revenue, and thus threw away a fulcrum which would have moved the Himalayah to within our reach. The Rajah is consequently more persuaded than ever of our foolishness and desire to take over his valued kingdom (of which we would not accept the gift). Is it not incredible that a man can be so weak as to fear the very power which placed him on his throne and to this day maintains him thereon from the being trisected, as Poland was, by the Goorkhas, Bhotanese, and Thibetans, any one of which would swallow him up in an hour? Lord D. has plenty of time now to think of the affair as I cannot go till October, the rains and the unhealthiness of the intervening valleys both precluding the attempt.

Six months passed before Sikkim, after repeated refusals, conceded a reluctant assent to the direct demands of the Governor-General. The chief expedition through Sikkim took place in the following year, albeit hampered by the obstructive devices of the Rajah's Dewan, which were successively overcome by Hooker's good-humoured firmness and amusing bluff.

But the partial permission for the autumn of 1848, followed by efforts to take away with the left hand what had been granted by the right, brought indirectly a still greater triumph. Thanks to the goodwill of the famous Jung Bahadur, Nepaul opened her eastern valleys to the traveller, and the Ghurka escort, disgusted by the petty machinations of the Sikkimese to prevent Hooker from ever reaching the northerly point at which he was to enter Nepaul, undertook to lead him the whole way through their own territory to the Tibetan Passes on the west of the Kinchinjunga group, through country never before and never since traversed by any European.

In the meantime Hooker was busy in other directions. 'If it were not for the Greenock-like climate,' he writes on April 28, 'this would be a very fine place, and I enjoy it much, for the vegetation is truly superb.' His new occupations were at first hindered by the necessity of completing the piece of unfinished work for his father, which he had brought with him from England.

This was the Niger Flora, of which he sent home the first part on May 18, the remainder on July 19. This was the only piece of work outstanding in regard to which he felt a personal claim; the rest could fairly be completed after his return, and so, when the way seemed clear for his journey to the Himalayan snows, he writes (September 12) with perfect unconcern:

> I saw that Lindley gave me a *touch* for travelling on my own pleasure while my *Flora Antarctica* is unfinished; to which I can say *Pooh!*

Indeed, to the end of his stay in India, he had no thought of writing a book of travels or working out his non-botanical observations. This he repeats to Wallich in 1850 as he had written to his father in February 1849, when sending him the Rhododendron notes and specimens he had brought back from the Sikkim-Nepal expedition. The future decided otherwise.

> Of them and of all my plants, MSS., and drawings, I beg you to make whatever use you think proper. The *Flora Antarctica* nearly broke my back; and except the Floras of New Zealand and Van Diemen's Land, I do not contemplate any other such great work. My present notion is to publish in the form of *Icones*, confining any large and costly illustrations to a few Natural Orders or Genera.[1]

In May, however, he took such opportunities as offered during the early part of the rains for botanical excursions near Darjiling. Without awaiting formal leave, he made

[1] For the success of the Rhododendron book, especially in India, see p. 326.

'a very favourite and interesting trip' by way of the cane bridge over the Great Rungeet, eleven miles away, into a deep, steamy valley admirably illustrating the successive zones of vegetation from temperate to tropical, that clothed the steep hillsides. The bridge itself was the British boundary; beyond lay Sikkim proper, where the Rajah somewhat ineffectively threatened punishment to any who guided a European, and where later Hooker's collectors going alone were maltreated by the Dewan's orders. But the inhabitants and even the Lamas, whose hostility had been represented as certain, were in reality most friendly. Indeed, on his second trip seven months later, the people brought supplies in embarrassing superabundance.

In this direction Hooker went as far as the junction of the Rungeet with the Teesta, and saw the mountains of Bhotan towering up over against him.

> The journey was, though not distant, a very difficult one, from the impracticable nature of the country, and had been accomplished by but one individual before; which is, however, mainly owing to the laziness and want of curiosity of the people, and the fact of the Rajah of Sikkim forbidding all crossing the narrow bounds of Darjiling. [Among his spoils were] three *Rhododendrons*, one scarlet, one white with superb foliage, and one, the most lovely thing you can imagine; a parasite on gigantic trees, three yards high, with whorls of branches, and 3–6 immense white, deliciously sweet-scented flowers, at the apex of each branch. It is the most splendid thing of the kind I have ever seen, and more delicate than the others.
>
> . . . I draw as many things as I possibly can, and send them to Falconer for transmission to you: the three first *Magnolias*, he tells me, are all new: two others I have not sent down: the 3 Rhododendrons are all drawn, and about 40 other plants, somewhat rudely; but they may give you some idea of the plants.

As to his various collectors:

> The Lepchas or mountaineers of Sikkim I like extremely. I have two men who collect fairly and climb trees à merveille,

and to-day have added two boys of 8 and 14 or thereabouts; one a very fine little fellow. Falconer has sent me up everything I asked for, including 3 Bengal collectors, regular *Haymakers*. I dislike the Bengalees very much; and these are lazy dogs, as all are. I shall astonish them to-morrow, when they will have to travel some 15 miles through these woods. One actually objected to carry the vasculum 6 miles, whilst a Lepcha carries 80–100 lbs. 16 miles on a stretch, and laughs all day long.

In the same month, May 19, he went further afield with his friend Mr. Barnes on what he considered the most interesting trip to be made from Darjiling. This was to Tonglo, a mountain 10,000 feet high, in the long subsidiary range dividing Sikkim from Nepaul, that runs south from Kinchinjunga, the then loftiest known peak in the world. Tonglo fronts Darjiling on the west, a dozen miles away as the crow flies, thirty by the path. The district was full of botanical treasures, the extra 1000 feet ascended presenting a total change in the Flora, but in the valley of the Little Rungeet the glories of the scarlet vaccinium parasitic on the trees, of the great white rhododendron named later after Lady Dalhousie, and of the tall magnolia with shining foliage that was to bear Hodgson's name, were sadly dimmed by the swarms of the large tick from the bamboos —'a more hateful insect I have never encountered'—and the persistent leeches such as had already been met with on the way up.

Unfortunately the bulkier things collected had to be left behind. Owing to the ceaseless torrents of rain, five of his fifteen men fell sick. Even the hardy Lepchas could not stand wet and cold together, especially on their poor fare of fern-tops, maize, rice, and whatever else they could get, from leaves of Solanum and nettles to fungi, 'which would give Klotzsch or Berkeley [1] the stomach-ache': in fact 'a vegetable must be very bad to be acknowledged poisonous by these people, who may come under Sambo's definition of the genus *Homo*, "an omnivorous tripod who [devours] all he can get."'

Still, what remained was 'a glorious collection,' making

[1] These were both distinguished mycologists.

a pile six feet high in the drying papers. 'If I can only succeed,' he cries, 'in getting these glorious things to Kew, how happy I shall be.'

As to the distribution of plants, these Himalayan valleys presented a striking parallel to the Antarctic. In the humid and equable climate of the latter, botanical orders which only reached lat. 30° or 40° in the northern hemisphere, reached Tasmania and New Zealand and even Cape Horn in 55° S. So in Sikkim, where it was not dry enough for the *Skimmia* in its native home to ripen the scarlet berries which light up our English gardens, some tropical genera pushed abundantly into the temperate zone, fostered by the damp and equable climate.

The general features of Himalayan botany he sums up as follows (May 18, 1848):

> In travelling N. you come upon genus replacing genus, Natural Order replacing Natural Order. In travelling E. or W. (i.e. N.W. or S.E. along the ridges) you find species replacing species, and this whether of animals or plants. Don't forget to send this to Darwin.

On July 24 (the extracts being given in brackets) and August 9 he writes:

> The rapidity with which the flowering season is advancing is quite wonderful, and I have accordingly doubled my establishment of collectors. I pay very liberally, often for trash, and they all like to bring me things. They are capricious and apt to run away if offended, but mine like me and I them, and such fellows will do anything for a master. I have always a horde of them in pay, at 8*s*. to 16*s*. a month. I have 18 at this present moment, for the plants are flowering and dying so rapidly that it takes *all* my energy to keep a good collection up. The papers too have all to be changed daily and dried individually over the fire—the rooms are so damp that hanging up to dry is no use. Everything moulds which is not kept *at* the fire. All my plants are on a circle of chairs immediately round the fender, inside which two Lepchas squat and dry papers all day long, in two rooms. [I am dreadfully badly off for paper, having used all that Falconer sent me up and all the newspapers (do you remember the

Bengal Hurkarus [1] in which Mrs. Mack's collections came ?) I can lay my hands on. This last fortnight I have got a glorious lot of things, such fine *Cyrtandreae* especially, and a good gale of wind helped me to many of the trees. Campbell too is as active as ever he can be, and I generally get two instalments, sometimes four, daily. I cannot possibly draw all I ought though I do my best to, and the poor Fungi are gone to the wall altogether. I cannot go 100 yards from the door without getting new things, to-day a new *Balanophora* [2] close behind the house, actually within a stone's-throw.

August 30 :—The rain it raineth every day, and the whole country between the foot of the hills and the Ganges is under water. . . . Such lots of rain was never seen nearer than the West of Scotland. Plants seem to enjoy it, for they are coming out and flowering faster than ever.

Besides the strictly geographical map already mentioned, a local chart was under preparation to show geographically the distribution of plants, 'a *Carte Géognostique* of the vegetation of this place from the plains to 10,000 feet (like Humboldt's of Chimborazo).' Notes on the agriculture of the Himalaya were being made for Professor Henslow. Loads of living plants for despatch to Kew were being sent down to Calcutta, where Dr. Falconer forwarded correspondence and repacked plants for the voyage. Many of these plants perished in the plains before reaching Calcutta ; the safety of the rest was threatened by the severe illness of Falconer at this juncture. But the supply was endless. 'The richness of this Flora is most remarkable and new things are brought to me every day. I dissect and sketch roughly the most important, including all the *Orchideae*.'

A great drawback during the first months was the absence of books of reference. In July, Falconer, in despair of an opportunity of forwarding them, took to sending them in small packages by post.

[1] A Calcutta newspaper.

[2] A curious root parasite of simple structure, without leaves or petals, related to the mistletoe, formerly thought to be allied to the Fungi. Hooker's paper on this order appeared in the *Linn. Soc. Trans.* for 1856.

A better opportunity, however, came before long (August 9):

> Falconer has kindly sent me four cases of books, soldered in tin, by *Post free* ! This is the only way of getting them safely now. They are De Candolle, Walpers, Kunth, and Royle. This week of *books and plants* has been perfect revelry. I find that my Rhododendrons are nearly all (perhaps they all are) new.

'My life here,' he tells his sister (September 28), 'is sufficiently monotonous to hear of, but far from so to me, my collections increasing very fast indeed, and never having a moment to spare.' Except for recording barometer, thermometer, wind and weather every hour, all the daylight hours were spent in writing and drawing and arranging plants. The plants generally came in at eight or nine in large baskets on men's backs. These Hooker always ticketed himself with the native name and any known quality or use, laying aside those he wished to draw and examine, and giving the rest over to be dried and the roots to be packed in moss. The perpetual wet forbade much going out. A recorded rainfall of twenty-one inches in July was perhaps nothing much for India,

> but it is like the difference between Glasgow and Edinburgh which I could never make Papa believe, that Edinburgh has more rain than Glasgow, though in the latter it is expended in a constant drizzle, in the former in a few downright showers.

Yet his health was perfect, 'living so regular a life in so salubrious a vile climate, far worse than Glasgow,' and 'here, in this dear delightful double-distilled Greenock fog, we know not what a headache is.'

Scottish recollections happily fill in the picture of Sunday morning at Darjiling which he draws for his sister Elizabeth (August 9, 1848):

> There is a church here but out of repair and the Parson, who is a visitor, gives service in a large room. This reminds me of Helensburgh, the majority of the congregation being made up of salt water looking people with faded bonnets and thick shoes ; very few people attend, including a *school*

of five children who really behave very well. What puts me most in mind of Helensburgh is the open doors and windows, the universality of fine weather on Sundays, the insects humming through the room, the stray bird, the leaves waving across the windows, and the irresistible attraction I feel to look out on the open valleys with huge mountains all round, the clouds chasing one another across the forest, and sunbeams dancing on the heavy masses of mist that keep floating along some thousand feet below us. The wind sighs the same sigh through the leaves that it used through the Limes at Row and these rustle in the same note. I see ripe blackberries too and small children gathering them, but don't see the Gare Loch and its boats; or smell the sea-weeds, no nor the tansy and peppermint, nor peat smoke of the new washed mutches and red cloaks—and above all, the Rev. Mr. Winchester, though a sober man enough, is far from a powerful preacher, indeed he may be called a powerless one, for you can't hear him three benches off, and his sermons, though better than Mr. Byam's, cannot keep my mind off the new trees and new weeds that grow up to the very doorstep.

In the same vein he wishes that Miss Henslow had ever been in Scotland so as to realise at a word that this rainy season was just like the climate of Dreepdaily, 'except that all the features are infinitely grander, the rains last longer, the mists are thicker, the fogs are more choking and the damp is more provocative of colds.' He gets up at six, but *hates* it, and equally hates going to bed at nights.

I have resumed my kitchen plan at Kew, of warming my back at the fire when writing and my feet when reading, during 'the sma' hours, ilka night.' Mr. Hodgson, who is in poor health, often sits up and reads with me, wrapped in a fur Roquelaure; now he is perusing Darwin's Journal, which I procured for him, and ever and anon he leaves off and battles with me upon some of the dogmas in *Lyell's Geology*, anent which we *pooh-pooh* one another's opinions very freely. Then we get to disputing on the course of a river, may be in High Thibet, and fight it out with old Chinese Charts and notes from various bad authorities. As the countries and rivers are utterly unknown to Europeans, it little signifies whether the latter *debouche* in the Arctic Ocean or the Bay of Bengal. Hodgson is a particularly gentlemanly and agreeable person, but he looks sickly; he is handsome, with a grand forehead and delicate, finely-cut features; when arrayed in his furs and wearing the Scotch bonnet and eagle feather with which it is his pleasure to adorn himself, he would make a striking picture. He is a clever person and can be wickedly sarcastic; he called Lord Ellenborough (the haughtiest nobleman in all India) a 'knave and coxcomb' to his face (true enough, though not exactly a fact to be *told* with impunity), and then squibbed his lordship; you must know that Lord E. had previously applied to Hodgson the sobriquet of an *Ornithological Humbug*, and had turned him out of his Residentship at Nepaul, because he had (by Lord Auckland's desire) clapped the Rajah into confinement. In short, Lord Ellenborough and Mr. Hodgson kept up a running fire, till his Lordship left the country. Happily, Hodgson lost no friends; but he lost by it his salary of £7000 a year, his Palace to live in, and the Insignia of the British Resident in the proudest court in India, and then withdrew to these Hills, on £1000, as a Retired Civil Servant.

It will be remembered how, in the early days of the Antarctic Expedition, botanists of the strictest school, like Sir William Hooker and Dawson Turner and Robert Brown, looked askance at divagations into other branches of science. Joseph Hooker not only possessed an energetic curiosity which overflowed by its very abundance into every branch of Natural History, but was convinced that the botanist as well as the traveller was incomplete without being also something of a geologist, a geographer, a meteorologist, and a map-maker. With a journey in utterly uncharted regions before him, he took pains to become a competent surveyor. Yet even then, after warmly thanking his father for ever generous help, he half apologises for spending part of his time on anything but pure botany.

October 1, 1848.

My solace is that you will not find that Botany has suffered by my fondness for other pursuits, without which no traveller of this exacting age is thought accomplished. I have gained great, though undeserved, credit here and no little help, by measuring the heights of the mountains and keeping up a good meteorological register. The Surveyor-General, who spent last season here, would tell no one what he was after; and the poor people who had shown him much kindness are very much disgusted. I keep no secrets, and if I cannot (and do not wish to) measure with the accuracy of a Surveyor, I do so sufficiently accurately for all practical purposes and at a very little outlay of time. With a pocket sextant and compass, lent me by the Deputy Surveyor-General (Capt. Thuillier, a most excellent fellow), I worked out in two hours the height of Kinchin from this place and made it 28,000 feet. Sinchul I have worked barometrically with no trouble at all, and make it 8653. Tonglo Mr. Müller and I have just worked out from the observations I took in May, and it is 10,009 feet.

So also a little earlier:

I have only seen the sun thrice this month so as to get observations. The time here was ¾ of an hour out, and my watch which you gave me before I went with Ross is the only good time-keeper here, so that all sorts of people send to me for the time. I spent one day furbishing up my surveying lore, so as to be ready for the *Terrae incognitae*, but I am wretchedly off for instruments.

Thus the rainy summer months wore away in busy employment, with alternate hopes and fears about the great journey to the snows in October. His plan, if this were permitted, was to spend a month there, and then, if at all successful, return again in May,

for I am sure [he writes on August 30] it will be better to work one part of the Himalaya well, from the Terai up to the Snow, than to proceed north-west towards the passes west of Nepaul, now so much better known [accepting the invitation of Major Thoresby, the Nepaulese Resident]. This, too, is the middle of the range, it contains the highest mountain, and so evidently differs in the Geographical Distribution of its Vegetation, from what lies East and West, that it presents

> the most advantageous point for research. The field is quite untrodden, and I hope to have 2000 species before I leave this year.

Were this impossible, his alternative was to take the journey to Assam and the Khasia Hills, which he actually carried out in 1850, visiting the tea plantations and looking out for a station adapted to the cultivation of guttapercha in Assam, and pushing northwards to solve the geographical riddle as to whether the Bramaputra was the same stream as the Tsanpo of Tibet.

As a last resort he managed to obtain a route to take him in five and a half days' journey to a village on the flanks of Kinchinjunga, and at first resolved

> to attempt it with or without permission, in the latter case with *very* small hopes of success, but every inch is botanising ground, and one direction is as good as another. . . . I have no hopes of penetrating into Thibet whatever, but no European has ever visited the snow E. of Kumaon and to do so here will be a feather in my cap.

Again and again he reiterates his intention, afterwards modified, of trying to reach the snows, even without permission, though this would sadly hamper his travelling, for his Lepchas would be kidnapped or fined or sold as slaves, for showing the way. 'My only requirements,' he reflects, 'are mountaineer servants, who have no property in Sikkim to lose.'

Should all these efforts fail, there was still a chance of reaching the snows in Upper Assam the following year, if he could time himself to be there in October.

The general political situation has already been sketched out in the letter of June 10, 1848. The present difficulty was in putting the right amount of pressure on the dependent Rajah who played at independence. Thus (July 19):

> Campbell and the Govt. are both anxious to forward me on; the Govt. won't order Campbell to send me without the Rajah's consent for fear of a war with China; Campbell won't run the risk of committing himself without an order!

He had already burned his fingers with the Government. But after a personal appeal from Hooker, Lord Dalhousie sent him 'an explicit statement from the Colonial Office of what it is conceived our relations with Sikkim ought to be.' In Dr. Campbell's hands this was a useful guide for negotiations—finally Lord Dalhousie in September addressed a letter to the Rajah with peremptory orders

> to give me full leave to travel to the Snowy Passes and to grant me every assistance. No one expected that his Lordship would do this; and considering how ambiguous are our relations with that crusty imbecile, and how much caution the carrying out of the object requires, it is the very strongest proof Lord Dalhousie could give of his true interest in my behalf.

To make the Government bestir themselves 'has cost me a world of pen, ink, and paper and the backing of very powerful friends.' Prudence, however, bids him add:

> Pray say little of these projects of mine; there are so many slips 'twixt cup and lip, and the objects to be attained do so fully jump with even my most sanguine expectations that I cannot venture to hope for perfect success.

Further he reassures his mother:

> No danger whatever will attend the excursion; a little plague and difficulty must be anticipated from the Rajah's innumerable petty headmen, and I am quite prepared to receive a great deal of insolence,—to put up with everything short of direct opposition.

No answer had come from the Rajah by October 1, but all was ready for a start by the end of the week, to Jongri at least (the village on the spurs of Kinchin already mentioned) if direct opposition were offered to the route east of the mountain and the Sikkimese passes into Tibet, on the manifestly untrue ground that Kinchin was a holy mountain, never visited by anyone, and that the Lhassa authorities must be consulted. It seemed certain that he must go alone, for illness or accident had laid up the only friends whom he could trust as travelling companions on such ticklish ground—Hodgson and Müller, Barnes and Campbell. Moreover, had the latter been uninjured, Lord Dalhousie forbade his going, lest, being an official, he should give a political aspect to the expedition. Dr. Hooker, he said, should act on his own responsibility alone.

To his Mother

October 13, 1848.

Everybody is solicitous to go with me; but I have refused all others, because I do not know them well enough to trust them; and having to bear all the onus myself, I should think it imprudent to risk taking any companion who might not be good-humoured and kind to the Natives, or willing to put up with insolence from the Rajah's people, should we chance to meet with them. Lord Dalhousie places great confidence in me, and the Rajah of Nepaul no less by granting me the first permission that any Englishman has ever received. Under all these circumstances, I shall do nothing in the peremptory way; for if anything disagreeable arose, I should be involving Lord Dalhousie in the necessity of vindicating me and avenging my wrongs, with fifty other troubles from which I should reap no advantage. I shall not therefore enter Sikkim, unless the Rajah consents. He has already committed himself; and my interference would do no good but harm.

To his Father

Darjeeling: October 20, 1848.

I wish you could have been with me this morning and seen the motley group of natives arranging with Campbell and myself the preliminaries towards my trip to the Snows, of various tribes, colors, and callings, such as one rarely sees any of, and still more rarely all together. I must, however, begin at the beginning and tell you that Campbell has at last wrenched a reluctant assent from the Rajah of Sikkim to my visiting his snowy mountains. In my last I informed you of his having returned a rude and flat refusal to Lord D.'s request in my behalf, as also of his having stationed 80 men at one pass and 25 at two others to intercept my exit from our territories into his, where his instructions were to capture my servants but lay no hands on myself; these Campbell insisted on being withdrawn, under penalty of dismissing the Rajah's representative (giving the Ambassador his letters, in short), and they were so. Campbell also gave the Rajah eight days to change his mind or have his conduct reported to headquarters with recommendations for condign punishment [i.e. by stopping the lease-money of Darjiling and annexing the Rajah's property at the foot of the Hills].

Ten days past and no word, when the Rajah's Agent, or Minister if you will (Vakeel is the technical term), was told that should no message arrive before the evening post hour, the letter to Lord D. should be sent. The answer was that advices had arrived to the effect that permission was given, provided Dr. C. would pledge his word that this should be my only visit and that a similar request should never be made hereafter. Such conditions were peremptorily rejected as not only derogatory in the highest degree but ensuring me the worst reception. They were again dismissed in disgrace to *read their advices again*, which they did and returned this morning with unconditional permission. This was followed by a long lecture on the impropriety of their conduct, the danger they had run in offending our Government, and wound up with a comparison of their conduct with that of an independent power, the Rajah of Nepaul, who had sent to Darjeeling an officer and guard to escort me to Nepaul, with instructions to provide me with carriage for my traps and food for my people.

All this was a curtain affair of course, as it would not have done to let the Goorkhas or others witness our scurvy treatment by the Sikkim Rajah's emissaries. The latter no doubt had their instructions from the first to deliver the rude refusal and if that answered the purpose well and good, if not to propose the other alternatives seriatim, and if defeated in all to give in with as bad a grace as might be.

This hard and disagreeable work over, we all met in the verandah and Salaams passed between myself and the characters to whom I should have liked to introduce you. First there was the Rajah's Vakeel, a portly, tall, and muscular Thibetan, clothed in a long red robe like a Cardinal's, looped across down the middle, and round his neck and down his shoulders hung a rosary. His face was not strongly Chinese at all, stern, grave, and stolid, thoroughly obstinate and impracticable; thin lips, a good chin, thin arched nose and narrow nostrils, high cheek bones and forehead, cold grey eyes and handsome brows; no beard or moustache, and a nut brown, but not bronzed complexion. His years must be above 60 and his hair was scant and grizzled. A stiff, black, small cap, with high brim standing up all round, rather set off the repelling look he maintained. Taken to pieces, he might be described as a funny mixture of the old woman, from his beardless face, the Lama priest from his dress and rosary, and a burly, well-to-do Landamman, deputed from some Swiss Canton to resist to the uttermost the demands of a dangerous neighbour power, unflinching under opposition and unscrupulous in makeshifts, always the bear, often the bully, and ever the sturdiest opponent of the overtures of his antagonist, even when designed for his own good. These qualities, together with an unblushing effrontery and consummate skill in fabrication and a large interest in the monopoly of Sikkim trade, rendered him a fit tool for the Rajah. Beaten at all points he has to give in, and there he stands, showing neither sulks nor smiles, just respectful enough to avoid censure and no more.

A real character stands at his elbow, a little old withered Thibetan, leaning on his long bamboo bow, simply clothed in a woollen robe, his grey hair floating in the wind, bowed with age, of mild expression, and stone blind. He is a Seneschal to the party, devoted to his country, and the Companion of the Rajah's deputations to the Political Agent of the powerful Government whose advances his master rejects. When he speaks, and this is very seldom (and as it is always in his own half Chinese tongue no Englishman can interpret it), the burthen of his story passes from tongue to tongue; he is evidently the oracle of the party; his placid looks and grey hair would lead me to confide in him and address him as Father, but I have a grim suspicion that his views narrow as his years go on, that he was bereaved of his best and brightest sense before our power showed itself in these hills, and that his crafty companions have taken advantage of this and done more than leave him in the dark as to our real power to punish, but wish to reward and encourage.

The attendants upon these, the Rajah's representatives (and their own, for, being a large sharer in the monopoly of the Sikkim trade, the Vakeel has more interest than his master in excluding strangers), were short, stout, thick-set Bhoteas, clad in purple worsted dressing growns, fastened round the middle by a belt, bare headed and footed, very dirty and ill-favoured withal.

Next conspicuous to these are my Nepaul guards, just arrived to *accompany* me to the Nepaul frontier and *conduct* me from thence; the Havildar (Corporal, I believe) is a small, fine-boned man, with little hands and small limbs and ankles, well-knit and active, of the Kawass tribe, who boast descent from the Rajpoots and are generally in Nepaul the slaves of the Rajah's body, sometimes soldiers and, more rarely, rise to the rank of gentlemen. He looks business-like and trusty, is very handsome, swarthy, with small moustache, broad forehead, bright open eye, good nose, handsome mouth, and small prominent chin. A pretty little turban sits nattily on his head, of black, woven with silver thread, and the number of his corps worked in silver in front, right over a red mark on his forehead which bespeaks his caste amongst the Hindus. His coat is a loose rover-like jacket of purple with silk braid in front, over a white under garment of cotton, open down the right breast and exposing his chest and long neck. A checked cummerbund is folded many times round his middle and over his nether garments, which are short, loose, and broad. What with his jaunty dress, careless air, and roving eye, he would pass for a sea free-booter (out of Cooper's novels for instance, but less mannered and theatrical and more real than the tricked out coxcombs of that author, who are the prototypes of Mr. T. P. Cooke,[1] rather than real fire-eaters).

The Goorkha Sepoys are immense fellows, stout and brawny, of curious cast of feature, heavy jowled and rather small eyed; they wear small linen skull caps over long carefully combed and jet black hair which hangs in heavy folds down the side of the head; they wear too scarlet loose jackets, very bright and gaudy, with a kookry stuck in the cummerbund and heavy iron sword at their side.

[1] Thomas Potter Cooke (1786–1864) served in the navy till the peace of 1802, and then took to the stage, being, as Christopher North put it, 'the best sailor out of all sight and hearing that ever trod the stage.' His greatest success was in the part of William in Douglas Jerrold's 'Black-Eyed Susan.' Another famous part of his was in 'Frankenstein.'

It would take pages to describe the various groups of bystanders: mild Lepchas in striped cotton, long naked limbed Goorkhas of model muscle and saucy air, Bhoteas of all shades of Chinese feature; Bhotanese, or subjects of the Dhurma Rajah, vieing with one another in rags, dirt, hideous ugliness and quaint ornaments, some deeply scarred with smallpox and the pits such receptacles of blackness that their visages looked as if peppered with duck-shot. Most have turned up eyes, very prominent cheek bones, projecting baboon mouth and large teeth; nearly all are of villainous countenance, of singularly low forehead and bad cut of head; the predominance of the animal propensities (fid. the phrenologists) being well displayed from the custom of clipping close the hair.

The Cis-Himalayan Bhoteas, whether of Sikkim or, worse still, of Bhotan, are as uncouth a race (short of savages like the Australian or Fuegian) as I ever beheld. A little sprinkling of Hindus and Mussulmen, chiefly our servants, with the above comprises the oriental population. Amongst them all were Mrs. Campbell's beautiful children, holding by our hands and as indifferent to the wild races about them as an English child is scared by the sight of an English beggar-man.

And now I daresay you will be ready to ask, what confidence I can expect to repose with remarkable prudence in such a gang—and this is easily answered. I take no money, and my plant papers and instruments are poor plunder. The people, though so averse to foreigners, do neither rob nor injure; were they inclined to, the Rajah's power over his people and his mortal dread of us would be a sufficient protection. Further I have the Nepalese guard before whom, for very shame, they must be polite and attentive, and in whom, as acting under the orders of their Government, the most implicit reliance may be reposed, for the Goorkha, when under orders and in confidential employ, is the soul of honor and of politesse too. Lastly, as they will not get a rap of pay till they bring me back safe and what they will receive then will be a fortune to each, they will consult their own interests as well as mine. So I expect devoted service from my guard, for it is their pride to devote themselves under such orders and auspices, companionship from what Lepchas I may take, passive obedience from such of the Rajah's men as may accompany me, perhaps a little obstinacy and presumptuous interference at first and insolence which I can better check with ridicule and exposure before the Goorkhas than by any other means. The Bhotea porters will keep one eye on me and the other on the Rajah's men and serve both masters if they can.

My great aim is so to conduct this attempt that it may be followed by another and to avoid suspicion. This will be difficult in Sikkim, and for the first few marches I shall make few or no observations, excepting of the barometer &c. in my tent, the only explanation a Bhotea can harbor of which is my desire to take the country. In Nepaul I may do as I like, the Goorkha having no orders to stop my observing; but in Sikkim I cannot knock a stone or pull a plant without disturbing the *Gods*, in other words exciting suspicion. I go, however, ostensibly as a botanist, and I will warrant that before two days are over every man jack of them will be collecting for me. I have always found frankness and kindness good policy with any nation, especially if combined with a reasonable amount of personal vanity, which I abundantly possess, and assumption of superiority and, above all, a liberally flattering opinion of the people openly expressed.[1]

The Rajah's people first offered carriers and porters, then withdrew the offer, which I am glad of, as the latter will be more my own people and have a double interest in behaving well; they, after some hesitation, give me a guide; he looks a good man enough and Campbell has seen him repeatedly. He is to accompany me to Nepaul too *if I like*, but this will depend on what sort of servant I find him.

[1] In the end the personal impression left on the Sikkimese by Hooker was remarkable. Twenty-two years later the country was again visited by a European, the botanist and traveller, Mr. H. J. Elwes, F.R.S. Even then, Mr. Elwes tells me, the Lepchas almost worshipped him. The learned Hakim, so friendly to his men and to the villagers, hale or sick, was remembered as an incarnation of high wisdom and kindly strength; and in 1908, after fifty-nine years, he was still a living memory (see the illustration which follows). As an observer, also, a high tribute is paid him by Mr. Elwes. Of all the countries in which the latter travelled, here only, whatever he saw, he saw with his predecessor's eyes. Hooker had noted everything that he himself found of interest: nothing was missed; places and objects all clearly described and promptly recognised. (See ii..125.)

I have no fear of managing one and all when the Rajah's own myrmidons are out of sight, for the natives like us and profit by our advance.

The present plan was to go five marches due north, to Jongri, then strike westwards over the spurs of Kinchinjunga, and thence north-west to the Nepaulese passes into Tibet.

I cannot tell you how comfortable I feel at the prospect of realizing the fondest dream I ever harbored as a traveller and botanist after all my toils with Lord D., tickling Campbell, bullying the Rajah. I have been pooh-poohed by one party, looked on as a visionary by another, and a very useful tool by a third, who say, you have not a ghost of a chance yourself of getting Government or the Rajah's permission, but you will prepare the way for a future. Lord Auckland, Campbell, Falconer, Hodgson, worst of all Sir Herbert Maddock whom Hodgson tried all his friendship (and they are most intimate) to move, all looked on with no hope and some of them giving me the comfortable assurance that my efforts would do good, though *not to myself.* Sir H. Maddock luckily went to Ceylon; had he got Lord D.'s ear it would have been all up; he has now returned to be President in Council in Lord D.'s absence.

Campbell has certainly wrought the battle well, with great forbearance and firmness, and is now as thoroughly devoted to me as it is possible to be. Mrs. Campbell is rummaging her larder and store-room for my comfort, making a veil for my face, providing me with fleecy hosiery, &c. Certainly Campbell has fought behind the Ajacian shield of the Governor-General, the tone of whose letters shows as kind an interest in me as determination to forward my aims, and C. has also had a heavy rowel in the shape of your teasing son himself. However I take your good motto and 'never look the gift horse in the mouth.'

Now I have written a famously egotistical letter; we bargained for unreserved correspondence and you see I fulfil my promise. I only beg that you will make no public use of this which holds out such bright prospects of success towards the snow in which, if I am disappointed, much chagrin will accompany my reverting to the contents of this same letter.

AND THEE AT LAMTENG.

"There was an old man there who remembers you extremely well, and even where you camped. He is still very hardy and active, and I send you a snapshot I took of him. He also sends you his best salaams. His name he pronounces 'And Thee.'" (From Mr. Charles E. Simmonds, June 12, 1908.)

I never mention Bentham, Harvey, Berkeley, &c., in my letters, nor have written to them; I still intend to, but know that you freely communicate all such intelligence as this is, and as from me. Also please send this to Darwin whom, as not being a botanist, you may forget. Best love to all.

Your most affectionate Son, JOS. D. HOOKER.

P.S.—The Sikkim authorities object to the Goorkha guard and are silenced by being told that they are my men and that I won't leave them in the lurch. This shows what I expected, that the presence of the Goorkhas is a grand check.

Hooker did not mean to be deprived of this lever against passive obstruction. Though more evasions followed, the sequel appears in a letter to Miss Henslow, October 26, 1848.

Whatever the Rajah's reasons may be for objecting to let these Ghoorkas enter Sikkim (and his fear may have some good foundation), he has acted with bad faith towards me; and he probably did so because he was aware that he could throw no *insurmountable obstacles* in my way, so long as I had a party of these Hill People in my interest. It is highly likely that the myrmidons of his Sikkim Highness had received orders to take me two or three days' marches by a wrong road, perhaps to where the rivers were impassable; then they would have shrugged their shoulders and said, 'We are as sorry as you can be, Sir, but what can we do?' And the consequent delays would cost me the season, etc. Meanwhile the Nepalese Guard came forward, offering to undertake the responsibility of conducting me to the Thibet Passes through their own country, if I chose; after which I might return by Sikkim, or by the way I went, according to my pleasure.

This offer was so handsome, and any intention of going through Sikkim (even if it were desirable or feasible) without this Nepalese Guard (which had been so promptly sent for me) would have been to put such a slight upon them that I instantly closed with the proposition, and am now all ready for the journey. I go due West from hence *to* and *across* the frontier of Nepaul, and then North to the Western shoulder of Kinchinjunga, and the Thibetan Passes. By following this course I shall occupy some days longer, and (what is of more importance to me) I shall lose the familiar landmarks of mountains etc. by which I should easily map my route, had I gone through Sikkim. I carry, however, a good time-keeper of my own and another chronometer lent me by Major Crommelin, by which I shall be able to take Longitudes with accuracy sufficient to determine my position approximately. As the day closes at 6 P.M. there is plenty of time to observe the stars, during the clear nights which I hope are coming; I say 'hope,' for October is called 'Darjeeling's Heavenly Month'; though it has been so rainy and cloudy up to the present time that I could not have started for the mountains, if permission had been granted, 4 weeks ago. Indeed the rains are not yet over: they are singularly late this year, which would have caused me heavy disappointment if I had been allowed sooner to travel Northward. The double evils of want of earlier permission, and of earlier fine weather, thus mitigate one another, on the principle, I suppose, that two Blacks *do* make a White, a neutral tint, at any rate.

On October 27 the party set out, fifty-six strong, including body-servant, collectors, shooter, stuffer, boys to climb trees and change the plant papers, and coolies, with Nimbo, the sturdy headman, and a Havildar in command of the escort, who carried additional weight of authority as being also tax-gatherer of the district through which they were to pass; returning to Darjeeling on January 19. It is interesting to note that the cost to Hooker was about £100. His friends pressed every assistance upon him. Campbell superintended the supplies for the men; there were personal stores from Hodgson, warm things from the Campbells; while

My friends, the Müllers, have rated my timekeepers, overhauled all my Instruments, furnished me with some capital tin boxes, and done more useful and necessary jobs for me than I can remember. They have also kindly promised to work out all my observations of Longitudes, Latitude and elevations, as I shall send them to Darjeeling. So you see I am admirably cared for, and have only the more to dread failure when so much kindness and trouble have been expended upon me.

In fact, until the positions of the chief places and heights were worked out so as to construct a map, he had but an imperfect idea of where he had been.

> During the greater part of my journey [he tells his father] I saw not a single known object, and had to observe with the sextant. No map contains the name of a single place which I have visited! That I was poking in and out over the western base of Kinchin is all I can affirm.

The line of route for ninety days finally showed the average daily distance covered to be eight miles—*one mile per hour!* Yet they walked full three miles every hour, so that two-thirds was wasted in the ups and downs and bends.

This and the similar chart made in eastern Sikkim, whence the passes led to Phari in Tibet, formed the basis of the carefully drawn map a copy of which appears in the 'Journals': a unique map of such value to the British officers of the Sikkim-Tibet Boundary Commission of 1903 that they telegraphed their congratulations from the front to the maker of it, who at the age of eighty-six was touched to receive this tribute to the work he had accomplished over half a century before.[1]

The first part of the journey was to follow the Tambur river northwards and proceed in turn up its west and east forks to the passes at the head of either valley, one thirty the other twenty miles to the west of Kinchinjunga. This great mountain, rising to 28,000 feet and continued in subsidiary crests all over 20,000, presented an impassable barrier of snowy peaks about sixty-four miles long, stretching between the western passes at the head of the Tambur, and the eastern passes at the head of the Lachen (Teesta), explored by Hooker in his second expedition. It was already late in the season,

[1] Khambajong, Thibet: 'Major Prain, Colonel Younghusband and officers Thibet Mission desire to send you their felicitations by telegraph from Khambajong and express their high admiration of that zeal displayed by you fifty-five years ago, which has enabled them to follow in your steps and has inspired them to emulate your devotion to science and to your country.' (See ii. 457.)

Major (afterwards Sir David) Prain, C.M.G., C.I.E., of the Indian Medical Service, was then Director of the Calcutta Gardens, and in 1905 succeeded Sir W. Thiselton-Dyer as Director of Kew.

for in the higher valleys the snow began to fall in October, and by the beginning of December, when Hooker approached the Wallanchoon pass, the snow lay deep on the last four miles of the track above the 15,000 foot level. Nevertheless he succeeded in reaching the divide, and from the col, more than 1000 feet higher than Mont Blanc, looked down into the forbidden land of Tibet. The still loftier sister pass of Kang-lachem to the east, however, was more heavily snowed up, and there the party did not ascend beyond 16,000 feet.

The next part of his plan was to return almost to the fork of the Tambur, and strike east, still through Nepal, towards the Kinchin group and eventually Sikkim. This involved crossing the huge ridges and profound valleys that successively stretch south-west and south from the Himalayan crest. But the pass over the third of these ridges, the Kanglanamo, was closed, and the inhabitants of the village at its foot had withdrawn lower down the valley. Thus he had to turn south forty or fifty miles till the alpine regions were left, and a snowless pass eastward into Sikkim presented itself, whence he could turn north again to the extreme flank of Kinchinjunga.

At this middle point of the journey, before turning north again, his solitude was most agreeably interrupted. Dr. Campbell, putting the final touch to his long-drawn diplomatic negotiations, was on his way to a personal interview with the Sikkim Rajah. After the complicated falsehoods that had been concocted to impede Dr. Campbell's progress, the friends were greatly tickled by the droll conduct of the Rajah and his court, who had found themselves compelled, after all, to go forth to meet him on the river, as the sole means of preventing his finally reaching the capital of Sikkim. On December 23 Hooker joined him at Bhomsong, on the banks of the Teesta, and shared in the formal interviews both with the crafty Dewan and finally, despite the Dewan's many subterfuges to delay or prevent this, with the Rajah himself, a *fainéant* devotee, half oblivious of mundane matters. Arrangements were made for Hooker's trip through Sikkim the following summer. The Dewan, indeed, as will appear later, organised secret obstruction to this; but the chief immediate result of the interview was the open friendship displayed by the Lamas and people of Sikkim.

> This man [the Dewan] and Campbell had become great friends, and he also became intimate with me. He was educated at Lhassa, and has very agreeable manners and personal address, but is the very most consummate liar and scoundrel in all political matters that you can imagine, and the coolest withal. He took me for a brother spy and rogue, and probably does so still. Next day we had an audience of the Rajah. He is a little, old, black man, of quick manners and eye, thoroughly Chinese in every thought and action, and *very sorry indeed* to see us so far into his country. We crossed the river on a bamboo raft; I wore a shooting-coat lent me by Campbell, my travelling cap and plaid; Campbell more respectable. We were received in a shed, fitted up so as to show off the Rajah to immense advantage, according to the taste of his poor self and people. The shed was hung with faded China silk; there was no furniture; we brought, at the Rajah's request, our own chairs; the leg of mine poked through the bamboo floor, and kept up a squeaking in a very high key. At the upper end of the little room was a high stage 6 feet! also covered with tattered silk, and over it a shabby canopy, under which the Rajah squatted, cross-legs, a little body swathed in yellow silk, with a pink, broad-brimmed and low-crowned hat on. Such an attempt at display was really humiliating! He never returned our salutes, but looked wistfully at us, and then at his courtiers, some dozen of very dirty fellows in silks (Kajis), ranged against the wall as mutes. The conversation was brief and trifling; it related chiefly to Campbell's insisting on having a responsible authority from the Rajah at Darjeeling. In the middle presents were brought, and white scarfs thrown round our necks, as a signal to depart, but we stuck to our seats in spite of all hints, and told him of my intention to visit again in spring the Snowy Passes to the east of Kinchin, and of how dissatisfied I was with the permission coming so late. He made no reply to all this.

After the interview the two friends travelled together till January 2, when Campbell was recalled by business. After two months' travel without a European companion, this ten days' comradeship with so good a friend stood out as a golden time in Hooker's journeyings. On January 2, 1849, he records:

> Here I bade adieu to Dr. Campbell, and toiled up the hill, feeling very lonely. The zest with which he had entered into all my pursuits, and the aid he had afforded me, together with the charm that always attends companionship with one who enjoys every incident of travel, has so attracted me to him that I found it difficult to recover my spirits. It is quite impossible for any one who cannot from experience realise the solitary wandering life I had been leading for months, to appreciate the desolate feeling that follows the parting from one who has heightened every enjoyment, and taken far more than his share of every annoyance and discomfort: the few days we had spent together appeared then, and still, as months. (Himalayan Journals, i. 332.)

After parting from Campbell, he turned north again to Jongri. This was a deserted yak post, never before visited in winter, consisting of two rude stone huts for summer travellers at an altitude of 13,000 feet on the great spur that runs south from the Kinchinjunga massif and divides Sikkim from Nepaul. Here he was on the veritable Kinchin, some fifteen miles as the crow flies from the actual summit 'whose grand snows rise on all sides on rugged granite precipices which have pierced the Gneiss and Mica-slate rocks, carrying them up in shattered peaks and cliffs to 20,000 feet.' Nearer along the massif stood the lesser giants, Kubra and Gubroo, the saddleback with a 25,000 feet peak at either end, and to the north-east the sharp cone of Pundeim dropping five or six thousand feet in a sheer precipice to the sea of glaciers below: the cliff, too steep to carry snow, showing a face of burnt red stratified rocks, so twisted and contorted as to appear like shot silk, permeated with broad white grains of the granite which caps the whole.

Here, till driven out by a prolonged snowstorm, he stayed three cold January days in his gipsy-like shelter, a blanket stretched for tent from the roof of his followers' hut, with a little stone dyke at the sides and a fireplace in front. The ground was frozen sixteen inches deep; to dig holes for the ground thermometers was a work of hours. Many of the mosses and lichens Hooker had last seen on the wild mountains of Cape Horn and the rocks of the Antarctic islands, and as on the Antarctic voyage, glacial terraces and erratic blocks suggested similar problems of ice action. Marching through snow from two to four feet deep among bushes was very difficult; as on his second, but not his third visit to high altitudes, Hooker was affected by mountain sickness as well as his men.

> The temperature fell to zero and it was bitterly cold. My Lepchas, several of whom had never been in the snow before, behaved admirably and not one uttered a complaint. At this elevation a few steps under any circumstances is fatiguing, and the glare of the new fallen snow in so rarefied an atmosphere gives soreness at once to unprotected eyes. I cut the veils Mrs. Campbell made me into little pieces for some of the party, others hung Yaks' tails over their eyes or pieces of paper, or unloosed their queues and combed the long hair over the forehead.

But the natives ascribed mountain sickness to another cause; namely, the Dwarf Rhododendrons:

> The scent (of resinous leaves) was overpowering; the Bhoteas attribute the headaches of these regions to them and not to the rarefied air. I think I can feel my head throb still every time I smell the plants in my collection.

Discomforts apart, the journey to Jongri was a great success. There was a rich botanical harvest on the way up, above the pines, ten species of Rhododendrons, one or two of them new; and lower down, forty-six species of ferns. Geologically it equalled in interest the Yangma valley, a remarkable glaciated valley on the west of Kinchin. 'I quite believe,' he exclaims, 'no two such spots have ever been explored in the whole Himalayan range.'

The trip wound up with a quaint episode. The homeward way led Hooker again to the Changachelling convents near Pemiongchi, the Lamas of which he knew from his visit on the outward march.

> They are re-ornamenting their temple very beautifully; the workmen come from Lhassa and the colors from Pekin.

To my amazement, I found myself on the walls, in a flowered coat and pantaloons, hat, spectacles, beard and moustache, drawing in a note-book, an Angel on one side offering me flowers and a devil on the other doing homage! I never laughed so much in my life, and the Lamas' artists were pleased beyond measure that I recognised the likeness.[1]

So, with the warm hospitality of the Lamas and four drenching days' march to Darjiling,

ended [he writes] my journey, without slip, accident, or the loss or hurt of a single man of my sometimes very numerous party. In Sikkim I have not spent an unquiet hour, except on the coolies' account, in the snow. I carried neither gun nor pistols, arms nor keys, and lost nothing whatever. From the simple people, Bhoteas and Lepchas, I have met every attention and kindness, and very pleased they will be to see me again, though, should the Rajah oppose, fear may deter them from coming near me; that I do not anticipate, however. A more interesting country for tourist, artist, naturalist, and antiquarian can scarce be found, and it was untrodden in any walk previous to my visit, and I have but flitted over the surface.

The only untoward incident at the outset of this march had been the unruliness of the fourteen Bhotea coolies, who plundered the stores, resisted their Sirdar and the Ghurkas, and finally made off on the seventh day of the journey, from the summit of Tonglo, their place being taken, after some delay, by a few well-behaved Ghurkas from the Nepalese villages. Then everything that could be dispensed with was sent back to Darjiling, and the reduced party went on its way.

This was troublesome for the moment, but not serious, and the note of satisfaction re-appears in the words:

I have not lost or broken a single instrument during my journey, though I have had 8 thermometers in daily use, 2 barometers, 2 chronometers, 3 compasses, a sextant, and Artificial Horizon. I consider this quite a feat—always remembering the roads to be of the worst, and that 50 men were bustling about me all day long.

[1] These drawings, unfortunately, are no longer extant (see ii. 471).

A few passages from the Himalayan Journals may be cited as bringing out personal impressions of the journey and the spirit of the traveller. Mountain scenery below the snow line is compared, as ever, to the perfection of our Scottish Highlands. In the Tambur valley is an old lake-bed, outspread under lofty hills. Through it

> meandered the rippling stream, fringed with alder. It was a beautiful spot, the clear, cool, murmuring river, with its rapids and shallows, forcibly reminding me of trout-streams in the highlands of Scotland.

Elsewhere the mountains rising out of the sea of valley mists are like the mountains by Norwegian fiords or Scotch salt-water lochs. A little lake, a rarity in these valleys, recalls the tarn at the entrance of Glencoe. We realise instantly the charm of the pool set in shining meadow greenery against the dark precipices beyond. It was a home-like delight to espy abundance of a common Scotch fern, *Cryptogramma crispa*, growing in the clefts of a rocky moraine under the Choonjerma pass, at 13,000 feet. High on the Wallanchoon pass, again, the same lichens coloured the rocks as in Scotland, and the dwarf rhododendrons and masses of a little Andromeda imitated a heathery hill side. Here, also, the magic of the familiar in the remote wilderness stirs the imagination :

> Along the narrow path I found the two commonest of all British weeds, a grass (*Poa annua*), and the shepherd's purse! They had evidently been imported by man and yaks, and as they do not occur in India, I could not but regard these little wanderers from the north with the deepest interest. Such incidents as these give rise to trains of reflection in the mind of the naturalist traveller; and the farther he may be from home and friends, the more wild and desolate the country he is exploring, the greater the difficulties and dangers under which he encounters these subjects of his earliest studies in science, so much keener is the delight with which he recognises them, and the more lasting is the impression which they leave. At this moment these common weeds more vividly recall to me that wild scene than does all my journal, and remind me how I went

on my way, taxing my memory for all it ever knew of the geographical distribution of the shepherd's purse, and musing on the probability of the plant having found its way thither over all Central Asia, and the ages that may have been occupied in its march. (Him. J., i. 221.)

Nor was imagination only stirred by Nature. It was equally moved by the diverse expressions of human aspiration.

The temple of Wallanchoon stood close by the convent, and had a broad low architrave: the walls sloped inwards, as did the lintels: the doors were black, and almost covered with a gigantic and disproportioned painting of a head, with bloody cheeks and huge teeth; it was surrounded by myriads of goggle eyes, which seemed to follow one about everywhere; and though in every respect rude, the effect was somewhat imposing. The similarly proportioned gloomy portals of Egyptian fanes naturally invite comparison; but the Thibetan temples lack the sublimity of these; and the uncomfortable creeping sensation produced by the many sleepless eyes of Boodh's numerous incarnations is very different from the awe with which we contemplate the outspread wings of the Egyptian symbol, and feel as in the presence of the God who says: 'I am Osiris the Great: no man hath dared to lift my veil' (i. 228).

It is interesting to note the traveller's full and careful method of observing on his march, and his scrupulous pains to avoid partial generalisations or the errors of the 'personal equation.' This method of recording observations, which left nothing to chance or the uncertainties of memory, is set forth almost parenthetically in the description of his descent from a Himalayan pass 16,000 feet high, when in the magical light of a young moon everything was bathed in beauty and imaginative suggestion, but all pleasure was lost in the headache and giddiness and bodily lassitude brought on by exertion in that thin air.

Happily [he writes], I had noted everything on my way up, and left nothing intentionally to be done on returning. In making such excursions as this, it is above all things desirable to seize and book every object worth noticing on the way out: I always carried my note-book and pencil tied to my jacket pocket, and generally walked with them in my hand. It is impossible to begin observing too soon, or to observe too much: if the excursion is long, little is ever done on the way home; the bodily powers being mechanically exerted, the mind seeks repose, and being fevered through over-exertion, it can endure no train of thought, or be brought to bear on a subject. (H. J., i. 247.)

As to overhasty generalisation:

The plants gathered near the top of Wallanchoon pass were species of *Compositae*, grass, and *Arenaria*; the most curious was *Saussurea gossypina*, which forms great clubs of the softest white wool, six inches to a foot high, its flowers and leaves seeming uniformly clothed with the warmest fur that nature can devise. Generally speaking, the alpine plants of the Himalaya are quite unprovided with any special protection of this kind; it is the prevalence and conspicuous nature of the exceptions that mislead, and induce the careless observer to generalise hastily from solitary instances; for the prevailing alpine genera of the Himalaya, *Arenarias*, primroses, saxifrages, fumitories, *Ranunculi*, gentians, grasses, sedges, &c., have almost uniformly naked foliage. (H. J., i. 225.)

As in other matters, so he sought for accuracy in drawing mountain scenery, with a deliberate endeavour

to overcome that tendency to exaggerate heights and increase the angle of slopes, which is, I believe, the besetting sin, not of amateurs only, but of our most accomplished artists.

Confessing that, as he did not use instruments to project the outlines, he could not pretend to have wholly avoided this snare (while the lithographer, alas, was not always content to abide by his plain copy), he is often careful to mention the angle subtended by lofty peaks in the distance, and the true slope on their sides. For, as he remarks (H. J., i. 347),

the vagueness with which all terms are usually applied to the apparent altitude and steepness of mountains and precipices, is apt to give false impressions. It is essential to attend to such points where scenery of real interest and importance is to be described. It is customary to speak of peaks as towering into the air, which yet subtend an angle of very few degrees; of almost precipitous ascents, which, when measured, are found to be slopes of 18° or 20°; and of cliffs as steep and stupendous, which are inclined at a very moderate angle.

CHAPTER XIV

THE SECOND HIMALAYAN JOURNEY

IT was now too late to proceed to the hills of Assam, where the healthy winter season would soon be over. This was small disappointment. The other mountains south of the Ganges, which had so charmed him the previous April, had lost all their attraction now that he had seen the veritable Himalayas. Moreover Hodgson laid stress on the simple fact that it was better to explore one district thoroughly than to wander. He resolved therefore to stay at Darjiling, where Hodgson's society and library, Müller's scientific aid, and Campbell's zealous interest, were strong inducements to a man who aimed at being something beyond a collector and tourist, and to follow up his success on the west of Kinchinjunga by an expedition to the east of it the next summer, completing the botany and sending home young plants and especially seeds, of which he writes to Sir William,

> I have done my best to give satisfaction. I stayed at 13,000 feet very much on purpose to collect those of the Rhododendrons, and with cold fingers it is not easy at the ripening season, December, to collect those from the scattered twigs, generally out of reach. (March 27, 1849.)

As to getting the seed of R. Dalhousiae, there was a further difficulty,

> for you cannot see the plant on the limbs of the lofty oaks it inhabits, except it be in flower, and groping at random in the woods is really like *digging for daylight*. . . . You must remember it is no light work to be the pioneer of these

fine things (April 2). I have obtained, however, plenty of young plants, and will send a tin case, direct, on my return to Darjiling (April 11).

The cold weather gave opportunity of a trip with Hodgson to the Terai in order to complete the botanical chain from the plains to the snows. Six weeks were spent in sorting and packing the botanical spoil from Nepaul; eighty coolie loads were sent down to Calcutta.

The most notable event of these intermediate weeks was

what I might call an *Angel's visit* from Mr. William Tayler,[1] the Postmaster-General for India, brother to Frederick Tayler the artist . . . a highly accomplished man and a splendid sketcher; and we became friends in a very few hours. . . .

The botanist among the mountains suggested an admirable subject for his brush.

He is pleased to desire my sitting in the foreground surrounded by my Lepchas and the romantic-looking Ghorka guard, inspecting the contents of a vasculum full of plants, which I have collected during the supposed day's march. My Lepcha Sirdar (which means Great man's Head man) is kneeling before me on the ground, taking the plants out of the box, that in his hand being a splendid bunch of *Dendrobium nobile*. He is picturesquely attired in costume, with a large pigtail. Another is behind me; the Ghorka Havildar and Lepchas, in their picturesque uniforms, are looking on, and my big Bhotea dog lies at my feet. On one side two Lepchas are making my blanket tent house, cutting Bamboos, &c. I am in a forest, sitting on the stump of a tree, with the Snowy mountains in the background; and a great mass of

[1] William Tayler (1808–92) was an Indian civilian who about this time was Postmaster-General of Bengal. His skill in portrait painting made him many friends; his caricatures some enemies. In 1855 he became Commissioner of Patna. His policy during the Mutiny had provoked great controversy, prolonged for many years, and an open quarrel with the Lieutenant-Governor led to his resignation, when he practised as a lawyer in Bengal till his return to England in 1867.

His brother Frederick (1802–89) was a water-colourist and etcher who enjoyed lifelong popularity in England, especially for his sporting and pastoral scenes. He was President of the Old Water-colour Society, 1858–71.

The Botanist in Sikkim

From the Picture by William Tayler.

Ferns and Rhododendrons, brought in by another man, are on the ground close to me.

My dress was the puzzle, but it was finally agreed I should be as I was when in my best, a Thibetan in the main, with just so much of English peeping out as should proclaim me no Bhotea, and as much of the latter as should vouchsafe my being a person of rank in the character. So I have on a large, loose, worsted Bhotea cloak, with very loose sleeves; it is all stripes of blue, green, white, and red, and lined with scarlet. Enough is thrown back to show English pantaloons, and my lower extremities cased in Bhotea boots. My shirt collar is romantically loose and open, with a blue neckerchief, which and my projecting shirt wrists, show the Englishman. My cap is also Thibetan, and only to be described thus: it is of pale gray felt, the upturned border stiff and bound with thin, black silk ribbon. On the top is a silver-mounted pebble, and a peacock's feather floats down my back. The latter are marks of rank. (April 25, 1849.)

The sketch, begun in February and finished during April on Tayler's later visit to Darjiling, was sent to England that Fitch might make a copy for Sir William. The copyist's practised hand improved to some extent on the workmanship; but in the interests of accuracy Hooker was constrained to write home (January 30, 1850): 'The stream of water and *fruits of Hodgsonia* which Fitch has brought into the foreground are doubtless improvements, though the latter are anachronisms when coupled with *Rhododendron flowers*, the one being the offspring of May and the other of September.' Later, a third version of the scene, more successful both in composition and in technique, was made from Fitch's water-colour by Mr. Frank Stone. From the former, which is in the possession of Dr. Charles Hooker, of Cirencester, the accompanying illustration has been reproduced.[1]

The big dog introduced into the picture was Hooker's faithful companion during his second journey to the snows till the unhappy day when, owing to his incorrigible habit of running on to the slippery bamboo bridges, he fell into a torrent and

[1] Mr. Stone's version belongs to Lady Hooker, Fitch's copy to Capt. J. S. Hooker.

was swept away. Kinchin, as he was named, is first referred to in a letter at the end of the Nepaul expedition:

> I have brought from the Snows a most grand Bhotea dog, about which I must write to dear Bessy, and a droll puppy of a breed which I hope will live in the Plains. The former is a huge and savage creature, but a faithful watch; he does not bite me, but has already so served three of my servants, chiefly at night. If you know a book called 'Youatt on the Dog,' and can refer to it, you will find a splendid wood-cut of this—'the Thibet Mastiff.'

The results of the Nepaul expedition being completed, from February 27 to March 24 he was in the plains. Happily the Sikkim Terai was free from the malaria, so deadly elsewhere, and he was able to reassure his parents, who would naturally be alarmed by the sudden death not only of his late companions, Mr. Williams[1] and his assistant on the Survey, who had imprudently camped in a most unhealthy jungle, but of his uncle and almost contemporary, Gurney Turner, who had entered the medical service of the E.I.C.

A reasonably good collection, as he modestly calls it, was the result, though in the densely wooded Terai 'the only safe way of botanising is by pushing through the jungle on elephants; an uncomfortable method, for the quantities of ants and insects which drop from the foliage above, and from the risk of disturbing pendulous bees' and ants' nests.' Geological research in dense tropical forests was exhausting, but he made many notes, including traces of inversion of the strata, as at the foot of other great mountain ranges, such as the Alleghanies and the Alps. By the Mechi river, the western boundary of

[1] The following is characteristic: 'If, as I fear is the case, the widow of Williams (of the Geological Survey) is left destitute—(she has six children)—there ought to be a small sum raised for her by the officers of the Geological Survey. I have written to Reeks about it, and requested that, if this be done, he would apply to you for £10 in my name; for during the two months I spent with poor Williams, he would not allow me to spend a shilling for board or travelling expenses. Reeks will only set down my name for £2 2*s*., and give the rest under a fictitious signature; for neither could some of my brother officers afford so much, nor are they called upon to give it by obligations to the deceased.' (To his mother, Feb. 1, 1849. Trenham Reeks, who died in 1879, was Registrar of the School of Mines, and Curator and Librarian of the Museum of Practical Geology.)

Sikkim, were 'reported Iron Hills'; inspection, however, showed that 'the Iron is, I believe, only Manganese, which will disappoint Mr. Campbell; but I have found a small (useless) seam of coal and vestiges of coal fossils.' Other observers had seen in the alluvial plains of the Ganges and the flat-topped terraces of gravel along the foothills the sure sign of a deep sea that in geologically recent times had washed the base of the mountains as they were gradually upheaved; Hooker himself confesses that he could never look at the Sikkim Himalayas from the plain without seeing in them the weather-beaten front of a mountainous coast, while the deep valleys he explored seemed essentially long fiords with terraced pebble beds and transported blocks such as could be seen on the raised beaches of our Scotch sea lochs exposed by the rising of the land.

For the rest, other picturesque episodes of the trip may be read in the 'Journals'; the elephant fair at Titalya, where Dr. Campbell joined them, on business as a buyer for the Government; the coolness of shooting the rapids of the Teesta after the heat and haze of the plains; the carnival at the young Rajah of Jeelpigoree's Durbar, with its battle, not of confetti, but of small paper bombs of red powder; the weariness of riding elephants, and the fierce storm of hail as they returned which cut to pieces Dr. Campbell's experimental tea garden and lay unmelted there for four days.

Now began preparations for the second and longer Himalayan journey, through eastern Sikkim. The plan was parallel to that of the former trip. As formerly they had ascended the Tambur river, so now the party was to follow the river Teesta to its head-waters; then ascend either fork to the pass at its head leading into Tibet. The western fork was the Lachen, its pass the Kongra Lama; the eastern the Lachoong, leading to the Donkia pass, under the great mountain of that name. These passes were far to the northward of the passes visited in 1848, for the barrier chain trends north-east from Kinchinjunga, and the line now taken was some fifty miles to the eastward. Thus it was expected that the direct route would take no less than twenty-five to thirty marches.

The journey produced wonderful results, but ended in a very unpleasant adventure. In the latter part of it, Dr. Campbell joined Hooker, and on their return both were seized and held as hostages for nearly two months while the Dewan tried to extort better terms in the treaty between Sikkim and India. The party had set out on May 3 for a three months' trip, but it was six months before the explorations were completed, and eight before the travellers returned to Darjiling on Christmas Day, 1849.

Hooker, travelling alone, would certainly not have been thus molested; but the chance of seizing the Political Agent was irresistible to the crafty Oriental, one of whose chief henchmen, moreover, had a personal score to settle with the Resident, who had caused him to be punished for the abduction of two Brahmin girls from Nepaul. For Hooker at first was reserved merely passive obstruction, triumphantly overcome by good-humour and patience, and the exhibition of the Rajah's formal permit and promise of assistance on the way to the snowy passes. The latter he was careful to obtain, despite the renewed shuffling of the Dewan, which would have left him with the poor alternative of a second visit to Jongri.

> As there are many rapid rivers to be crossed, and I must have relays of food, I cannot well venture without his permission. Though he cannot stop *me*, he may detain my coolies, and to remove the bridges is only the matter of *ten minutes*. Lord Dalhousie has again proffered his best services, and I write to him on the subject without hesitation.

Accordingly Campbell wrote a third letter to the Rajah, giving him ten days in which to make up his mind, and send formal permission and a guide.

This was effectual. By May 2 permission had come to visit the Lachen and Lachoong passes, and a guide, the same Meepo who had served on the former expedition, was to meet him a few marches ahead. It was a disappointment that, owing to a stringent order from the Court of Directors as to leave, Lord Dalhousie, however willing, was unable to grant immediate leave to Dr. Thomas Thomson, Hooker's old friend and fellow-student, the explorer of the North-western Himalayas, to join in this Sikkim expedition. He had, indeed, three months' sick leave which he was about to take at Simla, but his regular six months' leave was not due till the autumn. This he planned to claim immediately after rejoining his regiment in the Punjaub, and so share the final trip to Assam and the Khasia Hills.

A start was made on May 3, with a larger travelling camp than originally expected.

> They are 42 in all; 10 are soldiers, 5 are Hodgson's shooters, &c., 10 are Lepchas of my own, the rest Sikkim Bhoteas. Only two or three have ever been to the Snows, but all seem active, willing, and cheerful.

From his second camp he writes further:

> Everything promises happily for the success of this my present expedition, thanks to Hodgson and Campbell, whose kindness exceeds all I can describe. How far I may be able to proceed is very problematical, for the best collection of charts and routes will not reveal to me whither I am going. The soldiers inspire confidence in my people, and that is all I want. My own followers appear excellent fellows. To-day they accompanied me in a march which tired even my unloaded self, and though the weather is terribly hot, they uttered not a murmur.

The villagers everywhere showed themselves kind and civil.

> I have just been accosted by an enormously fat Lama, with a grand present of eggs, &c. The kindness of these simple mountaineers is very grateful, and their civil speeches quite *graceful.* They hope you will not fall ill, are sorry their roads are so indifferent, apologise for not bringing fowls (the priests say this) 'because they must not take life'—say they will hear of your progress in safety with pleasure, and hope to see you en route home again, to stay with them. A small joke convulses them with laughter, and the expected 'backsheesh' is always received with many thanks.

But official obstruction began with the first functionary encountered, to be answered, as always, with patience and firmness, seasoned with good-humoured contempt. The fellow declared he had no orders; the party must wait two days until word could be received from the Rajah. Confronted with the necessary permit, he apologised, but must mend the roads, and that would take two days.

> So [exclaims Hooker] these trumpery functionaries lie, cheat, and obstruct, and nothing but patience and cool contempt put them down. The moment I gather the contents of their long speeches from the preface, I cut them short with an answer which does not suit Bhotean idioms and fashions.

The personal difficulties on the journey may be measured by the fact that whereas to the snows was reckoned a matter of twenty-five, or for a heavily laden party, thirty marches, in the event it took eighty-three days, from May 3 to July 24, to reach the Kongra Lama pass. On May 5, the next hint of obstruction on the part of a friend of the hostile Lassoo Kajee melted away after the arrival of the Tchebu Lama on his way to Darjiling, though the latter, who was to prove himself a faithful friend, was formally commissioned to say that the Rajah had wished the expedition to be postponed on account of his son's death.

> Now [comments Hooker] as the Rajah had not spoken to his son for sixteen years, I doubt his sorrow. The period of mourning is over, anyhow, and, as I told the Lama, it was all one to me, if Rajah, son, and family were to die together—that was no reason why I should not travel through his country. He promptly apologised for his Master, and wrote an order (of what use it is the sequel will show) that I was to pass on unmolested, till I met a guide from the Rajah.

Five days later obstruction was renewed, but the tables were neatly turned on the obstructor. The Lama of Gorh, another underling of the Dewan's, having obstructed the roads and bridges overnight, officiously came forward as a guide, offering the choice of two roads. Hooker, all politeness, asked for the coolest, and at every obstruction, assured him that as he had volunteered to show the road, it was clear he meant to removal all obstacles, 'and accordingly I put him to all the trouble I possibly could, which he took with a very indifferent grace,' until fully discomfited by the arrival of the faithful Meepo with the Rajah's authority to proceed. Unfortunately the latter had never travelled the road, so that they were at the mercy of the guide he had brought with him, who was but a spy on both.

At Singtam, where he reduced his party by sending back the escort, he was delayed a day by the Soubah or governor, of whom he was to have much experience later, on the pretext of collecting food, which never arrived, and at Choongtam, where the Lachen and Lachoong join to form the Teesta, a full week. The motive was clear.

> The Rajah hopes, by throwing his Guide and party upon my resources, that he shall starve me into going away, and he has also followed up this scheme by sending a foolish old official to frighten my people; but the poor man cannot bear any degree of ridicule, and between laughing at his menaces and treating him with all kindness, I have fairly won his heart. I pay most liberally for everything I get; I give large presents to the Authorities and to the Convents; every day I heal the sick who come to me for advice and medicine; and nobody has received even a hard word from me, except in reply to the insolence of the Rajah.

It was more serious that the convoys bringing the promised supplies from Darjiling for the men, who required no less than 80 lbs. of rice per diem, were very late in coming, and when they arrived, brought only enough for eight days. That Campbell should not have fulfilled his promise of sending supplies regularly seemed at first incomprehensible, but it turned out that after the rains had begun on the 10th, the Dewan had taken care to leave the roads unrepaired; the journey was lengthened and the carriers had to consume part of Hooker's supplies as well as their own. Here, then, he stayed till May 25. His itinerary gave him six marches further to the snows, but two months were to pass before he reached the Kongra Lama.

It is worth recording, as an instance of his consideration for the people he was among, that he now resolved to forgo one of the most attractive parts of his programme, in the belief, afterwards dispelled, that the Sikkimese might suffer if he crossed the passes. Accordingly he tells his mother (May 24, 1849):

> It is my intention to proceed to the top of both of the Passes, without crossing, which the Rajah has forbidden; and though I dispute his authority to give such a prohibition, I cannot act in defiance of it and cross the Passes in secret. Thibet is the Headquarters of the Sikkim people's Church, and, if through any act of mine the Passes were to be closed, I should inflict upon the natives what they would consider a serious injury—namely, the *shutting of their Church Door.* It is most reluctantly that I give up the intention of crossing, especially as the Rajah's own order and other circumstances convince me that I could do so if I chose, and that no one has power to hinder me, for the first Chinese village is distant two days' marches on the other side of the Border. However, I have plenty to do on this side, and if by crossing I should throw any effectual impediment in the way of my Sikkim investigations, I should be a great loser by it.

At Choongtam he was forced to divide his party again, leaving there all but fifteen. Three marches further, at Lamteng, there was another week's delay and very short commons, meagre supplies taking twenty days to come from Darjiling over the bad roads. On June 23 came news that a large convoy had been driven back by landslips, but there was promise of another coming, so that on the next day he did not hesitate to move forward one march to Zemu Samdong, the bridge of the Zemu, a large tributary on the west bank of the Lachen. Here his guide, the local headman, or Lachen Phipun, alleged the Tibetan frontier to be. Not knowing which of these streams was the real Lachen, and having no crossing of a river marked here on his route, Hooker resolved to wait at least till sufficient supplies arrived, though both wet and hungry, learning the difference between a fowl and a chicken—'of the latter I eat bones and all, of the former I cannot.' Hunger, he also declared, made it a special martyrdom to science when, instead of eating a curious fruit called Gundroon, a polite present from the Rani, he put some aside to be sent to Kew.[1] Four weeks were spent up the Zemu, trying vainly to reach the head of the valley and the clearer Tibetan skies ahead, for the report of a pass in that direction was probably a deliberate blind. Large collections were made, for the grassy hills swarmed with rare plants, and were sent down the valley to be dried. Even so, the persistent wet destroyed much, and he laments to his father:

> Alas, one of my finest collections of Rhododendrons sent to Darjeeling got ruined by the coolies falling ill and being detained on the road, so I have to collect the troublesome things afresh. If your shins were as bruised as mine tearing through the interminable Rhododendron scrub of 10–13,000 feet you would be as sick of the sight of these glories as I am.

It was a rough time, but produced no ill effects, though

> a hole in the rock or a shed of leaves is very often my residence for days, and my fare is just rice and a fowl, or kid, eggs, or what I can lay my hands on—no beer or luxuries.

The great encouragement was that no other explorer had seen so much of the unknown Himalaya, or with results to be compared with his.

On the 28th and 29th came the Phipun's attempt to hustle him off with a rabble of threatening followers, which Hooker, supported only by his dog, Kinchin, entirely disconcerted by a show of unconcern, backed with plain speaking.

At the first alarm the coolie headman, Nimbo, Hooker's one courageous follower, took three lads with him down the

[1] Dispyros Kaki, Linn. The note by Hooker with the specimens in the Kew Museum is as follows: 'Fruit called Gundroon by the Bhotheas. Good eating dried in this state. Imported to Sikkim from Lhassa.'

valley to the drying sheds and rescued the plants from the marauders. Next morning, he tells Sir William (July 5):

> Sure enough, as I was sitting drawing on my bed, with a cup of tea on one side, it was 'Jenny Lass wha's coming!' and all the 'wild Macraws' were wending up the glen. Twenty of the most uncouth barbarians you ever set eyes on gathered at the mouth of my tent, dressed in scanty, tattered blanket kirtles, with long knives, long brass pipes, and long matted hair, bare-legged and bare-headed; they reminded me most forcibly of Scott's tales. I scarcely deigned to lift my head and look at them, but let them gather as they pleased, and then sent to ask what they wanted here. 'To speak to the Sahib.' I said they must report to me who they individually were, which they refused to do yesterday, and only gave insolence to my Sirdars. It turned out that every man was a Sikkim Bhotea and the Thibetans had all run away the previous night! I then sent word to the head man, that he must send every one of his rag-tag and bobtail away, or I would not speak to him either. This he did with some trouble, as a few were contumacious, and when he came to my tent I took him roundly to task for frightening my people, detaining my things, and giving insolence. Having rated him soundly, and taken all his answers down on a big sheet of paper, I sent him about his business, and have seen no more of the Bhoteas since! Can you fancy such fools! If you give in an inch it is all up; if you get the upper hand an inch, you may bully and swagger and knock them down like ninepins.

Intimidation having failed, dilatory tactics were renewed. On the return to the bridge at Zemu Sandong on July 1 letters arrived from Campbell and from the Tchebu Lama; conveying the Rajah's orders to the Phipun that he should aid the party. Three days later the Singtam Soubah arrived as conductor, with more commendatory letters and presents for Hooker from the Rajah. His secret business, however, was to starve the white man out, and though, after certain supplies arrived on the 11th, he led Hooker the following day one more march up the Lachen to the village of Tallum, he lingered there till the 23rd, alleging anew that this was the last point of Sikkim, and that Tungu, the next village, was in Tibet. All the villagers, down to the little children, were instructed to tell the same story. Tallum was the scene of the famous game of bluff, which convinced the Soubah that it was he, and not Hooker, who was being starved out.

Now the Singtam Soubah's instructions I also saw were to be most civil and draw me away; he represented the Rajah's affection for me as boundless; should I be but in a stream or come to hurt, nothing short of a Chait at Lhassa and annual worship could be thought of. The Rajah's anxiety on my behalf alone induced him to pray my return to Darjeeling, &c. &c. The more civil he was the more so was I, but I felt bound to assure him that my instructions were explicit, that I should wait where I was for orders from Campbell, which could not be before twenty days. He, knowing how short of food we were, grinned acquiescence, fancying he would soon starve me out. I in turn knew that the greedy old Rajah, by way of insuring his getting on with his duty, had allowed him and his coolies (sent to repair the road *back*) only six days' food.

Being camped at 11,500 feet, I had plenty to do, lots of new plants, and was as busy as possible every day and all day for nine or ten days. The Soubah visited me every morning and we had long chats; he is a fine fellow and has been in Lhassa, Digarchi, &c., and told frankly and freely all he knew, giving me most curious information. Talking one morning of the mountain chains, I asked him for a rude sketch of those bounding Sikkim; he called for a great sheet of paper and charcoal and wanted to make his mountains of sand; I ordered rice, of which we had sore little, and scattered it about wastefully; it had its effect, he stared at my wealth and, after bidding him good-bye (the custom always is you have to send your visitor away), I saw no more of my rice, which was ominous for his granary. Not long afterwards he volunteered to take me a ride to Tungu, which all swore was across the border. I agreed if the tent should go; he dare not let me. Why? It was in

'Cheen' (Thibet). Then I said I had given my promise not to go into Cheen, and would wait till my orders from Darjeeling came; he was nonplussed again.

Well, on the 10th day it pleased Providence to afflict the Soubah of Singtam with a *sore colic* so that he could not pay me his morning visit, and as I did not ask for him he took for granted that I was angry and dare not ask for medicine. This was owing to the quantity of wild stuff the poor soul had eked out his fare with. A servant came at night to tell me how bad his master was—'like to die,' he said, twisting his fingers together and laying them across the pit of his stomach to indicate the commotions of the Soubah's inside. I gave him a great dose at once and he was on his legs next morning looking woefully. He told me he had heard of 'Kongra Lama,' and would take me there if I promised not to stay more than one night at Tungu. I gave the same answer. Oh, he said, Tungu is not in Cheen. Is it in Sikkim then? Yes! Very well, we will all go to-morrow morning and I will stay as long as I please. There was no help for it, so he laughed acquiescence.

There is a triumphant ring in the first announcement of his success.

I have carried my point and stood on the Table-land of Thibet, beyond the Sikkim frontier, at the back of all the snowy mountains, alt. 15,500 feet.

He had not only defeated, but won over his old opponent.

We went to the Pass and into Thibet yesterday, the Soubah of Lachen, my arch enemy, the guide. He has made 100 rude apologies: the Chinese had threatened to cut his head off, &c., &c. I answer that an Englishman always carries his point, and that days, weeks, and months are all the same to me. He vows he will tell no more lies, not so much as that, hiding all but the very tip of his little finger. That now we are friends he will show me everything, and I must visit his wife in his black tent on the frontier. Now the tables are turned and the Bhoteas are as civil as they were before hostile and impracticable.

July 24, 1849, was the day of triumph. The day before

they had mounted to the high alp of Tungu, where friendly Tibetans from across the frontier were camped for the summer in their black horsehair tents, pasturing yaks. The journey to the pass and back was the best part of thirty miles. The ground was level enough for riding on the hardy Tartar ponies, stubborn, intractable, unshod, which never missed a foot among the sharp rocks, deep stony torrents and slippery paths, even in the pitch darkness of the final way back. Sorry-looking beasts, nothing could tire them, not even the sixteen stone burden of the Soubah. Hooker himself walked some thirteen miles of the way, botanising; but 'at dusk,' he confesses, 'I took horse, for alas! I am quite blind in the dark.'

Peppin, the Soubah, was as good as his word; going and coming they were most graciously received by his squaw and family.

> The whole party squatted in a ring inside the tent, the Soubah and myself seated at the head, on a beautiful Chinese mat. Queen Peppin then made tea (with salt and butter), we each produced our Bhotea cup, which was always kept full. Curd, parched rice, and beaten maize were handed liberally round, and we fared sumptuously, for I am very fond both of Brick Tea and curds.

Nature reserved an impressive setting for the last act of the serio-comedy. As they sat round Peppin's hospitable fire, a tremendous peal like thunder echoed down the glen. The men started to their feet and cried to Hooker to be off, for the mountains were falling and a violent storm was at hand. So for five or six miles they pursued their way up the river bed, shrouded in fog and deafened by the unseen avalanches that thundered down unceasingly from the great mountains on either side. Only the low hills which flanked the river fended off the falling rocks. Gradually, as they ascended, the valley widened; at 15,000 feet they emerged on a broad, flat table-land, and 500 feet higher reached a long flat ridge, where stood the boundary mark—a Cairn! This was their goal. The storm lifted its curtains. Beyond showed the blue and rainless skies of Tibet; behind, were revealed the two snow peaks of Chomiomo and Kinchinjhow, so named from its 'beard' of icicles, and between them the funnel-mouthed head of the valley up which they had come.

> Here [he exclaims], after three months of obstacles, I was at last at the back of the whole Himalaya range at its most northern trend in the central Himalaya, for this is far North of Kinchin-junga and Chumalari or the Nepal Passes I visited last winter, and opens on to the Thibetan Plateau without crossing a snowy ridge to be followed by other and other snowed spurs, as Kanglachem and Wallanchoon do. Here too I solved another great problem. There was not a particle of snow anywhere en route, right or left, or on the great mountains for 1500 feet above my position. The snow line in Sikkim lies on the Indian face of the Himalaya range at below 15,000 feet, on the Thibetan at above 16,000. I felt very pleased and made a rude panorama sketch on four folio sheets, very rude you may suppose, for the keen wind blew a gale and we were quite wet; above 15,000 feet too, I am a 'gone coon,' my head rings with acute headache and feels as if bound in a vice, my temples throb at every step and I retch with sea-sickness.

An hour and a half was spent on the Tibetan side, making observations. The letter tells how, in spite of the fire they made, 'my shivering Lepchas were numb and I gave them my cloak, going always well clad myself.'

Much as Hooker would have liked to stay for some time in the high alpine region of Tungu, the question of supplies forbade. The post took twenty days from Darjiling. Still, he stayed the rest of the week, exploring the high yak pastures and the glaciers, before setting out on the week's march back to Choongtam and plenty, 7000 feet lower, a weary march over roads in a terrible state with floods, landslips, jungle, and impassable places, and warmth attended with tropical discomforts.

> I think leeches are the worst; my legs are, I assure you, daily clotted with blood, and I pull my stockings off quite full of leeches; they get into the hair and all over the body. I cannot walk ten yards without having dozens on my legs;

they produce no pain but the itching and bleeding are troublesome; poor Kinchin can hardly walk from weakness, and he is blinded by the number hanging on to his eyelids, and his nostrils are quite full. (To W. J. H., Aug. 6, 1869.)

At Choongtam he rested ten days; then proceeded to complete his programme by starting afresh up the eastern stream, the Lachoong, to the disgust of the Singtam Soubah, who was still charged to accompany him, and longed to be back amid the comforts and the native beer of his own home. The unhappy man was also very lame from insect bites, and at the village of Lachoong (August 16) remained on the sick list, while Hooker, in unwonted freedom, made an eastward excursion to the unknown pass of the Tunkra-la, afterwards used by the British expedition to Lhassa. Of this cold, unsheltered spot and his botanical results so far he writes in a continuation of the letter to his father dated August 24th.

I think the botanical results of my little Thibetan cruise (which you may talk of) will astonish you, for number; not that they would have been increased by going further North; but I found what I so many years have only dreamed of, the remarkable change in vegetation that only occurs at the boundary of the mountains and plains, that prevalence of species and paucity of specimens which marks that curious zone where the perpetual snow rises 2000 feet [i.e. the snow-line is 2000 feet higher than on the southern side] on mountain faces opposed to the most sterile country in the inhabited globe. I am indeed more gratified with my Lachen journey than I can express to you, so long have all my friends here and at home thought the probability of reaching the Thibetan Plateau in this direction visionary. Campbell's and Hodgson's congratulations are extravagant. I am very pleased too to think that *any one may now go*, the egg-shell is broken; the intricate route once known and the nature of the impediments, it is easy to forestall the one and follow the other. Of the importance of its botanical results as to the Sikkim Flora you have yet no idea, nor had I till two days ago, when I returned from a long visit to another Pass of which nor I nor Campbell were aware and which took me to within ten miles of Phari and the Holy

Mountain Chumalari. I was four days away; it is amongst the main ranges East of Sikkim and leads to Choombi from this; though only of the same height as Kongra Lama, this, the Kankola, was heavily snowed, and indeed from 14,700–15,000 feet we were on snow the whole way. It took two days from hence to reach Tunkra; headache and fatigue prevented my botanizing much on the travelling days, therefore I camped at 15,000 feet and made a full Flora at 14–16,000 feet, wholly different from the Kongra Lama Flora at the same altitude.

Immediately above 15,000 feet there is far more rock and snow with vast piles of debris than anything else. This road is very rarely travelled, and then only by an occasional courier from the Rajah, when at Choombi, to the N.E. quarter of Sikkim.

Having no tent we slept on the ground, a great precipice our only shelter from the rain and snow. It was curious to waken in the morning and see the broad snowy faces of lofty mountains staring at you, the bright sunbeams dancing on their rosy peaks, and all within a few yards of you. Unfortunately the weather was extremely bad and always is so on this range. At sunrise it was invariably brilliant and clear, and I then hastily sallied out to a high place to take views, angles, and bearings. From such heights the prospect of the whole Kinchin group was superb beyond all powers of description; there was an exuberance of snow, and as the clouds of night rise and reveal peak after peak, with cliffs, domes, and tables of snow, it really conveyed the idea of a forest of mountains. At 8 o'clock clouds form, and before 9 A.M. every object far or near, is wrapped in thick fog, and you are fortunate if you can gain a glimpse of the sun with the sextant to make out your time and position. At 10 A.M. rain always commenced, and lasted with sleet or snow till sunrise of the following morning. Our camping ground was of course very cold, and the little sticks of firewood, for which we had to send down 2000 feet, were so wet, that with this, and the diminished oxygen of the air, it was very difficult to keep up a fire. I often think on these occasions of passages in your lectures, with keen appreciation of your tact and power in riveting the student's attention; how often do I remember your *Life* of Linnæus, and of what you have not realised for many a year, that it is

> The sweetest of pleasures under the sun
> To sit by the fire till the 'praties' are done.

Resuming his course up the precipitous valley of the Lachoong, he left Lachoong village on August 29, and in two marches mounted 6000 feet to Yeumtong, where a week was spent among the mountains. Here his long patience was further rewarded. On September 7 a new friend arrived in the person of his old opponent, the Lachen Phipun, who, having now, as he said, ascertained that the Tibetans were entirely indifferent, offered to act as guide still northward up the valley to Momay and the Donkiah Pass. Momay, at 15,362 feet, with its great yak pastures, the highest in Sikkim, proved to be an ideal place for observations of all kinds, and eking out two-third rations with what could be obtained in the village, the party stayed here till September 30. On the 9th they went to the Donkiah pass, 18,466 feet, and in order to obtain a still wider view over Tibet, Hooker scrambled up the mountain side another 1000 feet, an ascent made a second time when he revisited Donkiah later with Dr. Campbell. The climb eclipsed in altitude Humboldt's famous climb on Chimborazo, and this record of over 19,000 feet, as well as three peaks or passes of 18,500, held the field till the brothers Schlagintweit in 1856 reached the height of 22,230 feet on Kamet.

This stage of the expedition is well described in the following letter:

Lachoong River, Thibet Frontier (i.e. Momay): September 13, 1848.

From the top of the Donkiah Pass I had a most splendid view for 60 miles north into Thibet—first of extensive plains, dunes, and low rocky hills utterly barren and red from the quantity of quartz, tinged with oxide of iron, which form the hills north of Kongra-Lama; beyond that again, and as far as the eye could scan, were ranges of rocky mountains sprinkled with snow and of comparatively moderate elevation. From Kongra-Lama at 16,000 feet the view was wretched enough, but from hence, no language can convey an idea of the horrible desolation and sterility of the scene!

'A howling wilderness' is the only meet term; there was neither grandeur in the mountains nor beauty in the valleys to invite the traveller; in colouring, form of the land and mountains, and their composition and stratification, it strangely reminded me of the Egyptian desert. The rocks were disposed in horizontal strata, cropping out on the mountain faces and broken into low crags along their tops; not even lending fantastic shapes to relieve the eye. Range after range was like its fellows until, in the far distance, one range loftier than the rest, black, rugged, and heavily snowed in some places, shut out any more distant horizon. The whole landscape sloped N.W. and the ranges were East and West, so that I do not doubt the truth of the unanimous assertion of the people, that all the waters from north of my position and west of the Paniomchoo are feeders of the Arun which enters Nepaul far west of Kinchin-junga.

Very different from this dreary Tibetan landscape was the fantastic grandeu of the mountains hard by. There was a great amphitheatre of rock and snow under Kinchinjhow, walled in with precipices and an ice face of 4000 feet, 'a great blue curtain reaching from heaven to earth,' only fretted where 'icicles fifty feet long run along in lines like organ pipes'; its floor, two miles each way, 'a maze of cones of snow laden with masses of rock rising fifty or eighty feet—comparable to nothing but the crater of a stupendous volcano, where little enclosed cones of fire have been suddenly turned to ice.'

. . . What keeps me here is the very curious Flora, though not so rich as that of Kongra-Lama and the Thibetan plains. I have a set of most curious new plants from between 17 and 19,000 feet—Woolly *Lactuceae* and *Senecioneae* like *Culcitium*, *Gentians*, *Chrysanthemums*, *Saxifrages* of course, *Cyananthi*, and some very odd things. They are extremely scarce and require close hunting. Sometimes I get but one or two specimens of a kind, and poking with a headache is very disagreeable.

To-day I went up the flanks of Donkiah to 19,300 feet, amongst the knot of snowy peaks west of Chumulari, and such gulfs, craters, plains, and mountains of snow are surely nowhere else to be found without the Polar circles. Of course I have seen nothing to compare for mass and continuity with Victoria Land, but the mountains, especially Kinchin-jhow, are beyond all description beautiful; from whichever side you view this latter mountain, it is a castle of pure blue glacier ice, 4000 feet high and 6 or 8 miles long. I do wish I were not the only person who has ever seen it or dwelt among its wonders. Now I have been N., S., E., and W. of it, up it, down it, to 16,000, 17,000, and 18,000 feet; and every view enchants me more than another.

. . . I was greatly pleased with finding my most Antarctic plant, *Lecanora miniata*, at the top of the Pass, and to-day I saw stony hills at 19,000 feet stained wholly orange-red with it, exactly as the rocks of Cockburn Island were in 64° South [1]; is not this most curious and interesting? To find the identical plant forming the only vegetation at the two extreme limits of vegetable life is always interesting; but to find it absolutely in both instances painting a landscape, so as to render its colour conspicuous in each case five miles off, is wonderful.

[1] See *Himalayan Journals*, ii. 130 and 165.

CHAPTER XV

CAPTIVITY AND RELEASE

DURING the last weeks at Momay, as has been said, Hooker had again been happily relieved of the presence of the Singtam Soubah. Finding the situation unendurable, the wretched fellow withdrew to lower altitudes, uttering the gloomiest warnings against cold and famine and Tibetan interference. But on September 28 he returned to ask formal leave to go home, and brought the welcome news that Campbell, accompanied by the friendly Tchebu Lama, was on his way north through Sikkim, having been sent by the Government to seek a personal interview with the Rajah. His object was to cultivate better relations with the Sikkim officials, and to enquire into the breach of good faith displayed in the discourtesy and hindrances offered to Hooker. His authority to enter Sikkim, moreover, gave him the opportunity of learning something about the country which the treaty bound us to protect, yet from which we were so jealously excluded.

Leaving Momay, therefore, on the last day of September, Hooker hurried down to Choongtam at the junction of the rivers, and was joined by Campbell on the morning of October 4. Then, starting together on the 6th, they repeated and enlarged Hooker's previous trips, first up the Lachen to the Kongra Lama pass, then actually bluffing an entrance into Tibet, and following the upper Lachen round its eastern bend to its source in the Cholamo lake. This brought them to the Tibetan face of the Donkiah pass, which they crossed (October 19), and so completed the round by descending the Lachoong to their starting point at Choongtam, October 27.

Two letters to Sir William describe the happenings of this month.

Choongtam: October 3, 1849.

I arrived here late last night, having made three flying marches down from Momay Samdong to meet Campbell, who will be here to-morrow en route to Kongra Lama, as he tells me you are (ere the receipt of this) aware. I have been months stimulating him to the journey and with success at last. It is now six months since I have had any one to talk to, and now that the route is known and he has the Rajah under his thumb, I do not anticipate any difficulty. He had a most narrow escape for his life on the second day after leaving Darjiling: his pony slipped its foot in a most dangerous part of the road; feeling it do so he wisely jerked himself off, and the animal, rolling down the precipice, was killed on the spot!

I had hoped to make a very fine collection of seeds on the road down here, but it sleeted and snowed all the first day, and rained tremendously all the other two, which sadly impeded my proceedings. However, I did my utmost, and have ripe and good seeds of many very fine things, of which I send a few samples. I am now collecting seeds as fast and hard as I well can, and losing no opportunity.

The tardy advance of the whole flora is most remarkable, and many plants actually ripening their seeds, and uniformly past flower at 15–16,000 feet are still in full flower at 7–10,000. The reason plainly is, the further north you go the more sunshine there is. . . .

On the way down I passed an uncut maize field at 7000 feet—very high for the culture of that plant—and I stole several hermaphrodite heads. The villagers made an outcry at first, as they appear to know the value of the male panicle, but a sick woman turning up whom I doctored, gave me the run of the field as fee, and a pocketful of small, hard, tasteless peaches. . . .

I brought down three loads of 80 lbs. each of plants whose sodden state now keeps me hard at work. It is a very fine collection after all, with heaps of new and curious things from the Passes. The roads, mere tracks at best, were in a horrid state from landslips and deep mire, and I do wonder how my coolies made it out in three days, but they are all the best and most patient coolies you can conceive, never complaining. . . .

I have just had the big tin vasculum up from Calcutta, at which you shook your head so gravely. The Lepchas are charmed with it, and there will be a competition as to who is to carry it. You have not an idea how bulky the undried plants of these climates are ; the otherwise *very large* vasculum I use does not hold half, hardly one-third morning's collection. As to drying paper, you know I stow well, yet that ream of Bentall's paper does not suffice to lay in one day's collection, nor near it, if you take woody with other things. You may well wonder how I get on ; it is only by changing and drying papers every day. Bentall's is not nearly so good as the sugar refining paper I bought at Calcutta, and of which my stock cost £15. But after all good English brown paper is the best for all plants, as Mr. Brown always said. . . .

You will be glad to hear that I quite got over my headaches at great elevations and most of my other distressing symptoms, and I would not hesitate going to 20,000 feet if the mountains were but accessible so high. Still the lassitude is trying, and a sort of weight, like a pound of lead, dragging down the stomach, probably caused by over-action of the lungs straining the diaphragm, or diminished atmospheric pressure actually relaxing that organ and causing the abdominal viscera to drag heavily downwards. It is a horrid feeling.

Boiling point is a perfect nuisance at these elevations, and the Barometer is the only useful, accurate, or simple method. You must have a man to carry the wood and often the water too ; blowing the fire gives intolerable headaches, without blowing the best wood will not burn owing to the deficiency of Oxygen (i.e. rarity of the air) ; and if there be any wind (as there is sure to be) the temperature never comes up to the true B.P.

I have just had dinner (for which and all other mercies—including the safety of poor Campbell's neck, who writes affectingly on the subject and says he is spared a little longer to love me as a brother). To return to the dinner, it was a fine grouse tasting strong of Juniper tops, followed by the peaches, all I can say of which is, that if Loti were no better Plato (I think it was Plato ?) might have let his pupils eat their fill. A very large leech presented himself as the bell rang, to whom I did not refuse the rites of oriental hospitality, laying salt before him with alacrity.

The servant I left here has caught some beautiful butterflies and splendid beetles. I have rewarded him with fifteen shillings to buy a garnet-colored Bhothea cloak which is his [heart]-eating envy, and in which, with his long hair parted down the middle and beardless face, he looks like an auld wife at Kilmun Kirk. . . .

You will I fear think this a very childish letter, but really I have little news and can think of nothing but 'the Campbells are coming.' My little finger too is hurt and I cannot write much.

Lachoong (village): October 25, 1849.

What do you think—we spent four days in Thibet! in spite of Chinese guards, Dingpuns, Phipuns, Soubahs, and Sepas. It was a serious undertaking and required a combination of most favourable accidents, together with my previous acquaintance with the country, and a most indomitable share of resolution and boldness. Campbell has behaved splendidly, and diverted me by throwing all the sage precepts he sent me to the winds. He has frankly told me that he did not, could not, believe the real nature of the opposition and ill-treatment I had received; he had not been two days with me before he was storming and bullying right and left. The unfortunate Singtam Soubah, with whom at C.'s intercession I had kept such good friends, he gave no peace to, blackened his face, and sent him to the Durbar in disgrace.

On arriving at Tungu an hour after C. I found him at a drawn battle with the Phipun, my arch-enemy, and quite astonished that the ruffian cared no more for himself than he did for me, or the Rajah, or anybody else under the sun.

After fully weighing the possible consequences of breaking through the border and perhaps exposing the Rajah to menaces from China, &c., we determined to do it if possible, and we told the Border Chief that if he dared to oppose we would send a guard of Sepas from Darjiling to close the Pass. This threat, and promise of a present if we succeeded, got the man over, the Singtam Soubah (lord of all the district) being conveniently packed off in disgrace two days before. Our great ally was the Tchebu Lama, the Rajah's representative at Campbell's court, a man of intelligence and vigour, who had been dreadfully misused in Sikkim by the enemies of the English who surround the Rajah's park. This man we absolved from all participation and consequences, offering him an asylum and provision at Darjiling should the worst come to pass.

On the Border we were met by two Thibetan Sepas, who made a terrible row and endeavoured to stop us, without laying hands however on our bridles. They met us in *Sikkim*, swore that it was Bhota (Thibet alias Cheen), a lie of which we took advantage when really across the border. Then a terrible row was kicked up and the Cheen camp came out running after us with boots, matchlocks, &c. The Lama and Phipun both got frightened and implored us to stop for a conference, to which Campbell properly acceded, and I put spurs to my pony and galloped ahead on to the sandy plains of Thibet, determined to stay away all day and see what I could, for there was no good I could do by waiting with C., who could make no retrograde motion whilst I was ahead. Two Sepas started in pursuit of me, but Campbell kept them back with his stick till I was out of sight and of catchable distance. The elevation, 17,000 feet, was such that my pony was soon knocked up and I pursued my way on foot up the Lachen, at the back of Kinchin-jhow, over dry sandy stony dunes, with Carex, a little grass, tufts of nettles, Ephedra and a thirsty looking Lonicera? a few inches high. Proceeding N.E. from Kongra Lama I had long, stony, rolling mountains on the North and East, and to the South the stupendous snowy mass of Kinchin-jhow rose plumb perpendicularly from the sandy plains. Finding the country so traversable I thought it the best thing I could do to follow the Lachen to its source near the Donkiah Pass, as that would be our route *out if* Campbell should succeed in getting the coolies and *himself* past the guard, and because I had difficulty in making C. believe that I could and would guide him through the waste with compass and sextant if *he* only could and would *break* the frontier. Later in the day I arrived at Cholamo lakes, within sight of the Donkiah Pass, but my pony was so knocked up that I had great difficulty in dragging him after me. At the lakes I refreshed him with some tufts of green Carex and led him back, suffering much from headache as the sun was intensely hot, and a little exertion brings on headache at these elevations (nearly 18,000 feet).

Late in the evening I met Campbell's party, viz. the Lama and Phipun, looking for me; they told me that Campbell had gallantly pushed through thirty Sepas armed with matchlocks, that no hands were laid on him, but on our coolies (we had no Sepas nor arms), who of course were much frightened; that Campbell having shot ahead and I too being gone, he, the Lama, took on himself to point out to the Chinese officer that if either of us died for want of our tents, &c., it would be a terrible affair for the officer above all, who should have *taken us alive* rather than stop our men. The coolies were then allowed to pass on too, and came up at night suffering terribly from the dry heat, sun and dust and elevation. The Lama then went to find Campbell, who had mistaken the way towards Donkiah, and soon came in full of spirits and gave me a most ludicrous account of the mixture of fright and obstinacy and force the Chinese Sepas displayed.

In the evening the Chinese followed us, the Dingpun, or Lieutenant, riding on the top of a black Yak! surrounded by pots, pans, bags and bamboo bottles of buttermilk, a tent, blankets, &c., all bundled about his Yak, and he on the top of all like a gipsy on a laden donkey. He was a small withered man, in a green coat, with a gilt button on his Tartar cap; behind came the Sepas, enormous ruffianly looking men, dressed in blanketing, each armed with a pipe, a long knife, and a long rude matchlock lashed across his stern. These matchlocks are slung at right angles across the hip; they are very rude, long, with a pronged support or rest; the latter folds up with a hinge and projects like antelopes' horns beyond the muzzle. Such ungainly implements across their stern parts were comical enough looking. They marched in orderly, took no notice of us, and camped close by. We tented in a low cattle enclosure on the bare plain, burning Yaks' dung for fuel. The cold was intense and wind violent and dusty, sky brilliantly clear.

We determined to stay a day or two where we were, at

all hazards, and sent word to the Dingpun that we would condescend to receive him if he would visit us! next morning. This he did promptly, and we explained to him that it might be all very right and proper for him to obey the orders of the Lhassa Govt. and prevent (or try to) Englishmen passing from one Sikkim Pass round through Cheen to another, but that it was all stuff and we did not feel ourselves bound to respect their prejudices. Also we added that . . .

(Here the letter ends abruptly, the only addition is) Singtam, Nov. 1, Ripe Abies Webbiana, 3 packets sent.

The return to Choongtam prefaced the long planned treachery of the Sikkim Dewan.

Meepo, the guide, met them here, with orders to take them to the Chola and Yak-la passes in East Sikkim, a way leading over the same ridge (the Chola range) as the Tunkra pass already explored, across Chumbi, a wedge of Tibet running between Sikkim and Bhutan.

The road passed the Rajah's residence at Tumloong, and here Campbell desired an official audience of the prince. But although they were welcomed by the principal people and the Lamas as well as the populace, the meeting was prevented by the Amlah or Council, one and all relations or adherents of the Dewan, who directed them from Chumbi, where he was trying to stir up strife in Tibet.

On November 4 they left Tumloong for the Chola pass. This they ascended on the 7th, but were turned back by a Tibetan frontier guard on the plea of ' no road.' This guard was not only polite, but protected the travellers from the sudden insolence of a number of Sikkim sepoys who unexpectedly came up. No less unexpected was the re-appearance, lower down the road, of the troublesome Singtam Soubah, who had quitted them three weeks before, short of the Kongra Lama pass,—obviously ill at ease, and demanding a conference with Campbell; a conference naturally deferred till the evening's camp. Here was waiting a great party of Bhoteas, the rough, intractable element of Sikkim. They did not wait long. The night was very cold; the people crowded into the hut where Hooker and

Campbell were waiting. The latter went out to see to the pitching of the tents.

> He had scarcely left, when I heard him calling loudly to me, 'Hooker! Hooker! the savages are murdering me!' I rushed to the door, and caught sight of him striking out with his fists, and struggling violently; being tall and powerful, he had already prostrated a few, but a host of men bore him down, and appeared to be trampling on him; at the same moment I was myself seized by eight men, who forced me back into the hut, and down on the log, where they held me in a sitting posture, pressing me against the wall; here I spent a few moments of agony, as I heard my friend's stifled cries grow fainter and fainter. I struggled but little, and that only at first, for at least five-and-twenty men crowded round and laid their hands upon me, rendering any effort to move useless; they were, however, neither angry nor violent, and signed to me to keep quiet. I retained my presence of mind, and felt comfort in remembering that I saw no knives used by the party who fell on Campbell, and that if their intentions had been murderous, an arrow would have been the more sure and less troublesome weapon. It was evident that the whole animus was directed against Campbell, and though at first alarmed on my own account, all the inferences which, with the rapidity of lightning, my mind involuntarily drew, were favourable.

Soon the Singtam Soubah returned, 'pale, trembling like a leaf, and with great drops of sweat trickling from his greasy brow,' with the Tchebu Lama under arrest. He explained the seizure of Campbell as a political hostage, to be kept till the supreme government at Calcutta should confirm articles to which he should be compelled to subscribe. How would Campbell behave? What steps should Sikkim take to secure their end? Hooker refused to answer till informed why he himself was made a prisoner, whereupon the Soubah went away. Campbell was knocked about and tortured by twisting of the cords that bound him, especially by the scoundrel already mentioned who bore him a grudge; but he disconcerted the Soubah by declaring that whatever he might say or do under compulsion, the Government would not confirm it. The

Soubah's followers slunk away, and the Soubah himself left Campbell, who was then taken, much bruised, to his tent. 'It is *Tartar fashion* to catch and coerce a great man when they can,' and the Dewan had arranged for Campbell's seizure from the day he crossed into the country, three months before. But his tools were too timid, Hooker's popularity too great for arrest in the capital itself, where they were to be quietly detained unknown to the Rajah, till the Dewan returned from Chumbi. Here he had failed in his attempt to involve the Tibetan guard in his aggression, an attempt which drew down upon himself the anger of the Tibetan authorities when they investigated the affair next summer; while at the Chola pass the personal animus of his henchmen, delighted to outrun the letter of their instructions, created an impasse for which they were speedily disgraced by their master.

> The plan failing, they were utterly dismayed, having committed a gross outrage on Campbell's and my persons from which no imaginable good could come. The only course remaining was of course to trump up a new story and to detain us as hostages for no ill befalling them pending the Government's taking active steps for our release.
>
> Unfortunately they were so simple as to let out all their secrets *to me*, when trying to gain information *from me* by all manner of means, and over and over again gave me the Rajah's assurance that no fault whatever had been or could be laid at my door and that Campbell's offences were wholly political. Now, C. having Govt. sanction and approval for all his supposed offences, they do not know what to do, and urge our trespass on the Thibet frontier in the hopes that Govt. will commit itself and take up that grievance against us.
>
> This Dewan [writes Hooker, December 28, 1849] is an alien and universally detested; powerless except through his gang of Bhotean ruffians, who are runaways from their own land, and whom he protects, and who protect *him*. He is a man of some energy, and finds it easy to ride rough-shod over the simple and indolent Lepchas. He rules the old chiefs with an iron rod, monopolises trade, and is the

bitter foe of the English. All the summer he spent in Thibet, vainly trying to incite the Chinese to make common cause with him and drive me out of Sikkim and back to Darjeeling. This was the origin of his conduct to me at the Zemu river in May, June, and July. The Rajah is an old, timorous, and inoffensive being. The priests are all friendly, and hold Campbell and the British name in high respect; and the Lepchas are fond of us to a man, and would gladly transfer their allegiance to us if we would only protect them.

Force had first been used against Hooker to prevent him from giving help to Campbell; he was offered good treatment and presents, but refused such marks of respect so long as his friend was ill treated, and warned the Soubah of the consequences that must follow.

Writing in the first days of his captivity (November 12: the letter was not despatched till considerably later),' in the forlorn hope that this letter may reach England,' he tried to reassure his father:

My bonds are not very heavy, and I am under no apprehension either on my own or Campbell's account. I was seized in the hope of extracting information from me (by intimidation and otherwise) as to what course these stupid people should pursue. In this, I am happy to say, they have utterly failed; and I think they are so nonplussed, that they will not detain me much longer. Campbell is very strictly guarded. I am much better off; and have so very many friends among these poor people (to an evil faction among whose rulers this is attributable) that I hope and believe I can be useful. . . . I am altogether prohibited from approaching or communicating with Campbell, but he and I keep up a capital correspondence. My hand is so fatigued with copying out his Despatches to Govt., for I dare not send the originals by this opportunity, and sending a copy of my Journal for Hodgson to forward to you, that I can write no more. The said Journal H. will send you a copy of at once. I also so very much doubt this reaching you that I do not care to write much hereby. My old friend Meepo sticks well to me, and will I hope get this on to Darjeeling, where a demonstration from the military will effect our release at once. The Rajah has not fifty stand of arms, nor fifty men to handle them.

I have now to beg and implore you not to make a stir about this. I have never deceived you nor my Mother and entreat you to remark that all I say on the score of my position not exciting any apprehension of my safety, is strictly true, and to make it otherwise is mere romancing. I am allowed the free use of my instruments, plants, and books, and am busy and well occupied all day long.

I have heaps of letters written and writing, Bentham, Berkeley, Darwin, &c., but send only this by this chance.

After an interview with the Amlah, or council, on November 13, however, he was allowed, to his great satisfaction, to join Campbell, though they were both ill fed, and later horribly overcrowded, as unsuspecting messengers arriving from Darjiling were thrust into their narrow quarters; while their own coolies were starved or arrested.

The Dewan at last arriving from Chumbi on the 20th to find that his stroke had miscarried, professed anger and surprise. In sober fact, he had no conception how seriously the Indian Government would regard what he persisted in calling a mere mistake, which should be overlooked by both parties; Campbell's vigorous representations had their effect, and speedy release was promised; but a communication couched in mild terms arriving from Darjiling, where the real facts were unknown, complications with distant Tibet were feared, and an immediate incursion expected—to the great amusement of Sikkim spies—the Dewan was seized with a diplomatic illness, and nothing was done. Peremptory orders from Calcutta for their release were disregarded as not bearing the Governor-General's great seal, for Lord Dalhousie was in Bombay; and captivity, as shown by the following letter (received February 3), became more trying.

To Miss Henslow

December 2, 1849.

I am in great anxiety till I hear whether the report of Campbell's and my death has reached England; for we know that the Rajah purposely circulated the tale of his having compassed our destruction, and that it was believed in India. Now, we have, happily, no cause for apprehension, but every reason to hope that our captivity is drawing to a close.

My durance here has been somewhat of the vilest. Certainly the Sikkimites have left no way untried of making Campbell and me as wretched as possible. We are not allowed to stray ten yards from this miserable hovel in which we are immured, and we are debarred all correspondence and the power of laying our complaints before our own Government, or even before the Rajah. These people actually converse in lies,—they think in lies—and I verily believe that any appeal they may make to their own consciences is *answered by a lie*. Their utter mental degradation and distortion are inconceivable. I speak of the Bhotea authorities. The Lepcha population are a better set; they sympathise with us and show us many a little kindness by stealth. The Lamas, too, who are somewhat more enlightened than their rulers, are coming forward to a man, and representing to the Rajah the peace and comfort in which they lived under Campbell's sway; also that the Rajah is literally breaking his own head, for that when this outrageous conduct is answered, (as it must be and resented) by an appeal to arms, these people will assuredly come off second best. They have no muskets, their bows they handle very awkwardly, their long knives will be useless against Artillery. These warnings have already alarmed the Rajah, especially as we echo the same tale; he would be thankful now to be rid of us, but how to do so is the question! He has committed himself fatally by the violence used towards our persons; and as to the complaints he alleges against Campbell's public acts, the Superintendent, already appointed at Darjeeling, pursues, and will pursue, the same line of conduct, nor could Campbell alter if he would.

You would have been highly diverted by our schemes, especially for corresponding with one another; for Campbell and I were confined separately and debarred all communication. My Lepcha boys were so clever that we never failed to get little wisps of paper conveyed to and fro between us. Now that we are together we get on much better, and, although guarded, and closely watched by an ever-present spy, we never make ourselves unhappy.

Our only communication with the *Durbar* (Court) is through our spy, a truly odious being. He is perfectly made up of malevolence and falsehood, to practise which is his main occupation. He is a filthy squinting Bhotea, who drives away every one who comes near us, and causes our poor coolies to be flogged, when they approach the door to beg a little food from our small stock. We are, of course, more than civil, nay, we are *kind* to him, but he is equally untouched by our kind deeds and our remonstrances. Many a base scurvy trick he has played us and misrepresented our conduct to the Rajah, who treated us ill enough and starved both Campbell and me for the first fortnight; as he does our poor followers to this very hour. I suppose the evil *animus* this vile fellow (who rejoices in the name of *Toba Singh*) exhibits against us constitutes his recommendation in the Rajah's eyes. Happily neither he, nor any one here, can speak English, so my friend and I talk with perfect freedom, only using conventional names for persons. We call the *Rajah* Prince, the *Dewan* Butcher, *Toba Singh* Evil Eye, and so on.

Hodgson is our good angel now. Though his health almost imperatively requires him to go to the Plains, he stays at Darjeeling, in order to serve us by communicating with Government, threatening the Rajah, looking to the defences of Darjeeling, and comforting poor Mrs. Campbell and the few inhabitants who yet remain at the Station. The ostensible manager there is the brother of ——; he thinks (and is allowed by Hodgson to think) that he does everything, but he is a wholly inefficient person, and quite incompetent to stir a peg without the impulse, counsel, and correction of others.

From the middle of November, however, permission had been given to write to their friends, though even before their arrest, a whole packet of letters had been destroyed. Hooker accordingly sent a private account of all that had happened to Lord Dalhousie, then at Bombay, with instant effect. Troops were hurried up to Darjiling; an ultimatum despatched to the Rajah. Military force was a message the Dewan could not pretend to misunderstand; his vague promises of relief, his alternate boasts of the glories of Lhassa and peddling offers to sell them ponies cheap took on another tone. Propitiatory messages and gifts arrived from the Rajah and Rani, and the prisoners set out for Darjiling on December 8 under the charge of the Dewan, 'as slowly as he could contrive to crawl.' Messengers bearing Lord Dalhousie's despatch met them on the 13th, but still the Dewan, with his ponies and his merchandise, with which he yet hoped to do a roaring trade at Darjiling, loitered and talked and chaffered and allowed his bodyguard to make a parade of threatening the lives of his captives, till on the 22nd he halted in a state of hopeless vacillation within sight of Darjiling and its new barracks, twenty miles away, and shaken by the knowledge that the Rajah's peace offerings had been rejected. There was one last alarm. Nimbo, Hooker's sturdy Bhotea Sirdar, the special object of the Dewan's anger, had broken from prison, and with his chain still hanging to his ankle, had managed to reach Darjiling, and now threatened to lead a party to the rescue. Their attack would have been the signal for the murder of the prisoners.

Christmas Eve brought opportunity for a final stroke of diplomacy; the morrow was the great and only 'Poojah' of Englishmen, when they all met; it would be well to let Campbell join his relations and appease the exasperated soldiery. The Dewan, equally afraid to lose his hostages and to keep them, at last, with extreme reluctance and bad grace, consented. By 4 o'clock they were at the frontier, the bridge over the Great Rungeet, and by 8 safe in Darjiling, where, in addition to the rest, Hooker found his old friend and new travelling companion, Thomas Thomson, already awaiting him.

CHAPTER XVI

LAST DAYS IN SIKKIM

PUNITIVE measures against the Rajah were not very admirably carried out. Instead of the friendly chiefs being invited to Darjiling, the Rajah was bidden to come in and surrender, bringing the guilty parties with him, on pain of invasion. But when he failed to comply, and indeed to bring in the guilty was beyond his power, the threat was not carried out.

The army camped for some weeks on the north bank of the Great Rungeet, the Dewan with his handful of followers being on the hill not three hours away, and finally withdrew, while for penalty the fertile Terai lands, the British gift to the Rajah, were resumed, his pension withdrawn, and Southern Sikkim annexed. The fidelity of the Tchebu Lama was happily rewarded with money and a grant of land at Darjiling.

From his intimate knowledge of the country, Hooker was in a position to give sound counsel when asked, and to perceive, if he could not always correct, various false steps taken by the temporary administration; but he intervened as little as might be in matters which were not his proper concern, and his chief satisfaction lay in the eventual release of one of his men who was reported to have been murdered, and in the fact that thanks to his clear account of the affair, Lord Dalhousie acquitted Campbell of blame, and re-appointed him with wider powers than before.

For a short time the military preparations threatened to

upset Hooker's plans; his brief share in the abortive campaign appears in the following letter to his mother, dated January 31, 1850:

Before the time of the General and staff coming up here I was asked repeatedly whether I would go into Sikkim with the troops; I always say I did not wish to nor want to, but that if the General showed good cause for desiring it I would think upon it. Volunteer I could not and would not, being in another service and receiving pay from my own Govt. for very different work. Tom and I both went away from the station when the General was coming, but he had not arrived a day before he wanted me and sent the most urgent message through Campbell. I therefore returned about ten days ago, and found the old gentleman, Genl. Young, all in the clouds, as to carrying out his orders of occupying Sikkim with a military force. Meanwhile 14,000 men, Sepoys and Europeans, had come up with headquarters of one Regt., guns, a whole staff of officers, and nothing but the 'horrid din of arms' was to be heard.

Genl. Young is a very nice old gentleman and greatly obliged to me for my counsel, maps, and information, which settled him to march as soon as possible and take the Rungeet bridge. Both he and Mr. Lushington (the special Commissioner) begged me to conduct the troops which I refused except they sent me a written request specifying the urgency of the occasion, which I should forward to H.M. [Commissioners of] Woods, &c., and meanwhile take upon me the responsibility of acting with heart and good will under the General's orders. I objected on Thomson's account who had come so far to see me, and he was immediately put into orders for medical duty in the detachment (advance guard) with myself. This is a capital arrangement, for it gives him time of service in India instead of *leave* which he is now upon, and every hour taken off the time he will have to spend in India on his return after furlough is so much added to his life.

I went down with the troops the other day and took possession of the bridge over the Great Rungeet and camped some 500 men in Sikkim. As no further advance was to be made at once I returned to my plants at Darjiling, but expect to be summoned down very soon again now. No opposition of any kind was made to us, and I doubt if there will be any, so you need be under no alarm on my account. Under any circumstances it appeared to me so clearly my duty to undertake the service that I did so without any hesitation and have no fear for the result. Except Campbell and myself no one knows anything of the country, and hence the marching of the troops without good guidance would be most unadvisable. Campbell is so much the aggrieved party that he could not with propriety go to attack the Rajah's country; I, on the other hand, have no ill-will (nor has C. for that matter), the people, I know, are friendly to and fully trust me, they would far rather make overtures to me than to soldiers with guns in their hands, and with the heartiest desire and determination to bring things to a peaceful issue if possible, I do hope my presence may be useful.

The orders at present are to march to Tumlong and occupy the capital, for the Rajah refuses to give himself up or to offer any adequate concessions for his conduct. Many of the people I know from private sources are all ready and willing to come over to Darjiling, and only want our assurance that they will not be molested to grant a peaceful march to our soldiers. This they now have and appreciate. The Dewan has only thirty men to oppose us with and they will not help him, the Rajah has no army nor is he trying to raise one, so that he will probably flee at our approach.

It is said that the Rajah has sought succour from Thibet, and has received for answer that he has only got his deserts.[1]

[1] The expedition was abandoned, because the general, from his experience of the Nepaul campaign, reported the country as 'impracticable for British troops.'

In 1861 another punitive expedition was organised against the same Rajah for acts of violence and aggression on our territory. A staff officer engaged on this campaign wrote afterwards to the *Standard* (August 13, 1862) apropos of Hooker's military services:

'In 1859–60, on my way between Calcutta and Darjeeling, I studied Dr. Hooker's most interesting and valuable work, *Himalayan Journals*, which I found to be a most perfect staff officer's report, containing accurate information on every point that could be useful to the commander of an expedition, regarding hills, valleys, elevations, distances, rocks, soil, trees, vegetation, roads, rivers, bridges, productions, inhabitants, their character, climate, seasons, &c., and accompanied moreover by an excellent sketch map, which the government copied and furnished for our use.

'For the time that the force was in the field the work was as hard as has ever been performed by any force; but the rapidity of its movements and the

Dangers and troubles once over were characteristically treated as of small account, and in December 28 he writes to his mother:

> You see, by the above date, that I have, as usual, lighted on my legs and am safely escaped from the Rajah's clutches. Not that I think my own personal danger was ever very imminent; but the man who could commit one such rash and mad act (as the seizing and maltreating us), might be capable of doing what is really far more unlikely.
>
> The whole affair has been naturally exaggerated at Darjeeling, and so, into the Indian newspapers. My kind friend, Mr. Hodgson, especially, was possessed with the most dreadful alarm—due, I am well aware, to his intense solicitude on my behalf. He imagined all sorts of horrors, and attributed our capture to the *Chinese* authorities, whom he supposed to resent our having crossed into Thibet. He verily believed we should be carried into Lhassa—perhaps to Pekin, in a wooden cage—in short, he conjured up all sorts of chimeræ which, happily, did not enter our heads.

He concludes with a very light touch:

> I am dreadfully busy, as I need hardly tell you; and T. Thomson is an invaluable help. Hodgson says I am fat, and that my looks are a disgrace to the Rajah's prison house! Campbell is robust and rosy. The new baby is to be named Josephine.[1] It is very small and much the colour of blotting-paper, like all the little babies I ever saw; but some mothers' eyes have a property of neutralising that tint, as yours must have done, for you say I was a fair and white infant!

Similarly, to his uncle T. Brightwen, whom he thanks for a timely gift of new razors, 'now first used upon our truly

complete success of the expedition, which elicited the warm thanks and highest expressions of approval of the Governor-General in Council and of Lord Strathnairn, who was then Commander-in-Chief, were owing in a very large degree to the perfect information regarding the people and country afforded by Dr. Hooker's work, and which was not obtainable from any official source.

I am, sir, your obedient servant,

'G.'

See also ii. p. 183.

[1] He writes to his mother, April 27, 1850: 'Josephine was christened the other day, I answering all the responses I could in conscience, which does not include all the Church of England formulæ.'

patriarchal countenances,' he adds, 'though a close prison and heavy threats are *not pleasant*, still I fancy such books as Gonfalonieri's and Andryale's are indebted to a doleful imagination for much of their interest.'

Though for some months he confessed to being 'overwhelmed with Sikkim politics,' a return to his own pursuits was made all the pleasanter by the knowledge that his action was approved by Lord Dalhousie, who wrote him 'the kindest letter that ever gentleman penned,' and that, while the newspapers reflected on the conduct of all others concerned, he 'alone came off with high credit.'

For nearly three months he and Thomson were hard at work—'for hard work it really is'—preparing the collections to go home, filling up gaps where specimens had been lost, and completing the Sikkim flora by a visit to the foot of the hills. Thomson, fresh from exploration and botanising in the North-west Himalayas, was astonished by the magnitude of the collections, which by March 'form a huge mass, some 100 men's loads, and I am sure you will be pleased with them.'

> Altogether my collections are very handsome, though what with the Rajah's tricks and the horrible climate I have lost a great many of my large things, as Palm plants and fruits, &c., which were to have been dried whole for Museum specimens; these I am replacing as fast as I can, and Thomson being in the jungles, get on very well.

The latter was with the military, surveying the new boundary and choosing healthy positions for outposts. The forests continued to supply new plants. One budget contained seeds of 1000 species; others equally large followed. There were 100 kinds of woods, including all the Pines and most of the Rhododendrons.

> My specimens of Palms were each twelve feet long, and the new ones I am getting are as large, but the old ones almost all rotted though kept in a room with a constant fire during last rains.

There were also

whole specimens of *Rhododendron nivale* from 18,000 feet, the loftiest of all shrubs, and hitherto of any known plant,[1] but I have several species of plants from above that, curious half spherical balls of an Alsinea [2] growing in Thibet at 18,000 feet, like our old friend Bolax.[3]

Indeed the Himalayan heights were full of new marvels.

Donkiah is a wonderful place; 19,200 feet is the altitude of the Pass, and plants to 200 feet of top, Lichens to all but 20,000 feet. Wait till you see my colored sketch of Thibet. Jorgensen's works are moonshine to mine.

[1] This plant is described as follows in Hooker's Rhododendron Book:

RHODODENDRON NIVALE, Hook. fil.

Snow Rhododendron.

The hard woody branches of this curious little species, as thick as a goose-quill, straggle along the ground for a foot or two, presenting brown tufts of vegetation where not half a dozen other plants can exist. The branches are densely interwoven, very harsh and woody, wholly depressed; whence the shrub, spreading horizontally, and barely raised two inches above the soil, becomes eminently typical of the arid stern climate it inhabits. The latest to bloom and earliest to mature its seeds, by far the smallest in foliage, and proportionately largest in flower, most lepidote in vesture, humble in stature, rigid in texture, deformed in habit, yet the most odoriferous, it may be recognised, even in the herbarium, as the production of the loftiest elevation of the surface of the globe,—of the most excessive climate,—of the joint influences of a scorching sun by day, and the keenest frost at night,—of the greatest drought followed in a few hours by a saturated atmosphere,—of the balmiest calm alternating with the whirlwind of the Alps. During genial weather, when the sun heats the soil to 150°, its perfumed foliage scents the air; whilst to snow-storm and frost it is insensible, blooming through all, expanding its little purple flowers to the day, and only closing them to wither after fertilization has taken place. As the life of a moth may be indefinitely prolonged whilst its duties are unfulfilled, so the flower of this little mountaineer will remain open through days of fog and sleet, till a mild day facilitates the detachment of the pollen and fecundation of the ovarium. This process is almost wholly the effect of the winds; for though humble-bees and the 'Blues' and 'Fritillaries' (Polyommatus and Argynnis) amongst butterflies do exist at the same prodigious elevation, they are too few in number to influence the operations of vegetable life.

The odour of the plant much resembles that of 'Eau de Cologne.' Lepidote *scales* generally rather a bright ferruginous-brown, wholly concealing the ramuli, foliage, &c. *Leaves* one-eighth to one-sixth of an inch long, pale green. *Corolla* one-third of an inch across the lobes. The nearest allies of this species are *R. setosum* and *R. Lapponicum*, from which latter it differs in its smaller stature and solitary sessile flowers.

This singular little plant attains a loftier elevation, I believe, than any other shrub in the world.

[2] *Arenaria rupifraga*, Fenzl.

[3] *Bolax glebaria*, the Tussock grass of the Falklands.

The Rhododendrons by themselves claimed separate notice. The first part of the new book,[1] drawn up by his father from material sent home by him, had just arrived, following the eulogistic reviews, so eulogistic that they aroused Hooker's mistrust as well as his curiosity.

> To return to our *Rhododendrons:* I have further completed and copied out all the descriptions! together with a catalogue raisonné of the Indian ones known to me. It took me fifteen days' hard work, which I did most grievously grudge, and thought worse than my captivity, and assure you it needed all the stimulus of seeing, for the first time, the Book itself, to keep me on to the weary hackneyed Rhododendrons. As to the said book, it is above all notice from the like of me. The plate of *R. argenteum* likes me best; and that is not I think to be surpassed for drawing, perspective, colouring and portraiture, by Bauer's *Banksia*. It is a far grander and better book that even I expected, after all its panegyrics; and I am most heartily obliged to you for giving me the lion's share of the honors, which should by rights be as much your own as is the Victoria book.[2]

And he tells his mother of an appreciation from 'perfect strangers' which he confessed was very gratifying.

> All the Indian world is in love with my Rhododendron book, and extracts from my Tonglo journal, which I sent to the *Asiatic Society Journal*, have been praised in all the public papers. (August 8, 1849.)

The map of his travels was another labour to complete.

> I am so busy with my plants that I grudge working at the Map, and yet it *must be done*, whilst the materials and references to my note-books are fresh in my mind.

> January 23.

> Hodgson had got a map partly ready to send by this mail, but it is so very foul that both Thomson and myself

[1] *The Rhododendrons of Sikkim-Himalaya.* (Edited by W. J. Hooker.) 1849–51, 14 × 7, pp. 30, pl. with descriptive text. Fol.

[2] *Description of 'Victoria Regia,' Lindl., or Great Water Lily,* by Sir Wm. Hooker, 1837. M. D'Orbigny claims that he was the first to gather specimens in 1828 in the Province of Corrientes, in a tributary of the Rio de la Plata. Poeppig called it *Euryale Amazonia*, 1832.

think it better to retain it at present, and I will do my best to get one ready for February post, but really these are works of no ordinary labour, and I do dislike doing things in an inferior manner to what I can do them.

By the 30th, however,

I had just finished for you an excellent large map of my wanderings, but have thought it proper to give it to Genl. Young, who was all abroad as to how to dispose of the troops now marching into Sikkim.

July 18.

My map of Sikkim has been copied at the Surveyor General's office. Thuillier is greatly pleased with it. I have given it to the Govt. as they wished it, but Thuillier sends you overland either the original or a facsimile. Lord D. sent for it and expressed himself most kindly and flatteringly; it is the first and last of my performances in that line. As a topographical map I hope it will do me credit, it is as full as I could make it with accuracy, and I have the materials for working the elevations of 5 or 600 places over the surface, as also full ones for making it geological, botanical, and meteorological from the plains to 19,000 feet of elevation in one direction, and to 16,000 along the Northern, N.E. and N.W. frontiers.

After all this was not the last of the Indian map-making; in November he made a map of the Khasia Hills, which he visited during the autumn, and this

I finished this morning (Nov. 26, 1850); a very poor affair it looks. Thuillier will send it you with a copy, after he has copied it for insertion into the General Atlas of India. That finishes my *survey work*, I am glad to say. The work has cost me great time and labour, but I do not admit that it was so much time taken from Natural History, for I have had plenty of that too, as much as I could well put up with.

His experiences, shown graphically in the map, revolutionised current theories about the geography of the Himalayas, in which the veteran Humboldt was so deeply interested.

> I no longer regard the Himalaya as a continuous snowy chain of mountains; but as the snowed spurs of far higher unsnowed land behind, which higher land is protected from the snow by the Peaks on the spurs, which run South from it.
>
> It is singular that Thomson and I have, independently, arrived at precisely the same novel conclusions as to the great features of the Himalayan Range—its Glaciers, Geological structure and Epochs, Snow Line, &c.

For the rest of his travels in India, there were to be no more long months of solitary journeying. 'T. Thomson is with me at last,' he cries joyfully on returning from captivity. Thomson, like himself, was the son of a Glasgow professor; they had been fellow students. He too had travelled in Tibet, and had been a prisoner amongst Asiatics—one of the Ghazni prisoners in 1842. 'He parted from me in 1839, when we quitted England respectively for India and the Antarctic Ocean, and he was the first to greet me on my arrival in Darjiling.' He had fallen ill on his way to join his friend, for six weeks, but now, by the end of January,

> he has so wonderfully recovered that we walked together from Khasing to Darjiling, 25 miles, in 6 hours, uphill 3000 feet. Still he looks thin, grey and *very old*. . . . I cannot return the compliment when he assures me that I look fatter and younger than I did ten or eleven years ago (in 1839); for he is grown extremely like his father, and has literally quite as many white as black hairs upon his head.

Hooker's praises of him as 'a most pleasant companion, very clever, (he always was,) and generous too, devotedly fond of Botany and a famously hard worker, a regular Planchon for acuteness, but with twice the steadiness of character and none of the little Frenchman's crotchets,' culminate in the description 'the most valuable friend, certainly, I ever formed.' Their vast collections they proposed to work out together, when they returned to England; but even thus early a more ambitious scheme floated before them, and Hooker urges his father to engage a certain well-trained assistant for them, 'especially if you project a Flora Indica, at which Tom pricks up his ears with a will.'

To Hooker it was a great relief that the Borneo project had fallen through, after the death of Lord Auckland, who had arranged it. Apart from escaping that very unhealthy climate, there was a great advantage in having opportunity to complete a knowledge of Indian botany, albeit with emptier pockets. Thomson, however, to join in the expedition, had to sacrifice a year of his long-looked-for furlough; certain departmental friction was too strong to be overcome, and neither his recent illness, nor his scientific work, past or prospective, availed to let him count this period as Indian service.

The trouble in Sikkim at the first blush seemed fatal to the prospect of future travel so near as Nepaul. But good feeling was undisturbed. 'The Nepalese are so fond of Campbell and me that they even offered to come and rescue us from the Sikkimites' (January 2), and Lord Dalhousie continued to think the expedition feasible and did his utmost to bring it about. Jung Bahadur was passing through Calcutta on his way to pay an official visit to England. A meeting was arranged, and in the middle of March Hooker joined him and the Governor-General in hopes of receiving permission to start as soon as the weather served, in April. But though very friendly, Jung Bahadur was unwilling that Europeans should travel in Nepaul whilst he was absent and unable to protect them. Next year, certainly, on his return, but not this. Hooker, however, was unwilling for various reasons to stay out another year, though

> Lord Dalhousie entreated me, the last thing before we separated, *not* to give up the project . . . even offered me a companion, but I refused, saying that I would not choose to go with any one of whom I knew less than of Thomson.

Accordingly the alternative was adopted, of a journey to Assam and the Khasia Hills. As to Bhotan, 'I would not go there for the world, without 500 men in front of me and as many in the rear.' . . . As between Nepaul and the Khasia Hills, the botany of the former could not be very different from that of Sikkim, its chief interest being in its botanical geography and the éclat attending the traveller who traversed it from end to end; as to the latter, 'doubtless its vegetation is richer, though not so novel as that of Nepaul,' for it had been visited several times. (March 18, 1850.)

But the balance is finally struck with fair contentment (April 27) when a new actinometer and telescope had arrived and been tried.

> I could wish they were going with me to Nepaul instead of the Khasia Mountains! Still I really believe the latter country is the best in a botanical point of view both for my companion and myself, and it is certainly *far* the most practicable.

And the journey justified itself. He writes on August 8th:

> I have here the means of making extraordinary collections: had I remained in Sikkim, the same expedition would have procured no more plants.

Accordingly in mid-April he returned the weary five-days' journey from Calcutta to Darjiling, to make ready for the start, having

> been so much out and about Calcutta that I am very sick and weary of it. Greater kindness no man could receive than I have, but it is a killing sort of kindness that requires the compression into fourteen days of the good feeling of all Calcutta,

of which he says in a 'sadly idle gossiping letter' of April 6 to his mother:

> On the whole the society is more entirely agreeable than any I have ever mixed in. There is very little personal feeling shown, and there is much more real friendliness and kindness amongst the people than in your starched circles at Kew, where one feels far more *patronized* than shown attention to for your own sake, or from any desire of cultivating an acquaintance. Hospitality is here literally a ruling passion, and I am sure that I know twenty houses

in Calcutta, into which I should go unasked and be sure of a hearty welcome; indeed I may say I have been asked to be the guest of more than that number of families.

A few thumbnail sketches of character, mostly Indian, may be added from the letters of these days; the last, with its note of self-reproach for too easy condemnation of unobservant stupidity, is especially noteworthy.

I see by the newspapers that —— was married. I sincerely congratulate his family upon it; he is now provided for, and he had not talents for a profession, interest for a sinecure, nor industry enough for anything. I pitied him for his circumstances as much as I liked his really amiable disposition.

Mr. X. was a civilian and known as '*Jemmy Blague*,' the greatest liar in all India. His brother, Col. X., inherits the title, and says of himself that he killed so many Beloochees at Meanee, that Sir C. Napier had to stop him and took away his sword, when the gallant Colonel doubled his heroic exploits *with the scabbard!*

I have begun to like Capt. Y. in spite of his want of sense. He is a truly kind-hearted fellow and neither captious nor vain. When walking with me the other day, he mentioned that during three years of his childhood he had been stone blind. I was very much struck with this, and I felt ashamed of the harshness with which I had spoken of him. True I never dreamed that what I said would ever come to his ears; perhaps, too, if he had enjoyed the use of four eyes all that time he might not have profited by them; still, we really know very little of what we are doing when we pass harsh judgments upon others and condemn their conduct, and I felt tacitly rebuked for my want of charity.

CHAPTER XVII

TO THE KHASIA MOUNTAINS

THIS, the fourth and concluding expedition, lasted nine months. A start was made on May 1, 1850. A six weeks' boat journey took them down the (northern) Mahanuddy, an affluent of the Ganges, across the great delta at the head of the Bay of Bengal, then past Dacca, and by the course of the Ganges, the Bramahputra, and the Soormah to Chattuc and Punduah at the foot of the Khasia Hills. From this point first elephants and then an army of 110 coolies conveyed the travellers and their belongings to Churra, on the mountain tableland, on June 12. On September 13 they left for the eastern part of the plateau, and on November 17, having made an exhaustive collection, including 2500 species of plants and 300 kinds of woods, descended to Cachar, beyond Silhet. Lack of time and tribal warfare prevented entrance into the botanically unexplored valley of Manipur.

Cachar was left on December 2 for Silhet, where four days were spent, and Chattuc, whence on the 9th a fortnight of boat brought them to Chittagong. Here a botanical excursion was made to the north, and plants were collected apparently unknown since Roxburgh's time. But the higher hills were inaccessible, for the head-hunters were very active, and had taken thirty heads from one Bengali village the week before. Setting out again by boat on January 16 they reached Calcutta on the 28th, and leaving on February 7, arrived once more in England on March 25, 1851.

The first part of this eastward journey, what with the badness of the boats and excessive indolence of the crews, a thermometer ranging between 95° and 106°, and scenery destitute of all interest, was 'miserably slow and very uncomfortable to boot.'

We have been alternately winding through narrow channels, tossing in vast river beds, bumping on sandbanks, or lying, moored to cliffs of sand and mud, waiting for fair weather or calms. Scarcely a tree has been visible for days, and then came wretched cottages, accompanied by clumps of Mango and ghostly Palms. . . . The desertion of our crew compelled us to put into Pubnah for a few hours. You will scarcely believe it, but these people are so lazy and capricious, that our Headman and the crew actually *ran* away from their own boat, (a large covered luggage craft, 80 feet long) leaving it to be the property of nobody, (i.e. *our* property if we chose); so we had to hire other men at Pubnah, and brought it on to Dacca.

A fresh picture appears with the city of Dacca.

The dwellings of the English residents are truly magnificent, as much so as at Calcutta, with richer gardens and more beautiful prospects. The streets are open and clean, and this is literally the first Indian town I have seen where you can drive along the public ways without grievous offence to the nose. [The narrow-fronted native cottages of mud or plaited matting, running back fifty feet from the street, with their eaves dipping nearly to the ground at the corners, looked all roof.] In these hovels the famous Dacca muslins used to be worked: they were wonderful fabrics, of which they say that you could not see them when outstretched on the dewy grass, nor distinguish them from gossamer, when floating in the air. Aurungzebe reprimanded his daughter for appearing *en déshabille,* when she was really wreathed from chin to toes in one hundred yards of muslin. The manufacture has long been given up, or nearly so, but now there is a fitful revival, owing to the order given for the Grand Exhibition of 1851. For this, Dr. Wise [1] is collecting the article, materials and implements: the latter are

[1] Thomas Alexander Wise (*d.* 1889), appointed Assistant Surgeon in Bengal 1827, founded Hugli College and was its first Principal, doubling the work with the Civil Surgeoncy of Hugli 1836–9. Appointed Secretary to the Council of Education, afterwards Principal of Dakka College, retiring in 1851.

so simple that he justly remarks that it would require two natives to accompany them, in order that they should afford any degree of instruction to the public.

Dacca, which now has been restored to the position of a provincial capital, in 1850 presented 'the aspect of a tolerably well preserved and most extensive ruin,' still richly adorned, for

> all the houses are, or were, white-washed and stuccoed, much decorated, even the humblest; columns, friezes and arabesqued pediments, often extremely pretty, are everywhere seen; their ornaments strangely recalling what upholsterers and architects term Byzantine at home. I took for granted that this style was introduced by the Mussulman conquerors from the West; for Dacca rose to glory under Aurungzebe; but I am afraid that it is all borrowed from the ancient Hindoo Capital of Eastern Bengal, of which but a single street remains, twelve miles distant, and now buried in jungle. Certainly, I have neither met nor read of anything like it in India, for here there are none of the ugly variously constructed pillars, nor those of bulging form, or twisted like a rope yarn, which to my untrained eyes, seem typical of Hindu architecture. Nor are you offended with the gaudy colours, Peacocks, Elephants and vile deformities which appear on the friezes, capitals and every part of the Hindu temples. Grotesque figures are rare, and the running patterns and scrolls are elegant and quite similar in general character (so far as I can judge) to the Greek. The ruins of the more strictly Mohammedan buildings—Mosques and Tombs—are picturesque, and the damper climate does not accelerate their falling to dust, as in Western Bengal. Grass and climbers quickly bind and conceal the heaps of rubbish; while shrubs and Ferns spring from the shattered walls.

The Khasia mountains presented a great contrast to the Himalayas in other respects as well as in their small elevation of some 6000 feet. The long table-topped ranges were very precipitous, with roaring cataracts pouring over their scarped flanks, which rose from the plains like walls, the valleys receding in amphitheatres of cliffs. On the ascent from Punduah, the scenery was splendid, far more beautiful than any part of the Himalaya, and much more Brazilian in character; with groves of Areca Palm, fine rocks and a better mixture of brushwood and large trees than the complete forest of the Sikkim Himalaya presents. The vegetation was quite different, everything new to us.

The outstanding features were the heaviness of the rainfall and the abundance of plant life. At the very start they filled nearly a ream with the plants collected on their walk from Punduah before breakfast.

To his Father

June 21, 1850.

Scattered *Pandani* and the wonderful Stonehenge-like tombs of the natives are the arresting objects of the view; the former quite out of the places with which we associate their presence; the latter singularly in harmony with the moorland scene, whether as recalling the Druidical remains, or the erratic boulders of our own bleak open counties in England and Scotland, they are wild uncouth objects.

Of the weather 'most horrible' is the term, I believe, for all the time between May and October; we are considered to be singularly fortunate in getting out to any distance for seven days out of ten—for the first *three* it rained a deluge; and then the said clearance commenced, which is unprecedented in the memory of the oldest inhabitant.

Thick fog and torrents are the prevailing character, the rainfall equalling often in forty-eight hours the *whole annual English fall.* The statements are incredible, and I have set up my rain-gauges to see for myself; it is windy too, which is bad for me, as the rain gets on my spectacles and stops work; the damp is of course ruinous.

With half a dozen collectors at work and three good coal-fires burning in the drying bungalow, it was very hard work for Hoffman (his servant) and six men to get the plants changed and papers dried daily.

I had no idea [he continues] of the richness and variety of the Flora, nor can *you* ever have of the bulk of tropical plants, which, as I always say, puts the ordinary vasculum *hors de combat* in an hour; as to your notions of drying paper—80 lbs. is not a great collection for one day, and one and a half ream of paper to put them in (Tom adds 'at the very least'). Compared to the 4500 feet of Sikkim Himalaya (to which these mountains botanically answer), the latter is literally a poor botanising country; but again we have here no region like the 5–10,000 feet of Sikkim, nor of the Arctic of 10–17.000 feet.

Our collections, including those of this morning, amount to 1176 species, gathered since leaving Dacca; of which 800 were gathered since we quitted Punduah—this excludes all the species we found in these hills which we had gathered in the plains, and a great mass of un-numbered things out of flower. I am safe in saying that 1000 species might be gathered within five miles of Churra in a week.

Hodgsonia is in fruit and quite a different plant from the Sikkim one; so it is well you have stopped its premature début, as the confusion of plants and plates of Roxburgh's and mine would have been a terrible business. I have a fine fruit in spirits for you; it is not ribbed, and differently shaped.

Despite the rains and the limitation of local supplies, both friends kept well and hearty.

July 20, 1850.

Tommy Thomson and I get on capitally together—and pray tell Aunt Harriet, with my love, that he can still 'eat through anything' as well as your well-appetised son. We are getting on very comfortably here. Mrs. Inglis, of Churra, sends us *every day*, by the post which goes on to Assam, a tin with a fresh loaf of bread, pat of butter, and a muffin! We get plenty of fowls and eggs, and occasionally vegetables, but little or no milk: for these savages, the 'Khassya' people, though they keep cows, have a prejudice (*not religious*) against milk! I think this is almost a unique feature in the human race. We are extremely busy, as you may suppose, more so than we ever were before, and are making *enormous collections* of plants, but have much less time than we could desire for the microscope and examination, still less for drawing and none for other pursuits. The climate is cool and excellent; the thermometer is hardly ever up to 80°, or falls below 68°, at midsummer.

Darjeeling cannot compare with Churra—500 inches and more (i.e. upwards of 40 *feet*) of rain fell last year at Churra. I do not doubt that it is the rainiest climate in the world.

Nunklow: July 11, 1850.

Here Tom and I have arrived at our furthest North from Churra, all beyond this being very unhealthy. It is very tantalising to be stuck up here, literally within one day's horse ride of Jenkins,[1] whose dwelling at Gowhatty we can almost see; but the intervening Terai is deadly at this season. I have written to ask if he can send me an Assam native of tolerable cunning who will get me the *Palms* and *Bamboos* from the Terai. I have already thirteen species of *Bamboo* from Churra and ten from Sikkim: I believe those of the two countries to be perfectly different. Unfortunately they never flower, and I am determined with Tom's help, and by obtaining gigantic specimens, to describe them by habit, leaf, etc.

August 23, 1850.

What with Jenkins' and Simon's collectors here, twenty or thirty of Falconer's, Lobb's,[2] my friends Raban and Cave and Inglis' friends, the roads here are becoming stripped like the Penang jungles, and I assure you for miles it sometimes looks as if a gale had strewed the road with rotten branches and Orchideae. Falconer's men sent down 1000 baskets the other day, and assuming 150 at the outside as the number of species *worth cultivating*, it stands to reason that your stoves in England will still be stocked. The only chance of novelty is in the deadly jungles of Assam, Jyntea, and the Garrows. I am therefore not spending my money on Orchideae collecting but rather on Palms, Scitamineae, &c., which are more difficult to procure and not sought after by these plunderers. Oaks I will attend to, but they are most troublesome, as not one in a thousand is worth anything.

[1] Col. F. Jenkins (*fl.* 1833, *d.* before 1884) became Major-Gen. H.E.I.C.S. and Commissioner of Assam, the botany of which he investigated. He sent large collections of Assam plants to the Natural History Society of Cornwall. *Jenkinsia Acrostichum* was named after him.

[2] Thomas Lobb (*fl.* 1847) was a botanical collector for Veitch in India and Malaya. The genus *Lobbia* Planch. was named after him.

The vast extent of the collections and the amount of labour to be expended upon them at home appears from the following:

> Thomson's collections went home in April by the 'Wellington' in 28 boxes, directed to the India House. One box contains his books; he gave the whole collection to the India House,' being unable to pay the carriage of his own private ones, formed previous to the Thibet mission, *to Calcutta*. If Government do not do something, nothing can come of either Tom's or my collections; they cannot even be housed without. The collection you will receive (I hope *have* received) per 'Queen' will form at the outside one quarter of the bulk of what I shall have, and we are now packing in much larger paper layer over layer of plants to suffocation. How Bentham would storm, I often think, but we can neither afford paper, nor room, nor carriage. Luckily they are beautifully dried and all large specimens, but the separation will require great space, time, and unremitted labor.
>
> We left the hills on the 10th, and I had the pleasure of seeing all stowed safe away in a large boat hired to send all to Falconer's from Punduah. The dried plants in 70 bales are camphored and put up like bales of cotton in gunny[1] tight and dry. I could get no boxes. The woods, Palms, Bamboos, &c., are similarly put up, but, being very large, some 10 feet, they got a ducking going down the hill on men's backs. I hope none are injured and they had all dried when I followed them. Seven Ward's cases are full of Palms, Pines, a few Oaks and Larch, Nepenthes, &c. The Palms look splendidly; amongst them a new species of *Wallichia*, 20 feet high. There are also boxes with smaller things and bottles with fruits and flowers of more than 800 species of plants in spirits.
>
> As to the Calami and Bamboos, I ticketed them, wrapping the tickets up in folds of paper, but I doubt their surviving; and I do not see how they can be made available for the museum, except by Thomson or myself. The same may be said of the woods, tree-ferns, &c., which can only be worked up with the herbarium, and that will be a work of great time and trouble. I wish very much that the Government

[1] A coarse material used for sacking, made from jute fibre.

would give me a house at Kew for the collections, and a small salary to engage my working them up for the museum and public, and leave me to get a publisher who would illustrate, and over whom I should have some hold by having the offer of my Journal. I should greatly prefer this to having a grant for publication made to me. I shall never write well for profit, and would willingly give all my materials, scientific and popular, to the publisher, seeking no profit, but exercising a control over the amount of 'illustration' to be given to both Natural History and Journal.

Friends at home were more than ever eager that the vast results of the Indian labours should not be thrown away. Dr. Wallich offered his help if Hooker and Thomson should take up a Flora of India, joining with them as a preliminary in revising the Indian Herbarium at the Linnean. Hooker, in reply (June 12, 1850), tells of the hard measure meted out by the E.I.C. to Thomson, though he had lost all his splendid outfit and collections in the Cabul campaign, and of his own slender prospects from the Admiralty and the Geological Survey, the latter 'involving work he will not undertake again for the price,' though he hoped for some readjustment for the sake of a position he much liked.

To Dr. Wallich

June 12, 1850.

Other expectations I have none but a wife to maintain, and expensive appearances to keep up.

As to writing a book of travels, or working up my Geology, Physical Geography, or Meteorology—I have no thoughts of it.

Wealth I do not seek; but it is absolutely necessary that I be placed in unembarrassed circumstances to carry out the Fl. Ant. and Flora Indica; it were not expedient that I should have even the Geological Survey Work.

Reputation is a very fine thing, and Botany a very charming science, but neither will keep the pot boiling in that land of constraint and restraint—England—where my prospects are distraint for window-tax and poor-rates, if the Woods and Forests will not give me a barn at Kew.

My £400 here is, with prudence, equal to £800 in England, it has been more than that to me, but this year my expenses will be very great, nearly tripled. Had I my life to live over again, it should be in India—that, however, is not the question. I am homeward bound this cold weather to slap my empty pockets up and down Piccadilly, and sponge upon my friends at the Oriental for a dinner, since you *inhospites*, Athenæum, will not lay a plate for a stranger.

So here, my dear Wallich, is a good growl for you, after which I feel better, but not the less of a mule.

Thomson is the most good-tempered and -humoured fellow you can imagine, and no one can be more full of zeal and love of Botany, nor more willing to work; but the Flora Indica may go to Shaitan before we tax ourselves with such a responsibility under such wretched prospects.

To his Father.

It is easy to talk of a *Flora Indica*, and Thomson and I do talk of it, to imbecility! But suppose that we even adopted the size, quality of paper, brevity of description, &c., which characterise De Candolle's *Prodromus*, and we should, even under these conditions, fill twelve such volumes, at least; though excluding any word of English or not upon distribution, particular habitats, remarks on structure or aught else. About eighteen years of fair work would be needed, for I should not approve of any portion being so slightly executed as Decaisne's *Asclepiadeae*, Choisy's *Convolvulaceae*, and Alphonse De Candolle's various orders; and I further think that the plan of distribution is carried to excess. Our friend, Mr. Bentham, is truly the only first-rate Monographer of the present day. If therefore Thomson and I are to write a *Flora Indica*, we ought, I think, to be considered competent to do it all, or nearly all, except the *Cryptogamia*. That the East India Company will not come forward with money to aid the publication, you may rest perfectly assured. It may give Thomson military allowance, and he will be well content with that. It may also take copies, and by so doing, first raise up a Publisher, and then ruin him by distributing gratis copies to those who would, otherwise, be purchasers.

J. D. Hooker, at the age of 32.
From a Sketch by William Tayler, 1849.

i. 340]

Our Government may assist by granting me a small salary, or connecting me with Kew, so that I may have leisure to work, and thus it may stop my clamorous mouth; but neither our Government nor the E. India Company will give a sum, in any way proportioned to the work. What would a thousand pounds be, for a job, the labour of which must stretch over fifteen years? And I trow they will never award *both a salary* to me, *and money for the work*.

The question may be simplified by merely asking what is to become of my materials, MSS., and collections, on my return? I cannot undertake their arrangement, much less their publication, unless I am settled. If it be at all practicable, I desire to push for a house and small salary, attached to the Garden, and at once, because (firstly) Mr. Aiton's is now vacant, and (secondly) because the magnitude of my collections requires to be considered and accommodated. (Thirdly) because the money might now be granted as the continuance of an allowance hitherto enjoyed by a man, already in the service of the Government, and who has done his utmost to please his employers. They surely could never cast me wholly off, on my return?—(Still, there seems on other grounds an evident leaning that way)—But it must be surely remembered that I have hitherto received *nothing* in the shape of salary, and that every shilling has been spent in collecting and on travelling expenses. I do not much relish the idea of a Government Grant towards the cost of publication. It might only leave us in the lurch, as was the case with the *Flora Antarctica*. And supposing that Fitch's services should be no longer available—what sort of a predicament should I be in *then*?

The Admiralty, as you are aware, give me a salary and a grant, and the Woods and Forests, or whatever body may employ me, cannot (I should hope) do less. A salary would be far better for me than a grant as enabling me to work up my Journals; they cannot otherwise be given to the world. For such books as the work on *Rhododendrons* and its continuation, I shall grudge neither the plates nor the little trouble requisite to draw up the descriptions. But when such work is involved as the laborious publication of my Journals, of a systematic botanical work,—or of the scientific results of various kinds, arising from my travels, I must find myself placed, at least, in independence, before I can even begin. I already feel something of the Burchell[1] spirit, and nobody need be surprised if (the necessary and just stimulus being withheld) I should lapse into such a condition as his—so far as my collections and materials for publication are concerned.

[1] William John Burchell (1782–1863), the great explorer in S. Africa (1810–15) and Brazil (1826–9), published only a part of his S. African results. On his return with yet richer material from Brazil, the Prussian Government, it is said, offered him a handsome pension if he would settle with his collections in Berlin. This he refused, but his hopes of getting them published in England were bitterly disappointed.

CHAPTER XVIII

THE RETURN FROM INDIA

THE end of the Indian journey brought up the same problem as had arisen at the end of the Antarctic journey. What was the next step to be, and what arrangements could be made for the publication of the scientific results by the Government who had sent out the expedition? Government help, he held, might be given to working out research, but not to the endowment of researchers as such. As he puts it to his mother (August 8, 1849):

> Mr. S. is very clever, but one wants hard-headed, working men now-a-days, and Government pay should be doled out according to the amount of national profit, pleasure or advantage yielded by the science to the Public in general, and not to physiologists in particular, or philosophers. You need not apply this to me. I offer no excuse for myself and court no favour.

Hooker had always thought it proper to complete in India, apart from the voyage out or home, the three years for which his grants were allowed. That the last year was to be spent in India instead of in Borneo was in every respect good for him save as regards finance. If he was left to pay for his passage home the arrangement did not err on the side of liberality. He still received £300 from the Woods and Forests instead of the £400 for the two preceding years, but lost his full naval pay (£200), time of service and naval allowances, together with the free passage home which, under the Borneo scheme, would have been his with the rank of surgeon on board the *Maeander*.

Feeling that he could manage on his allowance, he had refused while in Sikkim to apply to the Indian Government for any grant in aid of his costly and laborious expedition to the snows, or to allow Hodgson to appeal on his behalf; but Campbell, before a similar disclaimer could reach him, had made representations to the Government, who generously granted him £100 to cover the cost of feeding his coolies, subject to the approval of the East India Company. However unwilling to ask, he was much gratified in accepting the proffered grant, and was free to spend the more on his collections and on scientific instruments. His total expenditure was £2,200; the official allowances were £1,200: the remainder was contributed from his own and his father's purse.

As for a permanency in the future, he had no wish to take up such a post as the directorship of the Botanical Gardens in Ceylon, offered to him on the death of his friend Gardner in 1849. Indeed his constant wish was to be settled at Kew. His father was short-handed. His former curator, Dr. J. E. Planchon,[1] had left him suddenly: 'Citoyen Planchon' or simply 'the Citoyen' as he was playfully nicknamed, Planchon of whom Hooker writes home amid the Revolutionary breezes of 1848:

> I hope Planchon won't be going to Paris now! He will be drawn (for a soldier) and quartered (not in Barracks), if he does not take better care. I doubt if the Republicans are so civil as were Napoleon's soldiers, who, at the battle of the Pyramids, gave the word, 'au milieu, les femmes, les ânes, et les savants.'

The little man, to whom the Hookers were much attached, was a paragon of botanical acumen, winning a second nickname from the 'ça touche' with which he invariably clinched a botanical argument; it was the highest praise to call T.

[1] He afterwards attained great eminence as Professor of Botany in Montpellier, where in his researches on the *Phylloxera* he discovered the only cure for this pest—namely, the grafting of the ordinary vine on the nearly immune stocks of American species.

Thomson 'a regular Planchon for acuteness.' But, with all his cleverness, he was, it seems, flighty and unstable, and he had unaccountably broken with Kew and the friends to whom he had expressed such gratitude and devotion.

> An assistant [writes Hooker to his father on Feb. 1, 1849] is now your chief requisite, and I wish I were at home to help you in this and other matters. It is the only drawback to my thorough enjoyment during my journeyings, that you should miss me in some cases where two pairs of eyes and hands, nay *two* whole heads and bodies are wanted.

And he urges his father to engage at once a man who seems suitable, using his Navy half-pay to secure him rather than lose the chance of an *honest*, careful, industrious man.

Soon after his return to England a long standing anomaly in Sir William's position at Kew was remedied. Though Director he had had no official residence in the grounds; the great herbarium, which was one of the scientific mainstays of the Gardens, was his private property. He had brought it with him from Glasgow; it was the one valuable inheritance he could leave to his son, and at his death was liable to be removed or dispersed if that son had not the means of maintaining it at Kew or elsewhere. Until 1853, for all its public utility, it was housed and maintained at his private expense in the 'three-storied, many-roomed' house at West Park, two-thirds of a mile from the Gardens,[1] 'a very pretty, genteel and comfortable residence' (in Dawson Turner's Johnsonian phrase), which, exclaims Hooker, 'has always been an incubus to me, so large in itself, while still your collections and Herbarium are outgrowing it!' These, with the study and artist's room, occupied thirteen rooms, while Sir William's expenses all along far exceeded his official salary.

At length, however, a change came about. First, the house in the Gardens belonging to Aiton, the late superintendent, fell vacant at his death in October 1849. It now

[1] The grounds of West Park, 7½ acres in extent, are now occupied by the Kew and Richmond Sewage Works.

was offered as the semi-official residence of Sir William, the thrifty Government, however, proposing to charge him £100 per annum as interest on the capital cost of the new Herbarium accommodation.

> All my Indian friends lift up their hands with amazement at it. . . . But it is an immense advantage that the Government can have it to declare that you put them to no expense, but that, on the contrary, you give them what interest they choose on their money.
>
> For some reasons [he writes home to his father, February 28, 1850] I shall regret West Park, a very pretty and nice place; and most of all I shall regret leaving it on poor Mamma's account, who will lose her pets of cows, poultry and pigs. Bessy will miss the garden, and I the wall fruit and the long gravel walk, which I have always cherished the memory of, for dear old grandpapa Hooker's sake. But really I never could endure the big house, without servants enough to answer the bells punctually, and in the rooms of which it was impossible that a dozen persons could be collected together with comfort. . . . I must add to the catalogue, the difficulty of getting to town from West Park, of sending to hire a Fly, or that perpetual trial to my temper, the waiting an hour for an omnibus, or the missing it (perhaps both), and in the rain, may be! The weary walk from our house to church, all in the mud, for Mamma, the want of any neighbour who can come and spend an evening hour with my sister, and my own midnight trudges from the omnibus, perhaps from Hammersmith, in case of my own staying at all late in town.

The plan dropped, till in 1855 another Crown house fell vacant by the death of Sir George Quentin, Riding-master to the family of George III. This became the official residence of the Director. It faced the Green and had its back in the Gardens. But it could not accommodate Library or Herbarium. Fortunately another large house close by was now available.

This was a house which had been purchased by George III. in 1818, at Banks' suggestion, to provide for a Herbarium and Library to be attached to the Royal Botanic Gardens. One of the rooms was already shelved for books. But the death of both in the same year cut short the project, and in 1823 George IV. sold house and grounds to the nation. In 1830 William IV. granted it to the Duchess of Cumberland for her life. From 1837, when the Duke succeeded to the Throne of Hanover, it was known as 'The King of Hanover's House.'

Now it reverted to Banks' purpose. Herbarium and library were placed here, and formally made accessible to botanists, while the Government assumed the cost of maintenance and provided a scientific curator.

True that from Sir William's first days at Glasgow, his botanical treasures 'had been open to botanists, as was its owner's hospitable table to visitors from a distance.' Nevertheless to the condition first proposed that the Herbarium should be thrown open to the public, while its owner paid for its upkeep and a curator, the younger Hooker demurred; Sir William had already halved his income by leaving Glasgow for Kew, and such a step meant surrendering to the Crown all private rights in this valuable property without adequate return. Possession of it, moreover, was an excellent lever to use in the gradual reorganisation of Kew from a semi-private to a wholly national establishment, the official home of botany with scientific and popular interests fairly adjusted, the centre to which lesser botanical centres should be correlated with due subordination. This transition clearly could only be effected through the Hookers, father and son, who owned so much of the material and were ready to enlist their unrivalled powers under the Government in the service of science and the nation.

Two important pieces of work under Government auspices now lay immediately before Hooker. One was to complete the botany of Ross's Voyage for the Admiralty. So far he had only published the Flora Antarctica in two quarto volumes (1847), with 200 plates out of the 500 for which an official grant of £2 each was made to cover the printer's bill. There remained the Floras of New Zealand, Australia, and Tasmania, and he made the usual application for his half-pay as Naval Surgeon to support him while completing this Admiralty work. This was granted for three years. Thus a great part of his labour was unremunerative. Nearly ten years later, however, he was agreeably surprised by a grant of £500 from the Admiralty in recognition of his 'zeal, perseverance, and scientific ability in his botanical services'; quite a new feature in his relations with My Lords Commissioners. His account of this appears in a letter to Huxley (1860), who responds: 'The Admiralty affair pleases me very much. It is only right and just, but still I think you may well be gratified. Justice does not always come in so pleasant a form.'

To T. H. Huxley

DEAR H.—My vanity will not stand the holding this back from you. I must confess to being amazingly tickled after twenty-two years' service of *sorts*, at receiving a handsome and spontaneous expression of unqualified approbation from my Lords of the Foul Anchor. I had made the application in due form for the small arrears (of three years' pay) that was due (for nine years' work); and just by way of not throwing a chance away, in spite of my wife's laughs, I sent a crackling cartridge of foolscap with a statement of the length and breadth of my works and pay. I said nothing of quality, the Navy being the only service in which I never saw a fellow do good by praising himself. I made no grievance. I used no influence of any kind or sort or description, nor did my Father. Washington[1] immediately took the matter up and sent me a dozen queries from my Lords; I answered all categorically, some three months ago—and lo the result!

His second great task was to work out his Indian collections. They had been made for Kew under the auspices of the Woods and Forests Department, which governed Kew, and unless worked out, arranged, and housed, were, like his Zoological collections on the *Erebus*, just so much labour thrown away. To this end he desired application to be made to the Depart-

[1] John Washington (1800–63), Rear-Admiral, entered the navy in 1812 and travelled much between 1822 and 1853: Secretary of the Royal Geographical Society 1836–41. He was engaged on the East Coast Survey 1841–7, and became F.R.S. 1845, Assistant Hydrographer and Hydrographer 1855–62. He was Hooker's companion on his Syrian exploration.

ment through his father for a continuance of the £400 a year originally granted him in India, and the tenancy, at whatever rent was usually asked, of one of the Crown houses hard by, unoccupied at the moment, where he could live and keep his collections, in close touch with all the materials for reference at Kew. It was surely the duty of the Department, whose commissioned officer he had been, to see that the work commissioned should be adequately completed.

This view of the case, however, his father was at first unwilling to adopt. However great Joseph's services had been, however deserving of later furtherance, the Department, he thought, had entirely fulfilled its duty by the simple grant of the sum originally asked for the Indian expedition. Anything more must be a matter of favour, not of due. Was the Department in arrears for the amount of its last year's grant? He offered his own purse instead; and prepared to make an appeal *ad misericordiam*, much to his son's misliking.

> All this [the latter writes to Bentham, April 2, 1851] is due to his excess of modesty; it is equally certain that he looks on his own Crown salary as mere kindness on the part of Govt. to himself, and that the fact of his liking his work and being willing or able to hold his post at half pay, would justify the Crown in cutting it down so much, should they wish to be just rather than liberal as they are in his opinion to himself.

Indeed it was rather a question of *himself* wanting aid, what with his broken health, the often trying Garden duty, and the extension of the Herbarium and Museum beyond his powers, while he saw 'the great accumulation of scientific objects which are gradually being consigned to oblivion in favour of showy articles.' But this was a subject which his son could not broach to him; it must be left to older friends like Bentham or Henslow.

But Sir William consented to delay making the application till he had consulted with these old friends; and meanwhile the presidents of the various learned Societies spontaneously deputed Lord Rosse,[1] President of the Royal Society, Robert Brown, the botanist, representing the British Museum, and William Hopkins,[2] President of the Geological Society, to press the Government on a matter of so much importance to science. By the following spring, just a year after his return, these representations produced their effect. The Department authorised the grant for three years, to the end of 1854.

Meantime in September he was in the act of moving into Aiton's old house in the Gardens when very onerous conditions were sprung upon him by the authorities. Refusing to be saddled with such a burden while his footing was still uncertain, he broke off at once. Furniture and all were taken away again.

> My collections [he tells Harvey a couple of months later] were turned out neck and crop of course—the dried plants into the Temple of the Sun, and the rest into the back shed of the Orangery ! where they are going the way of all parenchyma and pleurenchyma !

He finally settled in a house, now No. 350 Kew Road, belonging to Mr. Bryan, the Vicar, where the Curator, John Smith the elder, had spent his last years. Here he brought his wife, for at the beginning of August he had married Frances Henslow. Their engagement had been a long one, but this price had been paid deliberately. His position in the botanical world had to be assured by his great travels in India. Perfect confidence and rare strength of mind were needed to resolve upon a three years' separation within a few months of their engagement. But by birth and training she was able to help in his work, to share his aims, and appreciate the worth of their joint sacrifice.

Still, even after such sacrifice and achievement, his chosen

[1] The third Earl of Rosse (1800–67), whose laborious experiments for the improvement of the reflecting telescope culminated in the great telescope at Parsonstown, first used in 1845.

[2] William Hopkins (1793–1866), mathematician and geologist, nicknamed while tutor at Peterhouse 'the Senior-wrangler maker' : a teacher of Stokes and Kelvin, Tait and Clerk-Maxwell. He applied mathematical and astronomical tests to geological reasoning. Was elected President of the Geological Society 1851, and of the British Association 1853.

career was jeopardised for a time by this same lack of prospects. If he would exchange botany for mineralogy there was a vacancy at the British Museum to apply for, with salary and house: a firm establishment and tempting at such a juncture. Friends urged him to this prudential course. 'Shall I give up Botany and stand for Koenig's [1] place at B.Mus.?' he asks Bentham (September 3, 1851), adding ironically:

> To be sure I know nothing of Crystallography, Mineralogy, Chemistry, &c., but the Trustees are above such prejudice against a man who could wear a white neckcloth with ease, and take his fair share of their abuses with equanimity, which would be an all-powerful testimonial. I hate the idea of giving up Botany, but I am advised to try for it by Gray particularly and my Father proposes it.

The wiser counsel of waiting was, as has been seen, rewarded. Nevertheless in 1854, as the period of the departmental grant for arranging the Indian collections was drawing to an end, the same perplexities revived. Writing to Asa Gray [2] on March 24, 1854, he says, 'I sometimes think seriously of giving up Kew and living in London and writing for the press.' His family was increasing (his first child was born Jan. 1853, the second June 1854); his special work engrossing and costly; his only advantages, his father's Herbarium and Library, 'which are private and for which I am in no way indebted to the Crown.' Still:

> Pray don't think I am grumbling. I have had a long spell of pleasure as a purely scientific botanist, and it is time I felt some of the ills of my position. It does make me

[1] Charles Dietrich Eberhard Koenig (1774–1851) came to England in 1800 to arrange the collections of Queen Charlotte, afterwards becoming assistant to Banks' librarian, Dryander. In 1807 he became Assistant Keeper, and six years later, Keeper of the Natural History Department in the British Museum, finally taking charge of the Mineralogical Department. This was the post left vacant by his sudden death.

[2] Asa Gray (1810–88), relinquishing medicine for botany, became Professor of Natural History at Harvard 1842–73, and succeeded Agassiz as Regent of the Smithsonian Institution 1874. He was the first in America, in conjunction with Dr. John Torrey, Professor of Botany at Princeton, to arrange species on a system of natural affinity, whence he became a strong supporter of evolution as set forth by Darwin. His association with Hooker was not only that of scientific affinity, but of close and enduring friendship.

very anxious though, and were it not that my Father would feel my leaving the place, I would hang no more on in this suspense.

And in August he writes sympathetically to Bentham, who was suffering from similar qualms :

> If I thought you would be a happier man I would advise you to give up Botany; but you would not be so, and evil as our days are, whether they mended or worsed, it would be all the worse to you to have given up what is at least a wholesome and constant mental resource. I sometimes despond too, but as I was once told, 'I am limed to the twig,' and so are you ! Besides, you have a year's work for Cambridge Herb.,[1] and it would be dull work for you to drag through that as a termination to your Bot. career.

Sir William now made definite application for Joseph's appointment as assistant to himself at the Gardens, a very needful addition to the staff carried into effect in May 1855. In the preceding December, after his failure to obtain one of the Crown houses, Joseph Hooker had moved to a more roomy house at the top of Richmond Hill, No. 3 Montague Villas ; the new appointment brought him back to a house near the gates of the Gardens lately occupied by Mr. Phillipps. His wife, he tells Bentham (July 3, 1855),

> is not best pleased about it ; but I tell her she may spend the difference in fly-hire. As for me I am blazed or blasé (or whatever you call it in French) of change, and feel curiously indifferent—it is all out of one's lifetime ;

an attitude of mind parallel to that in which he had undertaken the previous move, proposing to take the house

> at or about the last moment, but being at present under a bad attack of Phytomania I am rather indifferent to all things in general, and my prospects in particular ; it is well I should be sometimes, for I am sure I feel worried enough when it does fall on my spleen.

[1] See p. 384.

Thus at length his own and his father's highest hopes were realised. Till Sir William's death ten years later, leaving his son and assistant obviously marked out as successor to the Directorship, father and son were settled together at the Mecca of botany they had created, united by strong affection as well as by a common work.

The culminating point of Hooker's scientific work during the decade is the Introductory Essay to the Flora of Tasmania, 'which in itself would have made Hooker famous,' writes Professor Bower.[1] This was published in 1859, just before the 'Origin of Species' appeared. Six years earlier he had published the corresponding Introductory Essay to the Flora of New Zealand. The difference between the guiding conceptions of these Essays, one in the middle, the other at the end of his first great period of systematic work, is a measure of the writer's advance in scientific theory, his long-standing dissatisfaction with the older view of fixity of species finding appeasement in the practical utility of the theory that species originate in variation.

He had long been the confidant of Darwin's views; had discussed and debated them with his old friend, providing botanical information, offering criticisms, citing instances and pointing out difficulties, suggesting his own solutions to problems which had vexed him ever more insistently as he more fully realised the fluidity of species, and the difficulty of establishing 'specific types,'—those abstract definitions, to which individual specimens should be referred, being as hopeless as the bed of Procrustes. On the main lines, at least, he was approaching conviction. The new theory, privately discussed, threw light on his own work if he was not yet, in the earlier fifties, persuaded of all its details; and he felt bound to avow publicly the change of view brought about by his later investigations. But Darwin's views had not yet been concentrated and expressed as a whole. A summary of them was given to the world in the 'Origin.' The sledge-hammer effect of this was still to be experienced.

[1] The present Professor of Botany at Glasgow.

Darwin and Wallace's[1] joint communication on Natural Selection was read before the Linnean Society in July 1858; the 'Origin' was not published till November 1859. The Introductory Essay to the Flora of Tasmania, appearing between the two, did not thus early proclaim Natural Selection as a proven theory and philosophic principle, whatever effect on his trend of thought Hooker confessed the publication of the 'Origin' might produce. He frankly employed the theory as a working hypothesis to see whether it did not explain the perplexing questions of botanical affinity and distribution better than its predecessor, which he had still accepted as the working hypothesis for the New Zealand Essay. Applied to the vast material over which his mind had ranged, the hypothesis 'worked' in striking fashion. So far as plant life was concerned, the Tasmanian Essay offered in advance a strong buttress for the 'Origin,' which dealt with life in both animals and plants.

Discussion of this progress in scientific views is most profitably postponed to a Darwinian chapter. For the present it is enough to bear in mind that the species question was constantly before him; and that while working on the ordinarily accepted lines until he could see more clearly, he was ready, when fuller conviction came, to avow openly his change of attitude.

With the publication of the Flora of Australia and Tasmania (1855–60) the Botany of Ross's Voyage was completed, the New Zealand Flora having been published between 1853–55. The next important work of this decade was the beginning of his *magnum opus*, the Flora Indica. The first year after his return in March 1851, 'slightly fatter, three years younger, and much stronger than when I left England in '47,' was mainly

[1] Alfred Russel Wallace (1822–1913), the joint discoverer of the principle of Natural Selection, gave up his profession as land-surveyor and architect to travel and study nature, visiting the Amazon with Bates, 1848–52, and the Malay Archipelago, 1854–62. It was from here that he sent Darwin in 1858 the paper which was read at the Linnean with Darwin's own, and led to the speedy publication of the *Origin*. Besides his two great books of travel, his most important scientific books are those on Geographical Distribution of Animals, Tropical Nature, Island Life, and Darwinism. He received the Royal Medal of the R.S. in 1868. Keenly interested in social reform, he wrote a volume on Land Nationalisation. He wrote also against compulsory vaccination and became a strong supporter of spiritualism.

devoted to getting ready the materials for the New Zealand Flora, so as to clear the field in part at least for the Indian work. Though the last boxes of his collections arrived in September and 'astonished' his father, to be followed immediately by Thomas Thomson and his collection, numbering twenty-five chests, it was not till March 20, 1852, that he wrote to Bentham:

> I have broken bulk with the Indian collections, done all the woods (about 500), Palms, Bamboos, and big things, and am all ready to plunge into the Haystacks, working in the rooms at Kew.

Some of his Indian results had already been published by the Asiatic Society of Bengal whilst he was in India. One folio volume with fine illustrations of the Sikkim Rhododendrons, edited by Sir William from his son's notes, drawings, and materials, appeared in successive parts between 1849 and 1851.[1] Another folio, a volume of illustrations of Himalayan plants from near Darjiling, chiefly collected by him on behalf of an Indian friend, Mr. Cathcart, was edited, with descriptions by Hooker himself, in 1855.

But now Dr. Thomson settled hard by and spent a great part of the next three years at Kew, completing his 'Travels in Western Himalaya and Tibet,' published in 1852, and working side by side with his friend at their common task. His masters, the East India Company, encouraged him to work with promises of reward if the work were satisfactory, but gave no immediate help. Nor was assistance forthcoming from the British Association. The nebulous hope of bringing out a whole Flora of India, however, took solid shape when, on the death of his father in 1852, Thomson came into a little money. This he promptly devoted to science, paying for the huge volume of 581 pages which he and Hooker brought out in 1855, and hazarding repayment from 'John Company.' The detail of this, the first and only volume of their Flora Indica, was so full that if the work had been completed on the same scale, it would have reached nearly 12,000 pages.

[1] See *ante*, p. 326.

Part of the plan was to find trustworthy specialists to deal with certain Orders. Thus Hooker writes in July 1852 to Munro,[1] the soldier-botanist, the 'wonderful grass-man,' who had been arranging the grasses in the Kew Herbarium, and who was keen enough to send home a collection of plants from the Crimea in the intervals of fighting:

> Bentham has already taken to preparing the Leguminosae Indicae. We shall ourselves commence with Ranunculaceae as soon as the collections are arranged, and beat about for assistance amongst good and true friends, printing for them *at once*, offering them copies for their labour, and selections from the complete collections in order of the extent and value of their contributions. What do you say to a *Graminologia Indica* with short, terse generic and specific characters, synonyms and a summary of the Geog. distrib. of the species, to be printed, published, and distributed gratis, to a certain extent, by ourselves as 'Munro's Gram. Ind.,' giving you 50 copies, and after distributing to all deserving public and private establishments, putting the remainder into a publisher's hands to sell? Such is our present idea of proceeding. Will you kindly think the subject over and offer any suggestions, not so much with reference to your doing the Grasses, as to the general principle? Great progress might thus be made towards a Flora Indica, by the serial publication of large Nat. Ords. and groups of small do. complete in themselves. We shall be very careful how we trust the materials to authors we have not satisfactory experience of.

But its completion was a task beyond even such energetic men. Time and opportunity were too scanty. Hooker was deep in other work. Thomson was bound to return to India. Enthusiasm did its best, and he had plunged eagerly into work, lightly proposing as a side occupation to index the Kew Herbarium, to Hooker's grim amusement. He was wholly in sympathy with the views of his fellow-worker.

[1] William Munro (1818–80) saw active service in the Sikh war and the Crimea, and held the West Indian command from 1870 to 1876. During the many years his regiment was in India he studied botany, becoming the chief authority on the Grasses. He did not live to complete his general monograph of the whole order of Gramineae undertaken after his retirement.

> Thomson and I [writes Hooker to Bentham, October 10, 1852] are not at all likely to quarrel about the limits of the species, which I hold that we should do if we were improper *lumpers* quite as much as if we were hair-splitters.

But the spade work was very heavy. By November,

> we have done a vast deal to the Malayan Flora, but not nearly got through the Khassya bundles. Thomson finds the arrangement of his *own* N.W. parts, which is not yet in Nat. Ords ! a much heavier task than he dreamt of. We are working steadily on, however.

But Thomson was constantly being called away by the claims of ailing relations ; his powers of persevering concentration had been sapped by much illness in India, and at the turn of the year 1853–4, Hooker writes in despondent mood to Bentham :

> He cannot work except under the very strongest stimulus, and every advantage being put under his nose,—it was so in India, there was no inducing him to study a plant though so keen and admirable a collector. . . . As to Flora Indica, I have no idea when Part I will be out, and between Thomson's excessive scrupulosity, his natural slowness, and his matchless *procrastination,* I see very little chance of its appearance under *x* months. The consequences of working by fits and starts tell very heavily, for it requires the same work to be gone over again and again. An immense introduction is nearly written, but also so by fits and starts that Mrs. Hooker has to go it all over, and it sometimes takes an hour to unravel a page of the MS. I have taken up the distribution of my own plants in earnest, and dropped Flora Indica altogether as hopeless under present circumstances.

Nevertheless the book, as has been said, appeared in 1855. It is described in a letter to Munro, November 8, 1855 :

> The first volume of Flora Indica is finished and consists of 2 parts, 280 pages of introductory matter, and as much of description, extending from Ranunculaceae to Fumariaceae ; it cost Thomson and me the best part of two years' hard labour and will, I hope, prove useful. We have a copy for

you, and I am half inclined to send it to the Crimea [Munro was then a Major and on active service], as if you are obliged or inclined to throw it away we can give you another. Thomson paid all the expenses of printing, publishing, and distributing, and I have offered the E.I.C. to continue and conclude it, if they will only pay at the rate of £200 for every 1000 species described, and I offer to get it printed and published free of all further expense to them and of any remuneration to the authors, also I would engage myself to stick to it for ten years at that rate. Hitherto they have given Thomson no reimbursement for any of his expenses, though he spent a year beyond his furlough at it upon no pay at all.

The financial fate of the book was very disappointing. It is recorded in another letter to Munro, December 21, 1856.

I am so disheartened about Flora Indica and the knavish conduct of the Court of Directors, that I have done nothing more to it; as soon, however, as I get Fl. Tasman. off my hands I shall return to it with zest; and devise some dodge to give John Company a Roland for his Oliver. You are aware, I think, that after paying all the expenses of the 1st vol. we put a merely nominal price on the 130 copies we put out for sale (after giving away 120), and that John Company, after refusing to subscribe for copies, or promote the work, or repay the authors, on hearing how cheap it was, *bought up* 100 *copies unknown to us*, which threw the work out of print, and left us £200 out of pocket, and our object defeated! I never was so sold in my life. I have begged and implored in vain that they give back the copies, and I have offered back not only the money but to give them gratis 100 copies of the Introductory Essay. As to poor Thomson, they will not give him 1*s*. for time or labour or expenses. Have not we a good *growl*?

The political sequel of 1857 of course precluded any scheme of tit for tat. Hooker enjoyed the grim suggestion that the dissolution of the East India Company was a retribution for this meanness as well as other more serious shortcomings.

After Thomson's return to India the two friends continued to work together, and from 1858–61 published in the Journal of the Linnean Society the 'Praecursores ad Floram Indicam: being sketches of the natural families of Indian plants, with remarks on their distribution, structure, and affinities.' But with Thomson's departure and Hooker's appointment as Assistant Director at Kew, the greater work was inevitably laid aside, and remained on the shelf for fifteen years, during which his only Indian work of importance was a considerable share in preparing Thwaites'[1] Enumeration of Ceylon plants (1858–64). But in 1870, the India Council was moved to take an interest in the matter, mainly through Mr. (afterwards Sir) Mountstuart Grant Duff,[2] with whom Hooker had some correspondence the previous year on Indian Forestry and Botany. The Duke of Argyll[3] also, Secretary for India, had scientific interests. Thus Hooker obtained support when he pointed out that the Indian Government had sanctioned the much needed Flora in 1863, but workers were wanted. The matter had slipped so entirely from official ken that the India Office could not even find the record of this official letter written six years before, and had to ask Hooker for a copy of it.

T. Thomson, the natural continuator of the work, was out of health, and in any case was bent on discussing details at impracticable length. There was no help for it; Hooker met the renewed interest of the India Council by assuming the responsible editorship, and with the help of a staff of collaborators made a new start. Twenty-seven years of further

[1] George Henry Kendrick Thwaites (1811–82), beginning life as an accountant, devoted himself to entomology and botany, especially the cryptogams, wherein his microscopic discoveries were ahead of his time. Most important was his determination of the algal nature of diatoms. For thirty years (1849–79) he was in charge of the Ceylon botanical gardens at Peradenyia, publishing an 'Enumeratio Plantarum Zeylaniae' (1859–64) which won him his F.R.S. He was also responsible for the successful cultivation of cinchona and other economic plants in Ceylon from 1860 onwards.

[2] Sir Mountstuart Elphinstone Grant Duff (1827–1906) was Under Secretary of State for India 1868–74, and for the Colonies 1880–1, when he was appointed Governor of Madras 1881–6. His series of Diaries contain many literary, personal; and political reminiscences.

[3] The eighth Duke of Argyll (1823–1900) was a vigorous Liberal politician and capable administrator who ultimately broke with his party over the Irish question. Between 1868 and 1874 he was Secretary of State for India. From his earliest days he was interested in science, especially geology, in which he did some original work; but his chief activity was as a polemical upholder of ideas left stranded by the progress of science.

labour saw the completion of the Flora of British India. This, he notes with regret, was conceived on a more restricted scale. It ran to seven volumes, published between 1872–97, containing but 6000 pages of letterpress dealing with 16,000 species. In the preface Hooker describes it as a pioneer work, and necessarily incomplete. But he hopes it may

> help the phytographer to discuss problems of distribution of plants from the point of view of what is perhaps the richest, and is certainly the most varied botanical area on the surface of the globe.

To complete the history of his systematic work on Indian Botany, let me quote from Professor Bower.

> Scarcely was this great work ended when Dr. Trimen died. He left the *Ceylon Flora,* on which he had been engaged, incomplete. Three volumes were already published, but the fourth was far from finished, and the fifth hardly touched. The Ceylon Government applied to Hooker, and though he was now eighty years of age, he responded to the call. The completing volumes were issued in 1898 and 1900. This was no mere raking over afresh the materials worked already into the *Indian Flora.* For Ceylon includes a strong Malayan element in its vegetation. It has, moreover, a very large number of endemic species, and even genera. This last floristic work of Sir Joseph may be held fitly to round off his treatment of the Indian Peninsula. His last contribution to its botany was in the form of a 'Sketch of the Vegetation of the Indian Empire,' including Ceylon, Burma, and the Malay Peninsula. It was written for the *Imperial Gazetteer,* at the request of the Government of India. No one could have been so well qualified for this as the veteran who had spent more than half a century in preparation for it. It was published in 1904, and forms the natural close to the most remarkable study of a vast and varied Flora that has ever been carried through by one ruling mind.

Such was the main channel of the enterprise; but the work overflowed into many subsidiary channels. Witness Hooker's numerous contributions on Indian subjects at this period to the 'Icones Plantarum' (Sir William's series of illustrations of remarkable and interesting plants), the 'Kew Journal of Botany,' the *Gardeners' Chronicle*, and the 'Proceedings of the Linnean Society,' two of these monographs being written in collaboration with Thomson.[1]

The work finally involved the arranging and identification of their vast number of specimens so that the duplicates might be distributed among other public and private collections. The heavy burden of this task finds a constant echo in the letters of these years, the more so as it was suddenly doubled. For, to quote the obituary in the Kew Bulletin:

> Before this work had been completed the Indian collections of Falconer, Griffith, and Helfer, made over to Kew from the cellars of the East India House, had to be dealt with in the same manner. The latter task had not been completed when Thomson departed, but another smaller though very important one was successfully accomplished. Besides the three collections mentioned, the residuum of the Indian Herbarium distributed by Wallich on behalf of the Honourable East India Company was also entrusted to Kew. The distribution of this great collection took place between 1828 and 1832; there was consequently no set of its plants at Kew. In this Kew did not stand alone; the herbarium attached to the Royal Botanic Garden, Calcutta, at whose cost and for whose benefit the collection had been brought together, was in like case. By a happy chance the friends were thus enabled to fill more or less satisfactorily a great hiatus in the herbaria of both gardens; a set, fairly complete, so far at least as the plants collected by Wallich himself are concerned, was made up and laid into the herbarium at Kew, while a similar set was taken to Calcutta by Thomson (who now succeeded to the Superintendentship there).

Thus in April 1857 Bentham is told,

> I am still struggling on with the general arrangement of the Herb. Ind. roughly into species and have only got down to Monopetalae. The number of sheets and specimens is frightful. I toil on and to little effect.

[1] See list of works, Appendix B.

In May 1858 he complains to Harvey of being appallingly behindhand with his work, and in June adds:

> I am working now extremely hard at these Indian collections, of which I am utterly sick. I expect another year will see them all arranged and incorporated in Herb.—and then comes describing.

In August, three weeks' enforced absence from the work had been such a gnawing anxiety that he could not think of prolonging it, since, there being no means of warming the distributing room, it was imperative to make an end before winter, lest it should drag on and cumber all the next year. By mid-November he came to the end of all he could do that year, namely 160,000 ticketed species. 'As for myself,' he tells Bentham, 'I am in statu quo, but considerably thinner, I am told.'

This was one heavy item. Then there was the Tasmanian Flora. 'I find it tremendous work,' and again (Aug. 8, 1859),

> this luckless Essay of mine has broken my back. I had no idea of the mass of material I had accumulated for it, or the time it would take to digest it; it is not half printed, and if I leave it in the present state for 2 months, it will take me many days to begin again, if indeed I ever could work myself up to completing it after such a break. I am daily working every spare moment at it, and have still several sheets to print and some to rewrite from the rough.

Then he was planning out the Genera Plantarum with Bentham, 'which I am deeply pondering.' His father's illness and prolonged absence in the summer of 1859 threw on his shoulders an accumulation of correspondence and all the work in the Garden, with the added responsibilities of looking after the erection of the large new Conservatory. Yet when his father did return for a few days there was no relief, for

> he now likes to consult me about *everything* he does, so that when he was here I had literally more to do than when he was away!

As a last touch, he was out of his old house and not yet in his new one, where the workmen were in possession. Much of this labour he had foreseen, but he had not foreseen its cumulative effect. Accordingly (August 8, 1859):

> I write till my fingers ache, tramp the Gardens and grounds till I am foot-sore, and go to bed at night to ruminate on the little I have done in the day. My wife presses me to go and join you, but with such a prospect before me I feel it would be folly or something worse, and the 'Genera,' which I am anxious to begin as soon as the V.D.L. Flora is off hands, would then be indefinitely postponed.

Staying alone all the summer at his father's house, for he had sent his wife and children to the Henslows', he reluctantly gave up the holiday he had planned to take with Bentham.

Meantime the 'Himalayan Journals; or, Notes of a Naturalist in Bengal, the Sikkim and Nepal Himalayas, the Khasia Mountains, &c.,' were published in 1854. These two volumes, containing together more than 900 pages of incident and adventure, as well as picturesque description and the most varied scientific notes, were 'dedicated to Charles Darwin by his affectionate friend, Joseph Dalton Hooker.'

The first edition met with instant success. A second, slightly abridged, followed in the next year with less good fortune. In 1891 a one volume edition was brought out in the Minerva Library, and was reissued in 1905.

The Journals ensured their author the highest reputation as a scientific traveller. The permanent results drawn from observations in so many branches of science have already been noted. His own view of it appears from a letter of thanks to Berkeley.

> I am greatly delighted with your hearty praise of my book. I did really take so much pains with it, and have for so many years looked forward to the publication of such a book, that I keenly appreciate the favourable notice taken of it by my friends and the public. To write a book of the sort, after travels of the sort, has been the pole-star of my life from earliest childhood, and now that it is really all over

and out I feel the great climacteric passed, and look back upon life after the fashion that people are described as doing after marriage, or the birth of their first child at latest, but as I do *not* after either of these occasions. I am greatly pleased for my wife's sake too, who took infinite pains with it, and but for whom it would have been a very differently rated book I fear.

Nevertheless, working out results in so many other directions proved a heavy distraction from his prime task in Botany, and he exclaims to Bentham :

Catch me at Quizzical Geography, Geology, and Meteorology again if you can ; they have afforded me much amusement and instruction and wonderful pleasure ; for I have always felt a keen pleasure in *practical* philosophy, tools and tables of logarithms, and now that I have said my say and added my quota to the heap, I think the wisest thing I can do is to leave it for work that is more expected of me.[1]

The one fly in the ointment was the extreme parsimony of the East India Company :

I have had a fight with them [he tells Bentham in August 1855] about discount upon the Himalayan book ; which would have left me out of pocket £30 by the copies they *did me the honour* of subscribing for, and I pitched them a letter that they could not say no to, telling them that they did not behave so in another case to which they were subscribing (Gould), and they were the only subscribers I had, public or private, who asked for 15 per cent. discount on their subscription. So much for my growls.

A variety of other occupations helped to fill up these years. Preparations for the Great Exhibition of 1851 were well afoot by the time of his return to England. His services were immediately secured as a Juror in the Botanical section and

[1] For this practical turn compare his description (to Berkeley the microscopist, 1854) of the Microscopical Society Soirée, 'where nothing short of a double-barrelled, revolving, etc., etc., instrument is thought worth notice. I saw some astonishingly pretty things, but the whole view is too kaleidoscopic for me. I never feel satisfied as to what I see if I have not poked at it previously with my own fingers.'

as editor of the reports, to see the whole series through the press, 'which is a great bore in some cases and very easy in others; there will be 1600 pages of it.'

This employment involved the tedious journey from Kew to town three or four times a week. His 'Report on Substances used as Food' was duly printed among the other reports that year; it was followed next year by his and Lindley's 'Report of an Enquiry into the best mode of detecting Vegetable Substances mixed with Coffee for the purposes of Adulteration.'

His own work as a Juror was honorary; for his work as editor of the reports he received remuneration, a grateful increase to his precarious income, albeit the time expended on the work ran to eight months instead of three, as proposed. As he writes to Bentham in July 1852, apropos of 'working very hard now at New Zealand Flora, the Garden and my Indian Journal!'

> Chicory versus Coffee report is gone in—Parsnips, Mangel wurzel, Beans, Acorns, Tan! etc., come next. I like the work, but that is the worst of me, I like anything for a change, and believe I should take to any pursuit with avidity (except drink and Wordsworth) that was put on me.

CHAPTER XIX

BOTANY: ITS POSITION AND PROSPECTS IN THE FIFTIES

HOOKER had long been conscious that something was wrong with the state of botanical science, in England especially. Physiology applied to plant life, as to animal life, was making fruitful discoveries. But systematic botany had almost exhausted the Linnæan and post-Linnæan impulse. The more nearly the Natural System of Classification initiated by De Jussieu and elaborated by De Candolle completed the cataloguing and classifying work along established lines, which seemed to be its sole remaining function, the more nearly it reached a sterile completeness. Schleiden in 1842 saw that Botany as an Inductive science must rest upon research into development and embryology. But these morphological studies with their comparison of structures which pointed to living lines of natural affinity, stood apart from systematic botany as a separate discipline. Though material was thus being laid up for a theory of descent, the doctrine of origins was still bound up with the traditional cosmogony. Research was cramped by the heavy hand of fundamental theory. It led seemingly to no promised land of science; no new vivifying principle which should reveal the clue to those perplexing problems in the affinities and distribution of plants, to which no rational and satisfactory answer was forthcoming on the old lines.

The search for novelties loomed too large; in the absence of good organisation between botanists, mere species-mongering had led to unspeakable confusion and overlapping. Observers had given different names to the same plant in different regions; their unco-ordinated observations tended to obscurity rather than light.

> What is to become of specific Botany I cannot think. I have only last week found out that the little Rhododendron anthopogon described by Don, Wallich, Royle, Lindley, Hooker and three times by Hooker-fil. is the very old Osmanthus pallidus—absolutely identical—not a variety even! I also took up the Indian Vaccinia and found that out of 16 species figured in Wight's [1] Icones no less than 9 were bad and old!

Man had not found what nature indeed had denied, a common standard for differentiation between species, varieties, transitional forms; nor an independent basis for that abstraction, the specific type, so useful as a label, so dangerous as a determinant. The very name conjures up the ancient logical battle between Nominalists and Realists; and the latter day Realists, perhaps unconscious of their intellectual affinities, were in the ascendant, upholding the existence of such types, the living approximations to which constituted species.

Full realisation of this state of things could only come through knowledge at once profound and far reaching such as Hooker's, uniting as it did the close personal study of entire floras and of the literature that dealt with them, representing every kind of region from the Poles to the tropics—the Antarctic, New Zealand, Australia, India, the Galapagos Islands, Aden, and the Niger, besides the botany of certain Arctic voyages, and much of Ceylon and the Cape. Only such intimate knowledge, ranging over the widest areas, could

[1] Robert Wight (1796–1872), M.D., of Edinburgh, entered the E.I.C. service and became a leading Indian botanist. He was early in touch with Sir W. Hooker, in whose botanical periodicals he began to publish his material when on furlough after 1831. At the same time he published with Arnott one vol. of his *Prodromus Florae Peninsulae Indiae Orientalis*. His later work in India included inquiry into the cultivation of useful plants and the charge of an experimental cotton farm, while at considerable loss to himself he published his *Illustrations of Indian Botany* with coloured, and *Icones Plantarum Indiae Orientalis* with uncoloured plates, numbering over 2000.

absorb and transcend the results of observation over lesser areas, with their comparatively clear demarcation of species. From such broad surveys came the gradual conviction that systematic botany was at once too artificial and too sectional to represent truly its professed ideal of natural grouping, being rigid and definite where nature proved to be plastic and variable. Only after dealing with thousands of specimens in the collections which passed under his scrutiny could he exclaim 'more specimens always break down characters,' i.e. destroy the rigidity of botanical definition and extend the fringe of individual variability. It began to grow clear that over a sufficiently large range every variety might exist between two allied species, and that where these intermediate forms had not chanced to be exterminated so as to leave the extreme forms in isolated contrast, it was impossible to lay down where the one 'species' ended and the other began.

But this upset the doctrines everywhere taught. Hooker, realising as no other botanist the difficulties involved and their reaction upon his science, divined in them one secret of the ineffectiveness he deplored in systematic botany. System, he saw, broke down at its widest extension. Unknown to its expositors, it had become formalised and abstract; it awaited a new interpretation to revive its powers.

Meantime, the same abstract formalism had invaded the lecture-rooms. All that could be done for the regeneration of botany was to improve the teaching of it, first, as has been seen, by setting examination papers which demanded a training not in simple memory, but in thought and observation; then by aiding in the preparation of the right kind of books for students and the right kind of lectures, in new organisation at the Universities and in the publications of the learned societies. His hopes take shape in a letter written to Huxley in the earlier part of 1856:

> My own impression is that we shall make no great advance in teaching Nat. Science in this country, except by some joint effort of Botanists and Zoologists who should pave the way by propounding a strictly scientific elementary system, —were this once effected we have sufficient command

over the public, as examiners in London, and as confidential advisers of examiners and professors elsewhere, to ensure the cordial reception of such a system. What with Henslow's Botanical School diagrams now in progress and Museum Types we have made a fair start, and if you do not occupy the field in Zoology some pitiful botcher or other will.

I am very glad that we shall meet at Darwin's. I wish that we could there discuss some plan that would bring about more unity in our efforts to advance Science. As I get more and more engrossed at Kew I feel the want of association with my brother Naturalists,—especially of such men as yourself, Busk,[1] Henfrey,[2] Carpenter,[3] and Darwin,—we never meet except by pure accident and seldom then as Naturalists, and if we want to introduce a mutual friend it is only by a cut and thrust into one another's business hours—it is the same thing with our publications; they are sown broadcast over the barren acres of Journals and other periodicals which none of us can afford to buy and then weed: if either the Linnean or Royal could be made to stand in the same relation to Nat. Historians that the Geological does to Geologists [&c.] great good would accrue,

[1] George Busk (1807–86) studied at the College of Surgeons and entered the naval medical service in 1832, leaving it in 1855 for purely scientific pursuits, chiefly microscopic work on the Bryozoa, and later, palæontological osteology. He became F.R.C.S. in 1843 and President in 1871, as well as serving on its board of examiners. For twenty-five years also he was examiner in physiology and anatomy for the Indian army and navy medical services. He did much public work as Treasurer of the Royal Institution, Hunterian Professor and Trustee, and Fellow of the Linnean, Royal, Geological, and Zoological Societies, receiving the Royal and Wollaston Medals, and was President of the Microscopical and Anthropological Societies, and edited various scientific journals. A close personal friend of both Hooker and Huxley, he was one of the nine friends who made up the X Club.

[2] Arthur Henfrey (1819–59) succeeded Edward Forbes in the botanical chair at King's College in 1853. His original writings, translations and editorial work did much for education and physiological botany.

[3] William Benjamin Carpenter (1813–85) was 'one of the last examples of an almost universal naturalist,' especially in the direction of marine zoology and deep sea exploration. His most notable work was in Physiology, his *Principles of General and Comparative Physiology* (1839) being the first English book containing adequate conceptions of a science of biology. His *Principles of Mental Physiology* takes first place among his researches into the relations between mind and body, including suggestion and the unconscious activity of the brain. He came to London in 1844, when he was elected F.R.S. and held various chairs of Physiology, and was Examiner in Physiology and Comparative Anatomy at the University of London, until elected Registrar, 1856–79.

but without some recognised place of resort that will fulfil the conditions of being a rendezvous for ourselves, an incitement to our friends to take an interest in Nat. Hist., and at the same time a profitable intellectual resort,—we shall be always ignorant of one another's whereabouts and writings. (The above is not English grammar but never mind that.)

The convivial plan was tried in the Red Lions [1] and has signally failed, as will any other that has no other aim but personal gratification of a kind that can but be got by dropping Science altogether, and admitting the rag-tag and bobtail of Literature and the Arts together with the dregs of Scientific Society. We want some place where we never should be disappointed of finding something worth going out for. A good Society well stocked with periodicals etc. answers these conditions and I wish we had one.

Ever your bore,
Jos. D. Hooker.

From the moment of his return from India the outlook was depressing. 'Botany,' he exclaims to Bentham early in 1852,

Botany is going down rapidly it appears to me; the Botanists die and take their mantles with them. Reeve [the publisher] talks seriously, almost positively, of giving up Bot. Magazine and Journal (Icones of course); [2] he hangs fire with my New Zealand Flora. I don't find *one single* Botanist started up since I went abroad; many are dead. Something it appears to me may be done by a combined movement in the Universities; is it a time?

It was little better in December 1856, when he writes to Harvey apropos of his reluctance to apply to the Royal Society for part of the Government grant in order to publish his researches, for being his own lithographer he would appear to seek pay for his own handiwork:

[1] The Red Lion Club, presumably taking its name from Red Lion Court, the depot of the British Association, was a dining club founded in 1839 which met during the British Association Meetings. A frequently schoolboyish jollity with no further aim or result made no appeal to Hooker.

[2] Sir W. J. Hooker's periodicals.

Botany is all going dogward through the desultory doings of its votaries. I have been for four years past much mixed up with Physical Science men, and have found much to admire in their way of doing business. They let no opportunity slip of getting all they can for the furtherance of their publications and observations, whilst Botanists stand by and depreciate their own efforts and studies. I wish I could get you here for six weeks and join in a general effort to lift Botany up in the scale of appreciated sciences.

And a month later he meets Harvey's reluctance to publish preliminary sketches in advance of the magnum opus on which he was engaged—Precursores to the first Orders of the Cape Flora, like Hooker's Precursores ad Floram Indicam:

We differ (you and I) toto coelo as to what we think good and bad. I suppose from your calling such diagnostic Praecursores of Cape Genera as I proposed your publishing 'fushionless stuff,' I am to take that as your verdict on the Praecursores ad Floram Indicam!! Now I daresay you are right as to the way in which the Praecursores are done, but I hold that such work, *if properly done*, is about the most valuable that can be contributed to Bot. Science. What the deuce do you call useful work, if accurate descriptions of the genera and species of very little known large tracts of the Earth's surface are not so? So 'fire away, Flanagan,' as your illustrious countryman Lever has it.

January 10, 1857.

DEAR HARVEY,—I assure you I was only in joke in pretending that you intended to snub the Praecursores, though I do assure you that they are not so good as you take them for: the complication of systematic Botany is so great that I make many important omissions, and I must say I am heartily glad that I am prefacing the Flora Indica (if it is ever to appear) with these less assuming attempts. Plenty of people point out omissions in such contributions who fear to plunge into a detailed work, or if they do to criticise its assumed learning.

On the other hand do you not undervalue the amount and kind of systematic Botany that you have to dispense?

You think that Praecursores of the first Orders of the Cape Flora would not be valuable, because of the little novelty, but I think you have the old error of preferring novelty to anything else. Where for instance can I go for a tolerably accurate notion of Cape species of *Ranunculaceae*? It will be a greater novelty to me to find in your Flora 3 *Anemones* reduced to 2, than to find them raised to 4, and in my idea it is a far more valuable fact, for reducing a bad species is far better than making a new. What I regret is to see so much good sound common Bot. information carried to the grave by the holders, because being insensibly acquired its real value is overlooked by them. I had the same difficulty in getting Thomson to supply sketches of the Tibetan and N.W. Floras for an Introd. Essay—he could quite see the value of my doing it—for the Sikkim Flora! So it is with the distribution of Southern Algae, and I do believe that I should do more good to Science by inducing you to give us a good unlabored essay on this subject than by attempting higher things myself or urging you to do so. Botany goes to the dogs from the prevalence of this mauvaise honte and false pride.

You are certainly far too hard worked, and I do long to see the end of some of your great labors. You should never work beyond 11½ P.M., and you should not poison *yourself*![1] The expense of that would be well reimbursed in otherwise employing your time. I think you should still lay out for gluer, and catalogue yourself, as these are very improving operations and easy ones on the whole, not demanding too much brain work or sedentary employment. With Wife's love, Ever your affect.

J. D. HOOKER.

Excuse this scrawl. Tim [the pet cat] bothered me most of the time.

In short, as he adds on October 23:

The besetting sin of the Botanists of the day is the *craving for perfect materials*; forgetful that these Sciences are all progressive, and our efforts but steps in the progression.

[1] I.e. himself apply insect-destroyer to his herbarium.

Another letter to Harvey (February 3, 1857) strongly repeats this appeal against the natural depreciation of his own familiar store of knowledge, and insists on man's duty of giving his formed ideas to the world.

I am quite prepared for what you say of your work, it was always my case on first venturing, nevertheless you have done a great deal already and will soon fall into the way of it. A few steady weeks at Systematic Botany in the Herb. wondrously renovates and reinvigorates one I find, and when weary of desultory head work, I find the Herb. a great relief.

As to your publications I would urge you to think now of putting together some of your ideas and facts on wider branches than purely descriptive. I think that this becomes a duty after a certain time of life with those who keep such subjects before them—too much of our dear bought experience dies with us, and the pursuit of careful descriptive Botany rather renders us too timid about striking out into generalities that are the product of years of insensibly gained ideas. I express myself abominably and write as I think, but I am myself urged on all hands to treat some branches of Botany in a larger manner, and as soon as I have completed my rough lists of Indian and of Australian plants I intend to make them the data on which to establish some attempts to estimate accurately the relations of numbers of genera and species in given areas with climate and elevation, the relations numerical of genera to orders, of number of species in globe, etc., etc., in short to bring to book upon absolute data (tolerable as far as they go) certain principles now vaguely enunciated on no fixed data at all. This you could do for Southern Algae and connect their migration with ocean currents and temp. of Ocean, not in detail, nor upon exact data, but upon fair data, and be they good or bad you are the only one capable of doing it, and it will take any other man many years to come up to your capability and opportunity. Heaven knows I dread *my* subject and feel enough my own incompetence, but the work wants doing, nobody else has the opportunity, and it is in my position of life as clearly my duty as any moral obligation can well be. Others can and will work up species, and I have no right to withhold the result of my personal experience in generalising on these subjects and in handling them so long as I think myself and am assured by my fellow Botanists that the attempt on my part is called for. These, however, are not matters for a week or a month; but shape a course towards them and you will find a wonderful mental relief follow, when distracted with 'choses à faire.'

Thus amid the fluctuations and discouragements of the outlook for pure Botany, Hooker found that to take stock of his ideas and marshal them in the Introductory Essays to the Flora of New Zealand and the Flora Indica was a re-invigorating process. The synthesis meant new force, new interest. To Bentham, who was in Paris for the Exhibition, he writes in July 1855:

The Flora Indica Introd. Essay is going ahead. Henfrey is shot and proposes altering his whole system of Botanical instruction at King's College! my chers confrères the geologists shrug their shoulders and do not half like it, and H. Watson is going to review it in the Phytologist.

I shall be amused to hear what they say of the Introd. Essay in Paris, mind you tell me. I have frightened them here out of their wits, and some of them thank me for the presentation copy with a frigidity that delights me. Hitherto Botany has been dull work to me, little pay; no quarrels; an utter disbelief in the stability of my own genera and species; no startling discoveries; no grand principles evolved, and so I have a sort of wicked satisfaction in seeing the fuse burn that is I hope to spring a mine under the feet of my chers confrères, and though I expect a precious kick from the recoil and to get my face blackened too, I cannot help finding my little pleasure in the meanwhile.

Before long, however, a better era for Botany seemed at hand; a more cheerful strain is apparent in a note to Henslow (January 6, 1856) apropos of his son George's career:

Keep him to Botany if you can, but not to the exclusion of other scientific pursuits, drawing, &c. I am well sure that there will be openings and good ones for accomplished Botanists ere long, and I cannot fancy a more agreeable, fairly profitable and useful life than that of a scientific man who is really attached to his pursuits.

The same note is sounded in correspondence with Harvey, who, a month before (October 1856), had returned from his three years' cruise in the Indian Ocean and Australia, and had been elected to the chair of Botany in Dublin[1]:

[Nov. 1856.] You know that I am not a sanguine man, and yet I can see that you have in yourself, with an unembarrassed life, abundant resources for a fair income, and I am sure that you have resources in your collections and previous career for continuing the life of a pure man of science, with honor and profit to yourself and to the lasting benefit of science. I would much rather see you the Curator of Trin. Coll. Herb. on £100 and free of all Lectureships whatever than hampered with even the Botanical.

The serious matter was that to the Botanical chair at Dublin various duties had been attached, seemingly 'pluralities without sinecures,' as Hooker defined them, and especially the duty of lecturing on Natural History at large, for was not Botany a part of Natural History? Hooker, backed by his father, strongly urged the inexpediency of taking up a Zoological Professorship in any shape at all, joint or disjoint:

[Nov. 25, 1856.] I cannot say that I at all stomach your Zoological lectures and duties, not from any aversion to Zoology or to your joint Professorship, so much as because it will involve all sorts of other minor and major zoological inroads upon your time. You talk of lecturing on *Invertebrata* as if they were nothing; do just read Huxley's lectures in the *Medical Times*; they are admirable, though in saying so I feel like the old Scotch wife who said, 'Ae, it was a grand discourse, I couldna understand the ane half of it.' By Jove, the whole science seems to be so changed to what I learned, and the literature of any one such small Order as Annelida or Rhizopod or Cestoid worm! so overwhelming, and the new facts so revolutionary, that I cannot fancy any

[1] See p. 400.

but adepts mastering the Invertebrata. Of course you can give an elementary course on these things such as they were, but so much science and philosophy is now expected from a professor, that I would rather you could confine yourself to Botany.

Embodying his friend's arguments in a letter to the authorities Harvey obtained relief from this anomaly, and was able, as Hooker put it, to settle down to a quiet Botaniphilus' life. The letter of November 25 continues:

> You ought now to take the highest position in Bot. Science and regard the aspiration thereto as your destiny. You are loaded with honey and your calling is science, and you and I should have no thought but to make ourselves useful to Science, without fear of personal failure. The less we think of ourselves the better so long as we are no burthens to our neighbours. Bentham's unselfish love of science always charms me, he has never a thought of personal aggrandisement in money or honor; but indeed we have both of us lived under the highest examples and happiest influences in these respects. My Father, Bentham, and Thomson are such a trio as we shall never see again. Except Faraday and Darwin I know of no others in the walks of science so pure and disinterested, except perhaps Asa Gray in America. I am getting prosy, however.

More than once during this period the necessity of lecturing nearly fell upon Hooker. In 1851 it was proposed to appoint a Professor of Botany to Kew, to lecture in London, and Prince Albert suggested him for the post. But such a proposal did not fit with the real position of Kew or of its Director. Hooker, being 'pumped,' answered frankly that work on his Indian and Southern collections would put lecturing out of the question for himself; that making such an appointment to an establishment having neither Library, Herbarium, Secretary, nor Museum-keeper was putting the cart before the horse; and indeed, so long as his father was supporting the establishment in these points out of his private purse or energy, appearances must be deceptive. Rather call in the services of good outside lecturers.

In 1855 a fresh lecturing scheme was suggested in connection with Hooker's appointment as Assistant at Kew. Kew ought to justify its scientific endowment by giving the public of its science as well as its pleasure walks. At the cost of his personal inclinations, Hooker was ready to help the development of Kew by focussing public opinion on its national character; but the official world would have none of it.

Similarly he tells Bentham (January 1854):

> The Royal Institution are pressing me very hard indeed to lecture for them. I refused on the grounds that it was wholly incompatible with my duty to Govt., whereupon Faraday writes offering to go to Ld. J. Russell[1] and get me the Govt. sanction. I have refused definitely again, and added that were any application made to Lord J. it would be to appoint an assistant to my Father. The offers were most kind and flattering and too pressing—it is always excessively disagreeable to refuse such invitations, however little inclined one may be to accept.

It was at least the fact that if lecturing in London exacted too heavy a toll from the Director's working time at Kew, Kew was too far from town for a London audience. The only stimulus to public interest that followed was the opening of the Gardens in 1857 on Sunday afternoons as well as week-days. He tells Bentham on June 1:

> My Father remonstrated and my Mother is in a sad way about it, as you may suppose. For my own part I had no wish for it and on private grounds oppose it, as probably disturbing the only quiet day I get in the week; but on the other hand I consider it a wise and beneficial measure in a public point of view, and therefore feel that I have no right to complain.

The consolidation of the scientific side of the Gardens took a long step in advance when Bentham in 1854 presented to the nation his great herbarium and library, valued in cash

[1] At that time President of the Council.

at £6000.[1] Bentham, moreover, left Pontrilas and settled first at Kew and later in London; saw to the final arrangements of his herbarium, and continued his own botanical work, more especially the monumental Genera Plantarum in collaboration with Hooker.

This accession both weighted the scales in favour of Kew as against the other and in many ways less suitable centre of botany at the British Museum, and offered a new factor in the problem of the ultimate destination of the Hooker collection. As to the status of the Herbarium he tells Harvey (January 21, 1857):

> We have no funds for buying plants; my Father pays himself for all appertaining to the Herbm. as of yore, and calls it his own. We should hardly dare to ask for money to buy Cryptogams, as the Herbm. is upheld ostensibly for the naming of the Garden plants, and we are not yet in a condition to throw down the gauntlet to the British Museum. We have just drawn up the Garden Report and pitched it very strong about the uses of the Herbarium as a scientific adjunct to the Gardens.

With the death of Robert Brown in 1858 the question came to a head. Ten years before, the Parliamentary Commission had determined that on Brown's death they would abolish the Botanical Department; and, Hooker confesses, 'every reason for doing so then is redoubled in force since,

[1] In the Memoir of his father, p. lxxx, J. D. H. writes: 'This was second to my father's alone in England in extent, methodical arrangement, and nomenclature, and was placed in the same building. Its formation was begun in 1816, in France, where and in the Pyrenees Mr. Bentham collected diligently; but its great expansion by the inclusion of exotic plants dated from his introduction to my father in Glasgow in 1823, when the friendship between the two commenced which remained undisturbed for forty-two years. From that date the two botanists may be said to have hunted in couples for the aggrandisement of their libraries and collections, sharing their duplicates, Mr. Bentham giving my father the preference in all cases of purchase, &c. The one great difference between their aims was, that the former confined his herbarium to flowering plants, whilst my father's rapidly grew to be the richest in the world in both flowering and flowerless plants. The offer of this gift was prearranged with my father, who with his wonted disinterestedness put aside the obvious fact, that its acceptance would greatly diminish the value of his own herbarium and library, should the Government ever contemplate its purchase.'

and endless others added.' The Hookers were summoned to meet the Trustees of the British Museum on the subject of the Botanical collections coming to Kew.

> Brown [he writes to Harvey] leaves everything to Bennett except the fossils, which he gives to Brit. Mus. if they will keep them with the plants; if not they are to go to Edinburgh. The Trustees will put Bennett in Brown's place and keep their collections at B. M., but whether Govt. will not insist on the Brit. Mus. N. Hist. collections being turned out of the building is quite another question. My idea is, that eventually all the Nat. Hist. will go to Kensington Gore but the plants, which will come here.

That the collections should be moved from the dust and grime of their cramped quarters at the British Museum was certainly an excellent thing; the zoologists wished the zoological specimens to go to a new museum in Regent's Park, close to the living animals in the Zoological Gardens; the botanists were agreed that the botanical collections should be merged in the greater Kew collections, instead of maintaining an independent existence. But Natural History carried little weight in the House of Commons, and was very slightly represented among the British Museum Trustees, Geologists and Physicists especially having been appointed to this body owing to official interest in the Jermyn Street Museum. Thus in the eyes of working men of science there was great danger ahead lest the collections should be handed over to the charge of the non-scientific Science and Art Department, and that at South Kensington science and the interests of research should be subordinate to exhibition as a popular show.[1]

[1] The surplus from the Great Exhibition of 1851, amounting to £213,000, was invested by the Commissioners in land at South Kensington. Here a Museum of Art was established, the nucleus of which consisted of exhibits purchased by the Government. To these others were gradually added, such as the collections from Marlborough House, the Sheepshanks collection, and so forth. In natural sequence proposals followed for the transfer bodily to the same centre of other institutions and museums that received Government support, especially those connected with scientific instruction. For in 1853 the Science and Art Department was detached from the Board of Trade by the amalgamation of several minor establishments with the School of Design, under the Secretary of the latter and the indefatigable Henry Cole (afterwards K.C.B.), himself the chief organiser of the Great Exhibition, and reorganisation

We know to-day how amply science, in the persons of the late Sir William Flower and his successors, has fulfilled the scientific mission of the Natural History collections at South Kensington. The germ of this success lay in the movement set afoot by Hooker and Huxley to amend and strengthen the influentially signed memorial that laid the case for science before the Prince Consort as head of the Kensington Committee.

The two friends joined forces on what Huxley called their 'permanent Committee of Public Safety' to watch over what was being done. Huxley, who professed himself 'thoroughly roused,' eagerly enlisted the support of the progressive among the scientific and the scientifically inclined among public men and editors of the Reviews, and as for the attitude of the Laodiceans in science he writes with cheery defiance:

> I don't think it is necessary to trouble one's head about such opposition. It may be annoying and troublesome, but if we are beaten by it we deserve to be. We shall have to wade through oceans of trouble and abuse, but so long as we gain our end I care not a whistle whether the sweet voices of the scientific mob are for or against me.

A few passages from Hooker's letters may be quoted:

To T. H. Huxley, 1858

> My present impression is that a compromise may prove to be the best thing—anything to keep out of the K. Gore people's clutches—and that if we could only satisfy ourselves that the Nat. Hist. would certainly be moved we should without delay apply for a building in the Regent's Park, near the Zoolog. Gardens, so arranged that vast sufficient Galleries should be filled with enough Birds and Beasts for the public to gape at *daily*, with parallel private side galleries where Naturalists could *daily* work (and where

was the order of the day. Finally the Government ended its partnership with the Exhibition commissioners, and became sole owners of the Kensington site.

A familar nickname for South Kensington and all its works sprang from an interim iron building erected in 1855, unjustly supposed to be from Cole's designs; it was popularly known as the Brompton Boilers, or shortly 'The Boilers.'

the public were *never admitted*) and where the specimens would be arranged for work and not for show. . . . Proximity to the Zoological Gardens and its live beasts and birds is however, I fear, the only pretext that could be offered for not accepting the K. Gore offer.

The real secret of our anxiety is, not that the separation from Art at Gt. Russell Street would be injurious, but that we would lack support as a National Museum of Nat. Hist. except we huddled our collections under the wing of art. This gives our cause a bad look.

I do truly say that we at Kew do not want the Brit. Mus. Herbarium here at any price; it is no use to us, and if it be the means of breaking up the Brit. Mus. Nat. Hist. collections, or withdrawing support from them, I shall deeply regret its coming here; but as an honest man I must say (with every working Botanist) that it is for the interests of Botanical Science it should come here; it would take 22 years and as many thousand pounds to make the B. M. Herbarium anything like ours here, and there are no men to do it. Besides which, a working herbarium cannot be kept clean enough to work with in London; it must, if worked with, be exposed for hours daily to dust by great portions at a time.

So far as the Bot. Department is concerned the Trustees are in an awful fix, and my opinion being clearly that they should clean, poison, and stop adding to the Banksian Herb. and the Govt. should take my Father's as the National Herb., keep the plants at Kew and increase it so as to keep it as far ahead of all others as it now is, I am far too deeply personally interested in the matter to take any prominent part with decorum.

I am further for having at the British Museum a Botanical collection, illustrating *Plant life* such as Henslow could best plan and develop, and for which perhaps our friend Lindley or Henfrey would be a highly qualified keeper. It should be as popular as Bentham suggests in every respect, but also as scientific in its details and completeness as the most profound vegetable Physiologist and Anatomist could wish. This would cost little, be very instructive to the Public, and useful to men of Science. It would be unique, there would be nothing like it in the world. I had often planned such a thing for Kew, but we are still young, and have far too much to do to complete what we have on hand.

Were a Herbm. not necessary to Kew, I would say at once let my Father's go to the B. M., but it is impossible to work scientifically a garden of 20,000 to 30,000 species, and name the hundreds of things sent to us to name, without a first-rate Herbarium and Library here, as good as ever the B. M. ought to be made. The seeds sent are often to be known only by the accompanying dried specimens which go into the Herbarium, and the latter becomes in a thousand ways an indispensable adjunct to the Garden and reciprocally (by being the depository of the plants once cultivated in the Garden) an integral part of the establishment, and a record of its progress and efforts, its successes and failures as a horticultural establishment, all quite apart from its scientific uses.

The offer of other botanical collections to Oxford and Cambridge, neither of which was enthusiastic, had already given opportunity for pushing the cause of science in the older Universities, where it was still of small account. The Fielding and Lemann collections were on offer, but there were difficulties to be overcome. Thus 'The Fielding Herbarium,' [1] he writes to Harvey in January 1852, 'is to be offered to Oxford upon conditions of good keep, accessibility and extension: terms which I think Oxford won't agree to.' Moreover the question of extra-mural Trustees and their duty after the collections had been accepted was a thorny one, alike to Sir W. Hooker, who had been nominated, and to the University as legatee.

I cannot help thinking [he writes to Bentham, Feb. 5, 1852] that these Legacies may be the means of instilling new life into the Universities; the conditions being reasonable. A proper representation backed perhaps by P. A. [Prince Albert] as Chancellor, with the offer of such a Herb. as Fielding's or Lemann's, should do wonders, especially as, in future, a

[1] Henry Borron Fielding, a country gentleman whose health prevented him from taking any active share in scientific life, devoted himself to botany. He purchased Dr. Steudel's herbarium in 1836 and the Prescott collection in 1837, bequeathing his entire herbarium and many books to Oxford on his death in 1851.

Botanical Fellowship or two might be insisted upon, from whom the Professors should be chosen. Rooms and £50 a year should do a great deal for a Herbarium, supposing it to have the superintendence and zealous curatorship of a working Professor, such as Henslow would have made before he got his Father's living, or as Berkeley might now.

Though there was at first no very reassuring answer from friends in either University, affairs straightened themselves out. By March 16 Henslow is told that

> Oxford is inclined to behave much more handsomely than we anticipated, offers £1000 for a building, £50 and a good suite of rooms for a keeper, and £25 for annual increase—constant accessibility to the public without a Master of Arts or any other drawback.

On Bentham's advice Mrs. Fielding withdrew some of her conditions; the gift was accepted, and before long a curator was found in the person of Maxwell Masters,[1] of whom Hooker wrote to Harvey:

> We are hunting for a curator for Hb. Fielding. I hope young Masters will get it, a fine lad ætat. 20 who has just finished a most distinguished medical education at King's College and took medals galore—is son of Masters, nurseryman at Canterbury, and early passionately attached to &c., &c., &c., &c., &c., &c. It is only £50 and two rooms at present and worth no one's having but a scrub's, or a man who will take zealously to science and trust to providence for a future competence as a Botanist. I have a great idea that a *good* Botanist and good Herb. would advance science greatly in the Univs. Daddy cannot see it somehow, but I had Masters out to dinner yesterday and the old Gent. takes to him—a mere scrub or half educated man would lower the position of Botanical Science in the eyes of ignorant bigoted Oxford (I hope I do not offend your High Church ears),

[1] Maxwell Tylden Masters (1833–1907) was a pupil of Edward Forbes and of Lindley at King's College, and Sub-Curator of the Fielding Herbarium. After standing unsuccessfully against Henfrey for the Chair of Botany at King's College in 1854, he took up general practice, but lectured on Botany at St. George's Hospital and edited the *Gardeners' Chronicle* after Lindley's death in 1865, besides writing many botanical monographs.

a well educated and passable Botanist would be tolerated for his own sake, but a really zealous ditto, well educated *elsewhere,* and commanding the respect and esteem of men of science in general, must I should say force a proper appreciation of Botany in the University.

Similarly a personal conference between Hooker, Henslow, Lemann [1] (who was preparing to break up his collections and distribute the fragments where most wanted), and the Cambridge authorities, established the other collection at the sister University. As he tells Bentham, who arranged the Herbarium:

> Henslow scouting the idea of valuing the species or specimens because they were uniques has told well, and proved to the Dons that such collections have other and a higher value than old china. I must say they express themselves liberally and well.

[1] Charles Morgan Lemann (1806–52), M.D. Camb. 1833, F.L.S. 1831, F.R.C.P. 1836, collected in Madeira 1837–8 and at Gibraltar 1840–1, and presented his Herbarium of 30,000 specimens to Cambridge University. He wrote, but did not publish, a Flora of Madeira. The genus *Carlemannia* was named after him by Bentham.

CHAPTER XX

SCIENCE TEACHING: EXAMINATIONS

THOUGH neither lecturing nor teaching in person, Hooker found a useful educational lever put into his hand by his twelve years' examinership. In the autumn of 1854, thanks, he presumed, to the influence of Sir James Clark,[1] he was appointed to examine in botany the candidates for the medical service under E.I.C. He was already examiner to the Apothecaries Company, and writes of the special standard in the papers set by him in a letter to Huxley.

> I should certainly give a very different examination to the E.I.C. candidates to that for Apothecaries' Company Medal. The latter, you see, is competed for on Bot. grounds solely, by 'all England,' and should be a right good tough affair in my opinion, and very different from a Pass, or Matriculation Examination. It was not to be expected that you should have answered half the questions. I did not expect one candidate to answer 2/3 of them, but just see. There was *only one* question that no one answered and that because misunderstood: and three answered nearly all. I had 6 men, and by far the very best men I ever tackled; there was not one bad paper, and the first three were excellent—the worst answered 2/3 of the questions (better or worse). You may remark that I did not put one catch-question, or one that did not involve general principles. There was not a man amongst them

[1] Sir James Clark (1788–1870) began as a naval surgeon, and after successful private practice abroad and at home, became Physician in Ordinary to Queen Victoria on her accession. He served on various Royal Commissions, on the Senate of the London University and the General Medical Council. Without adding much to science, he possessed considerable official influence.

that had not studied plants for himself. I had also another object in my paper, which was the leading men to study plants rather than books. Every one but Henslow thinks my questions dreadful because nobody thinks of them. You must also remember that they had 8 hours; and that my object was to give questions requiring *thought* rather than *memory*. What does Busk say to them?

Continuing the subject, he writes on September 12:

Sir C. Wood [1] has written me a powerfully flattering letter, asking me to accept the Examinership! This is rather good after my name has been battledored and shuttlecocked in the medical papers for the best part of the month as I am told, for I have not read them yet.

God knows there was no jobbery in my election. Of course I graciously accept; and of course I get thanks for the same, from this pink of politeness who seems a regular official Mantalini with his 'demnition sweetness.' What are Busk's ideas on the subject of the examinations? I have long held that the Army, Navy, and E.I.C. examining good passed men of the Royal Colleges is a piece of the most confounded impertinence. As to the Navy Examination we know what that *was* and I suppose is; it has always appeared to me that the said services should seek from the Colleges men proved by them to be first-class in their profession, and then let the Examiners of the services examine for accomplishments and qualifications essential to shed lustre on the service and improve it. I am going to talk over this subject with Paget [2] to-morrow, but of course shall take no initiative and am rather groping my way in utter ignorance than anything else. The success of my Apoth. Co. examination has put new ideas into my head, and convinces me that even in Botany men at the examinations are rather to be expected to exert their reasoning faculties than their powers of memory. If we only reflect we shall see that the Oxford and Cambridge honours papers, and even high class examination and pass

[1] Sir Charles Wood (1800–85), created Viscount Halifax on his retirement from public life in 1866, had been Chancellor of the Exchequer under Lord John Russell from 1846–52, and in 1854 was President of the Board of Control and from 1859 Secretary of State for India.

[2] (Sir) James Paget. See *ante*, p. 25.

papers, are of their kind far better tests of the intellect expended in the attainment of the subject than our Medical Examinations are.

The outcome of his ideas on these examinations is summed up in a subsequent letter to Sir C. Lyell:

October 26, 1869.

I was one of the four who, at the request of Sir C. Wood, originated the system of competitive examinations for the Medical Officers of the *Indian* Army, which produced most extensive and important reforms in the Medical Schools (after they had abused us well for our pains!); the system was extended thereafter to the British Army, and now to the Navy, for twelve years I examined twice a year, in all branches of Science! I did not retire till I was appointed Director here, when the fees of the Examiners were immediately doubled!—*post hoc*—I cannot say *propter hoc.*

It was a very arduous and poorly paid duty. Paget, Busk, and Parkes[1] were my coadjutors.

For the next six years the letters contain constant references to these examinations. They meant a bout of hard work in January and July, with, say, 600 foolscap sheets to look through as a first step. Experience showed the frequent lack of good preliminary teaching and of any single system of teaching. In 1855 we read of twenty-eight candidates for thirty places, of whom six were ploughed, 'they were excessively badly taught, in Botany especially'; in 1857, forty-three men for twenty-two places, again showing much ignorance, while in 1858 the men are on the whole better. But he was sometimes in despair over the answers given, and writes to Harvey at Dublin, July 14, 1859:

I am examining at India House and ask a man what the value of Duramen is in contrast to Alburnum, and he answers that Policemen's batons are made of it! Guess his country.

[1] Edmund Alexander Parkes (1819–76) was the first organiser of the Army Medical School, and the founder of the science of modern hygiene, especially military hygiene. As an army surgeon he served in India for three years, returning to London in 1845, and became Professor of Clinical Medicine at University College in 1849. In teaching and in physiological research he was equally distinguished.

> If I had asked him the economic value of Rosaceae he would have quoted Shillelaghs! Another told me that the freezing point of water was 50° below zero, and another that the boiling point was fixed by filling a thermometer tube with boiling mercury! What are your Colleges of Surgeons about? Some of their licentiates are *consummate ignoramuses.*

Nevertheless he was convinced of the value of Botany in medical education, writing to Henslow in 1855:

> I wish very much you could afford half an hour to think over the subject of 'Botany as a branch of education and a means of mental culture specially adapted to the early education of Medical men,' and send me a few notions on the subject. I am preparing a notice of the mode of conducting the Botanical Examinations for the E.I.C., and want to drive it into the heads of Medical men and students; that it is not with the hope that the Botanical knowledge obtained will ever be of the slightest direct advantage to the man in practice that it should be taught, but because a right elementary knowledge is necessary to the right understanding of the Pharmacopœia, Hygiene, therapeutics, Mat. Med., etc., and especially because the mental training of a good elementary Botanical or Nat. Hist. course is the best means of becoming skilful in diagnosis of diseases and of developing his ideas. I am, however, a bad hand at expressing my ideas in mental philosophy and yet would like to do it properly.

Thus he was the more bent upon establishing good scientific teaching and reasonable examinations. He is consulted by Henslow in 1855 as to the papers the latter is setting in the Tripos at Cambridge, and later by Harvey on the corresponding papers set at Dublin. In querying various points he says to Henslow (March 15): 'I am no scholar, but sometimes do instinctively sniff out a clumsy expression, and in this case certainly did not know a good one.' In another case, criticising the wording of a sentence, 'I do not doubt you mean right, but it appeared very wrong on the paper.' He also urges Henslow not to use a descriptive term which had already failed to win general acceptance among botanists.

The following undated letter to Henslow further illustrates his difficulties:

> Better not recommend *books* except perhaps to advise the study of such a thing as Lindley's 1*s*. pamphlet on descriptive Botany, which is quite unique, and I think the men should be told that it is best to work upon the Candollean system of Orders. I should not recommend any other of Lindley's works, or indeed any works as *works*: and the 1*s*. pamphlet only as indicating a *method of working* that will certainly meet the exigencies of the Examiners.
>
> I find yearly the difficulty of having to do with men who have never been taught on any system, or all on different systems. I feel the difficulty of recommending books, but I see in the present condition of the Science and its Professors, the necessity of indicating *a method* both of working and of arranging the Nat. Ords. To make the *book work* depend on the coaching up a particular author's work, as Babington[1] proposes to do by Lindley's Elements, would be fatal to any good examination.

The proper method of examination is further dealt with in a letter to Harvey, who had just been appointed Moderator in the College examinations at Dublin.

> [March 24, 1857.] What is a Moderator-ship? Steam or sail? I like your programme of it, but do, I beg, insist on their demonstrating characters both on *dried* and living specimens of Brit. polypet [alae] and see that their knowledge is founded on sound Morphological laws, as studied by themselves on the plants. Henslow has just issued an admirable *dried plant* Examination Scheme, write and ask him. You are quite right to stick to elementary knowledge of British plants, and however much you change your subject never lose sight of the principle of keeping within the limit of what

[1] Charles Cardale Babington (1808–95), botanist and archæologist, who succeeded Henslow as Professor of Botany at Cambridge in 1861, was especially enthusiastic as a field botanist, and his *Manual of British Botany* in successive editions from 1843 onwards brought the subject from the Linnean stage into harmony with continental progress in systematic and descriptive botany. His lectures, however, did not expand with the new developments of botanic teaching in histology and physiology, and his detailed descriptive work, such as the *Synopsis of British Rubi*, ran to an extreme of analysis in basing new species in minute differences.

> they ought to know *practically* and well, and of so conducting the examination in Physiology (when you take that as a change) that it shall include Morphology and the Natural Orders. Do stick to the motive that Botany is a knowledge of *plants* and do not budge one inch from that. I am quite convinced that one of the greatest evils done to science is the fashion of making men learn solely or chiefly matters of which they can have no practical knowledge: their education is thus a forced one, the honors they get are not for the kind or amount of knowledge which enables them to make their way on afterwards, and they have been thus led to form a low estimate of the only useful branches, and they do not like to hark back upon these afterwards; and are deterred from going on with the science for ever after. The whole subject of education in Science is being better appreciated now that the German school is falling into disrepute.[1]

The writing of good handbooks was as essential to the progress of Botany as the elaboration of a satisfactory system of lecturing.

Bentham's 'Handbook to the Flora of the British Isles' (published 1858) was a great step in advance, and a letter to the author while still at work upon it strikes a confident note (February 16, 1854):

> I am rejoiced at the progress of the British Flora, and regard its appearance as a new era to British Botany. The public are really prepared for a change radical and complete. Your Flora must appear as a Precursor. I shall keep your letter in the hope that you will work out such remarks as you embody in it for a good sound introduction to the book. After all it is doing far more good to publish a Flora that will set people on the right way to know plants for themselves than one which aims to tell them everything about them. I would announce boldly my aim as the desire to put people on the right track and not to supply them with what they ought to find out for themselves.

Next came Henslow's work in elementary teaching of botany. John Stevens Henslow, who was born in 1796, and was therefore eleven years junior to Sir William Hooker,

[1] Compare the reference to Heer's lectures; p. 402.

had been Professor of Botany at Cambridge since 1827. His chief interest was not in systematic botany, but in the life history and geographical distribution of plants; his great distinction to have been the pioneer of practical teaching in England and the inspiration of those who came under him. As a keen observer, he knew the value of learning through one's own observations and discoveries. The average lecturer taught the students in the Medical Schools to learn botanical facts by memory; Henslow led his students to discover their facts by their own dissections of plants, and demonstrations from living specimens. Teaching by things, not words only, he made his subject alive, and on the same principle, arranged the public galleries of the Ipswich museum to be a connected demonstration of types, not a 'raree show' of curiosities.

> I am extremely glad [Hooker writes to him, May 10, 1856] to hear such good news about your class-men, and hope that you will turn out a Botanist or two amongst them. Pitch into the Dons and bigwigs.

The enthusiasm he awakened among his University students was renewed among the village children of Hitcham, to the living of which he was presented in 1838. Here, every Monday after school hours, he gave them lessons in botany, simple, accurate, intensely interesting, combined with systematic dissection of specimens and the making of local collections and observations. These village lessons were the source and pattern of the excellent nature-teaching now so widely diffused. The enthusiasm of the children, the lasting effect in interest, attention, character-building, were most remarkable.

He was gradually putting together the MS. for a projected book of Village Botany, which was left unfinished at his death in 1861, but formed the basis of Professor D. Oliver's[1]

[1] Daniel Oliver (1830) came to Kew at the invitation of Sir Wm. Hooker, and while working at the Herbarium found time to prepare and deliver, without fee, lectures to the foremen and gardeners of the establishment, 1859–74. In 1864 he was appointed Keeper of the Herbarium and Library, a post he held until 1890. He succeeded Lindley as Professor of Botany at London University (1861–88) and received the Royal Medal 1884, and the gold medal of the Linnean 1893. He was editor of the first three volumes of the Flora of Tropical Africa, one of the great Colonial Floras projected by Sir W. Hooker. Oliver was both right-hand man and close friend of J. D. H., with whom his 'omniscience' was proverbial.

'Elementary Lessons' (1863). He also designed a series of botanical diagrams, with explanations, for use in the National Schools, then under the branch of the Board of Trade known later as the Science and Art Department. These diagrams were prepared at Kew, and Hooker writes of them to Asa Gray (March 29, 1857):

> Fitch has just completed a most magnificent set of 9 Elephant-folio plates with illustrations and analysis of about 50 Nat. Ords. and genera designed by Henslow, and superintended by your humble servant. It is done for National Schools under Board of Trade.

These met with skilled appreciation in wider circles also. 'I find your diagrams,' he tells Henslow, 'greatly admired in Dublin. Harvey was copying them out in grand, and they had a very good effect'; while another letter remarks, 'I like your little explanatory book; it will, I hope, do great execution at the schools.'

In 1858 also:

> I met a Rev. J. T. Graves[1] at Dublin, a Fellow of Trinity Coll. Dublin, Mathematician, a man of renown in these parts who has been employed by Govt. in enquiring on Endowed schools and other Educational matters. He is immensely strong on your point of teaching the science of *Observation* to all men, especially to the young of all classes, and he has reported the same to Govt. in perhaps the very words you would have used.

In formulating this scheme of teaching and condensing it from his naturally more diffuse oral style, Henslow gladly sought the help and keen criticism of his son-in-law. The following letters illustrate Hooker's own sympathy with such a plan, his insistence on the need for the pupil's perfect understanding of the 'hard words' and definitions which form the

[1] John Thomas Graves (1806–70), a great mathematician, whose correspondence gave stimulus and suggestion to his friend Sir William Rowan Hamilton in his discovery of quaternions. Called to the English as well as the Irish Bar, he became Professor of Jurisprudence at University College, London, in 1839, and from 1846 was a Poor-law Inspector for England and Wales under the new Poor-law Act.

indispensable tools for scientific teaching, and for accuracy in the use of them, and—striking personal note—the happy freedom with which two friends could speak their minds to each other.

Many thanks for the perusal of the enclosed, which I like very much indeed—I have made a few pencil suggestions.

The term systematic Botany is a bad one, but there is no better in ordinary use; it hence wants a little amplifying upon to show that that branch is more than classification. Morphological is the right, in contradistinction to Physiological, but not adapted to your purpose. Few people appreciate the fact that Syst. Bot. is the exposition of the laws upon which plants are *formed* as well as *classified* naturally—somehow they do not.

Have you read Huxley on Methods in Nat. Hist.?[1] How do you like it? I very much.

My pencil remarks on your sheets are only suggestions. I like the whole thing very much.

December 12, 1854.

My dear Henslow,—The enclosed seems very explicit and clear; I have no suggestions to offer but a very few verbal ones. Would it not be as well to put all the technical terms in italics, it seems to give them weight? Under *Flowers*, I have put a pencil through 'through arrest of development'—as I think it is rather questionable and at any rate will be canvassed. Can we say that the Papaveraceae, having 4 petals and only 2 sepals, is through an arrest? this order being formed on a binary plan quite as normally as other Dicots are on a quinary. If we hold this to be an arrest of development, we must also consider the Monocots to be ternary through arrest—or reason in a circle. The fact is we call 5 the normal number, simply because it is prevalent: and by the same token 5 being prevalent in phaenogams as a whole, the Monocots which are in the minority are as much entitled to be considered arrests, as are Papaveraceae.

Under Gymnosperms,—'*an unfolded scale*' is very ambiguous, the said scale never was folded; but if you say

[1] On the Educational Value of the Natural History Sciences. An address delivered on July 12, 1854.

that hypothetically it was so, then you had better say 'an unfolded leaf.' I have suggested 'flat or concave' with 'unfolded' in brackets.

I do not at all agree with the terms Milkworts, Tutsans, etc., as English equivalents for natural orders, seeing that the same name more often applies to the genus only and most properly. Mallows are Mallows, and their family or order, the Mallow family or Mallow order. Mallow-*worts* means nothing—wort not being a recognised equivalent of any value, generic or ordinal. I think that by introducing such terms you lose all the little point English names have and gain nothing whatever. What is a wort? in English surely not a *tree,* to justify Mast-worts, more especially as mast is an equivalent to wort, in one sense. Wort I believe means *weed* or *herb.* I am still all for Crowfoot family (or order), Mallow family, etc., etc.

You will have a little difficulty to adapt a good name for all, but any genus contained in this family will be right, whereas the introduction of wort is wrong in grammar and more wrong in science. Let one of your pupils ask you to explain why you say an Oak belongs to the *Beech* family, or *Nut* family, or *Hornbeam* family, or any other contained genus you may adopt, and you can explain at once, rationally, and shew that the name conveys definite information—but what conceivable excuse have you for calling a nut a *mast-wort*! wrong in sense, in English, in sound, and in science. I think such terms are a retrograde step in the progress of sound elementary education. 'There then,' as Willy[1] says. It would be further exceedingly important to designate the Nat. Ord. in English, by the same genus or term as the Latin ordinal name is derived from—thus 'Cruciferae,' and 'Cupuliferae' = 'family of cupped fruits,' and 'Primulaceae' = 'family of Primrose.' You could thus explain both the Latin mode of giving ordinal names, together, and save much complexity and loss of time and of no little confusion too to young ideas, the only explanation needed being that there is *no* English inflexion that answers to the Latin '*Primulaceae*'—in English it must be expressed by the word order or family affixed or postfixed. Better than all this would it be to tell them that they can no more

[1] His small son, now aged two.

dispense with the word *Ranunculaceous* than with *perigynous* if they are going to progress in Botany, but if they are going to learn only a little, they had better take the English generic name and add 'Family of' to it. It appears to me essential that you should not throw a word or termination away.[1]

[February 1855.] I have gone over the accompanying very carefully, but fear it will hardly answer the purpose. It appears to me (but I may very well be wrong) far too laboured; too much is attempted to be taught by each sentence, they are hence too long and involved; there is a constant wandering from particulars to general Laws; and a great many too many words just a little too difficult for beginners. To be so philosophical it should be in aphorisms, for you cannot be clear, concise, and learned too, in a conversational form. My own impression is, that it would be better to make the demonstration of the Bean first, simple, clear and to the point, giving no words except the simplest. I object to 'axis,' 'relative,' 'modification,' etc., when *superadded* to the necessary and unavoidable technicalities; each of these, though familiar to us, being a subject of thought, to the 'village school,' before understood.

Having demonstrated the Bean, etc., you might then go over it again and another dissimilar plant along with it, and explain how the buds form, and the leaf buds give place to flower buds and how the leaves become floral whorls, how simple leaves become compound, how petals unite, etc., etc., but I am sure no pupil can learn all these things at once.

You are so much accustomed to teach with specimens and pictures, illustrating every point and making everything clear, that you perhaps forget how much of these advantages you lose in a book; and how necessary it is to be extremely simple in diction and in separating your kinds of information. In short I doubt if you will succeed in teaching the uninitiated young structure and morphology *at once*, which you here attempt. I further doubt your being able to do a book of this kind piecemeal. It is a most difficult task the writing down to the capacity of ignorance. I know it by experience; you must weigh every word and prune and clip every

[1] This is a rooted objection, repeated emphatically in a letter to Harvey, July 1858: 'I hate the whole system of English names. Why is not Myosotis and Epilobium better than Mouse-ear (of which there are two), or Willow Herb, to which there is as good an objection?'

sentence to the shortest, consistent with perfect lucidity. It requires a short severe study and some little regular attention.

Fanny has been looking over parts of it, and quite agrees with me that the words underlined in pencil will be so many stumbling-blocks to village school children and even higher class ones. In short the whole is not only too scientific but in too scientific language.

[March 3, 1855.] I am extremely glad to find that you have not taken umbrage at my severe criticism on your little book MS. I am always severe and often unreasonably so, though I do not think I was so in that case. I have often thought that it is impossible for a really highly educated man to write a good book for the ignorant, except he be checked by another; to write down to a low capacity, or low standard, is of all things the most difficult. Your present plan is excellent and will, I should say, answer perfectly if you will rigidly resist all temptation to digression, long sentences and giving more than *one idea*, or fact, to be mastered at a time. I made large allowances in your MS. for Leonard's copying, and am fully aware that the lesson was to be learnt by the developing plants, and therein lay another difficulty, it would be impossible to arrive at a general accurate idea of 'the plant' by such protracted means, and it is by giving such a general idea of all the main parts and their relations, as rapidly as possible, that we must begin. In your MS. there is far too much to be learnt of each organ to allow an ordinary intellect to grasp the whole at the end of the first lesson. You talk of a return to collect 'scattered ideas'; now these said *scattered ideas* are what of all things I would avoid the possibility of the pupils acquiring. The *first* acquired knowledge should be systematic and definite. [An analysis of eight Lessons follows.]

I doubt your doing with less than these viii Lessons, but I do not doubt your doing with far fewer words than you imagine. Fanny says that your diffuseness is your snare; I say it is of all clergymen, and of all those who are much in the habit of writing for the public, with no mentor or critic to check them, and whose time is their own in the rostrum. I never read or heard a sermon that I could not weed of half its words to the greatest advantage *of the reader*, mind you, I do not say to the *hearer*, though I think I could almost add that too. To write well and concisely is a rare acquirement, and the pulpit being beyond criticism, clergymen almost invariably become diffuse and verbose. In too many cases words are thrown in to fill up the time allotted to the discourse, partly because the clergyman has other more important duties and in many cases because he has often nothing new to say on his subject. Be all that as it may, I would avoid in the book the diffuse style that is so well adapted to lecturing and demonstrating, and be as sparing of words and concise as is consistent with an easy style. The *aphoristic* will hardly do for a school book, I fear. In lecturing on specimens you cannot so well cloud your meaning by words, or weary by repetition, because the *fact* demonstrated is visible and tangible; repetition impresses it on the mind, verbiage gives time to the audience —but in a school book it is quite different; here the fact is not visible or prominent; you have to impress an idea or image and repetitions and verbiage take the mind away from it. Contrast Faraday's[1] lectures and his writings, and they are models for each, but no styles can be more dissimilar. Your MS. was more a lecture in writing—and this is a lecture *on* writing—but I really am interested in the book and feel my own incompetence to such a task so keenly, that I cannot forbear doing everything I can to put you on your mettle. You *were* an admirably clear writer; perhaps 15 years of a country living has not tended to develop the faculty. You have all too much your own way in lectures and the pulpit; and write your weekly allowance for the pulpit with nobody to pull it to pieces. Do not fear bothering me with questions. I like them from you.

I return your MS. with some suggestions. I like its plan very much, the only apparent defects (and which would probably be much reduced if read in print) are the attempt to explain too much as you go along. *Facts* are one thing, the *rationale* of them is another; and I doubt if you help the bonâ fide beginner much by mixing causes with effects. The beginner *must learn by heart* a certain number of

[1] Michael Faraday (1791–1867), who, starting as Sir Humphry Davy's assistant, became the greatest discoverer in pure experimental science, was proverbial for the personal magic of his lectures, especially to the young.

> *definitions*, and those you do not put before him categorically. Many men have many minds and my mind always revolted at having to read up a long yarn about a word, whose meaning alone in a tangible form I wanted at the time. My own plan would have been to have left much of what you say in the first part to a chapter on Morphology. I think too that by using too many words and attempting too much simplicity, you involve the sentences and mask their meaning. I did honestly try hard, and for the life of me could not understand your definitions of Hypogynous, perigynous, etc.

A similar letter to Asa Gray on the appearance of his excellent 'Elements of Botany' (March 30, 1857) re-enforces these points of view. Some loose definitions are criticised, but the chief one desideratum was an Introductory Chapter 'written in the same lucid, simple, and still accurate and sober style,' introducing the beginner to some of the more leading ideas in a practical study of plans—telling him what to look out for, and giving examples of them. He must insist also on certain definitions being 'absolutely and unalterably impressed on every pupil's mind and at their fingers' ends.' A glossary at the end is not enough.

> It is true that 'Organs,' 'Morphology,' and most of these terms, not all, are defined in the Glossary, but ten to one the pupil will go through and through the work and be unable to define 'Anatomy,' 'Organs,' 'function,' 'type,' at the end of it!
>
> The definition of Physiology is rather loose, is it not? 'The Science of the Forces that determine the action/play of functions.' Your term 'the way it grows' (act of growth) is development, which is not physiology but a branch of morphology. Physiology is Physics + Chemistry. It is true that bad Botanical definers class ovule, growth, and such things under Physiology, but if so then aestivation, vernation, and every other phase of development comes under Physiology.
>
> A little might be said on the great advantage of Systematic Botany as a means of schooling the mind (as good as Mathematics) to habits of close observation, accurate defining, and

diagnosis. Some of our greatest lawyers and medical men have pronounced Systematic Nat. Hist. as an admirable training for medical and legal enquiry, in sifting evidence and disease, etc. etc. Also Syst. Bot., i.e. the Nat. Ord., should be the prominent goal for the beginner, as they are the expressions of the Morphology, Structure and all other attributes of plants. Classifying plants is further an exercise of the reasoning faculties, always bringing memory and judgment into play, and we all know 'Memoria augetur excolendo.' An Introductory Chapter of this kind would invite many thoughtful pupils to think for themselves, and give a dignity to the study that teachers would appreciate. These hints, if worth anything, may help you to a new feature for a reprint.

Another thing must be impressed at the present day,—that Botany is a knowledge of *plants*—that Physiology, Anatomy, etc. etc., are one thing, but Physiological, etc., Botany quite another. Also that in examining in Botany the teacher should never go beyond what the pupil has a practical knowledge of. Botany is a Science of Observation, and the present plan of examining pupils in what they have coached or crammed up is ruinous. They are disgusted at finding that after taking an honor in Botany, when they want to progress in the Science, they have to go back to the Elements. If teachers understood this, they would themselves see the necessity of learning. Tell them that a child with a buttercup could make out whether Torrey[1] or Gray knew most of Botany, but that neither Torrey nor Gray could tell which of two children knew most of plants by examining them on what they had only read. Reading without observation on the Sciences of Observation is most destructive. The difference between the modes of teaching required for the Natural Sciences and Moral Sciences, etc., has never yet been properly put, and until it is, all hopes of getting the Nat. Sciences introduced into Elementary Education are illusory.

Allowing for the difference of aim between a handbook and a course of lectures, there is a close parallel between these

[1] John Torrey, M.D., LL.D. (1796–1873), was born in New York, and became a pupil of Amos Eaton, pioneer of Natural Science. In 1818 he took his medical degree and practised as a doctor, but devoted his leisure to botany and mineralogy. He published a Flora of the North and Middle Sections of the U.S.A., 1824, and a Flora of New York, completed 1843, &c., &c. Professor of Botany in the Medical College, and at Princeton College, and was also State Botanist.

criticisms and the advice given in a letter dated February 3, 1857, to Harvey, who in November 1856, being newly appointed to the Botanical chair at Dublin, consulted him as to the best scheme of lecturing.

The essence of this advice, based on experience as examiner, is to give the students a moderate amount of matter, very thoroughly; teaching through mind and eye and hand, first by clear explanation of fundamentals with three or four examples of each, and exact definition of essential terms; next by big diagrams keeping these chosen examples and exact definitions always before the men's eyes, then by teaching the men to dissect and draw, examining them with specimens, as Sir William Hooker used to do, in the second half of each lecturing hour.

> If ever I lectured on Botany to Medical students and others, I would not give half the matter others do.
>
> Whatever you do, strive to be *under* the mark in amount of what you teach, and over it in well illustrating what you mean.
>
> Never forget that the men have had no elementary training, and come to you absolutely unfit to take up the study of Botany, and keep the elements always in view.
>
> Use as few terms as you possibly can, never using one in two senses, or two for one purpose. I never get a man who can give me a straightforward answer as to what a *seed*, a *fruit*, or an *ovule* is. [The answer is given in a] sort of unsystematic, illogical fashion, showing that those who know what a seed is have no precise notion of it.

As to the ever repeated insistence on the men knowing perfectly the definition of terms employed, such as analogy, affinity, homology, species,

> if any one objects, tell those who know them that they need not look at them, but that in a recent London Exam., out of 45 members of the 3 Colleges of Surgeons examined, not 5 could give a logical, accurate definition of any 5 or more of these terms, and many of *none*! and that without them a right knowledge of any branch of Nat. Hist. is impossible.

> Explain that the philosophy of [the great divisions of plants] can only be understood when they know what a seed and its germination is, an axis and the arrangement of its parts, an ovule and its ovarium.

The course being for medical students:

> Illustrate as many Nat. Orders as possible by Medical plants, showing the drug but *alluding* only to its preparation and uses.
>
> Finally, the less preparation you personally make, except in the way of diagrams, &c., the better; be certain that he who has read up for an elementary course is either unfit to give one, or will fly over the heads of students.

Of existing handbooks, he remarks that Lindley's, dating from 1830, 'are capital as guides, but antiquated,' and 'Henfrey's rudiments not bad,' but the work of another popular writer

> the worst I know, containing every fault elementary books can have, loose, inaccurate illogical, bad English, without distinction of what is useful and useless to the beginner. . . . Impress on the men the folly of attempting to go beyond [these] elementary books except with specimens in their hands; and in conclusion din for ever into their ears that the principal Nat. Ords., properly studied and rightly understood, are the exponents of all branches of Botany, embrace a knowledge of all, are the application of the results of all to practice, and are synonymous with 'Botany' in its highest signification.

Finally:

> I have been talking a good deal about lecturing, since I wrote to you, with Huxley, who has come to absolutely identical conclusions, and is going to alter his course accordingly at the Govt. School of Mines; this *entre nous* at present. He and I have often talked over the subject, and he is quite of my opinion that the present mode of teaching is worse than useless.

The contrast between the old style Botanist and the new was forcibly brought home to him when in July 1862 he paid a visit to Oswald Heer [1] at Zürich and heard him lecture to his pupils.

> All I can say [he tells Bentham] is that if he is a type of the old school of German Bot. teachers, I do not wonder at the Physiologico-Microscopists, Okeno-Schleidenists, carrying the day; for any more dull and dreary exposition of Genera and species I never heard, with no specimens in students' hands, none in the lecturer's, no diagrams, no pictures, no nothing. It opened my eyes to the real facts of the great battle between the systematists and Physiologists.

The great change in English botanical teaching, when it came at last, took shape under Huxley's inspiration. He it was who revolutionised biological teaching in 1872, making his students study the chief types of animal life not merely through lectures and books and specimens prepared by other hands, but from their own observation and dissection of the actual objects, under the guidance of himself and his enthusiastic lieutenants, Michael Foster [2] and Rutherford and Ray Lankester. From animal to vegetable biology was but a step. While Huxley was away ill in 1873, a similar course in botany was instituted with equal enthusiasm by another

[1] Oswald Heer (1809–83), Swiss investigator of fossil plants and insects. Educated at the University of Halle, ordained minister 1831. He went to Zürich in 1832 and lived all his life there. He studied medicine, but soon devoted himself to botany and entomology. In 1834 he became Privat-docent and was the first Professor of Botany at Zurich 1852, and in 1855 the Polytechnicum there. His first publications were on fossil entomology, 1847 and 1853; and his first paleo-botanical paper in 1851. He passed the winter of 1854–5 in Madeira. His *Urwelt der Schweiz* was published in 1865 and his *Flora Fossilis Helvetiae* in 1877.

[2] Sir Michael Foster, M.D. (1836–1907), the physiologist, after a brilliant career at London University, was for some years in practice with his father at Huntingdon. His career as a teacher of physiology began in 1867 as prelector, 1869, professor at University College, London, and Fullerian professor at the Royal Institution. In 1870, after acting as Huxley's assistant, he migrated to Cambridge, first as prelector at Trinity College, then 1883–1903 as professor in the chair founded for him by the university. He became F.R.S. 1872, and biological secretary R.S. 1881–1903; President of the British Association and K.C.B. 1899; M.P. for London University 1900–6. A close friend of Huxley, he carried forward his method of teaching, and edited his *Scientific Memoirs*, 1901. His chief works were a *Textbook of Physiology* and his *Lectures on the History of Physiology*. He was the joint author of *Elements of Physiology* and of *Embryology*.

of his lieutenants, Professor Thiselton-Dyer, afterwards Assistant and successor to Hooker at Kew, himself a student of the physiological botany which had made such strides in Germany, as well as 'knowing plants' after the fashion of the older botanists.

Hooker's own excursions into botanical physiology enabled him to realise the vast importance of this, as an educational influence, as technical training, and as a guide to the true relations of plants as determined by descent and kinship. But to his mind, with its encyclopædic knowledge of specimens, there was one drawback to this insistence on the study of structure and function. 'You young men,' he once exclaimed to Professor Bower, 'do not know your plants.'[1]

His appreciation of the change which ten years had brought about is well shown by his advice to a botanist, then working abroad, who had been trained in the old school, not to stand for a botanical chair then vacant in England (1884):

> My impression is, that it would not suit you, without indeed you have kept up a knowledge and practice of Physiology, minute anatomy, and chemico-phytology, and indeed physico-phytology, which now form the staple of the Botanical teaching, and above all of Botanical examinations in this country. Botany is no longer a knowledge of plants, but how parts of plants 'come about' and what they do! you begin with yeast, moulds, &c., and the higher you go the less you know of the whole plants and the more of their 'inwards.' There is no question of the high scientific value and interest of all this, but the outcome of years of it may leave a man in utter ignorance of any plant bigger than the Torula and Mucor he began with. Botany of this sort is the study of the laws of life, the highest of any: but to pursue it requires a special education; and to teach it, a special practice; and I do not know if you have had either. I have not. It is most necessary for the modern physician and surgeon; it is the gate through which he enters the study of

[1] Apropos of the knowledge of plants and their uses possessed by the old field botanists, Mr. Elwes tells a story of how he and Hooker and Berkeley the mycologist were lunching together, when some new pickles from the West Indies were placed on the table. Berkeley alone, with his knowledge of Materia Medica, was able to identify the ingredients.

his profession; this sort of botany, in this respect, plays the same part in modern medical teaching, that the botanical course which taught the Natural Orders, &c. did of old.

The botanical teaching of my day was the Student's first schooling in diagnosis, and it taught him medical botany, and the origin and history of drugs. Now, diagnosis is taught clinically, in a way it was not in my time, and a knowledge of drugs and their origin is left to the druggist; and botany is made the introduction to organic chemistry and physiology in the application to the problems of life in health and disease.

Our careers are very different from this, and you are making your mark in yours; would it not be better to stick to it? or only to leave it for something in the same line?

CHAPTER XXI

SCIENCE ORGANISATION: SOCIETIES, JOURNALS AND REWARDS

Though the organisation of Science at the Universities and other centres of education was important, more important still was its organisation through the learned societies, partly as meeting places for scientific workers, partly as providing the means of making scientific results easily accessible through their publications. Where these were inadequate to the necessities of the case, established journals of literary repute might be taken into alliance, publishing a scientific column regularly, or, in the last resort, a Review entirely devoted to Science might be set afoot. How heavy a burden such non-original and administrative work imposed on very busy men was to be learned from experience.

One conclusion to which it pointed appears from a letter to Huxley in the spring of 1861, when Bentham, who with characteristic modesty never claimed to be more than an amateur in botany, was proposed as President of the Linnean, a post he held from 1861 to 1874.

Kew: Wednesday.

> You know my prejudice against professional Scientifics being Presidents of these heterogeneous bodies: and in favour of independent men who make a bond of union between Science as represented by the Society and the outer world —and who if really Scientific, are so as amateurs. Bentham is one such, and for the life of me I cannot find another at all eligible on the whole list.

On the other hand the methods of the societies which combined Science with 'Society' and lionised travellers before making very sure of the value of their reports, were as repugnant to Hooker as they were to his friend Huxley. The present generation can remember the laughable explosion of the de Rougemont boom which took place at a meeting of the British Association: a much more notable personage with a tale of tropical exploration and hunting and discoveries in natural history provoked a furore in 1861, followed by a storm of criticism which has never been definitely settled, the most balanced opinion being that very probably what he said was substantially true, but that no less probably his so-called experiences, which were not borne out by subsequent reports from local collectors, had merely been gathered from hunters on the coast.

> The man [writes Hooker to Dr. Anderson,[1] July 7, 1861] is a victim of Murchison's lionizing system: an unscientific bad observer is raised to a first-rate scientific geographical lion, and *after* that has to write a book to justify all the fuss made about him. The poor man is honest enough in purpose, but is dizzy with all that has been done to him and unable at any time to write—he exposes himself awfully of course.[2]

But this Leonine Heresy was not without a medicinal value.

[1] Thomas Anderson (1832–70), botanist, M.D. Edin. 1853, entered Bengal medical service in 1854. Director of the Calcutta Botanical Garden, organised and superintended the Bengal Forest Department 1864; left an incomplete work on the Indian Flora.

[2] In November 1862 Hooker received a letter from Gustav Mann, the Kew collector at Fernando Po, saying that he had been across the country described by this traveller, and that his accounts were all unreal. Mann himself suffered under another 'lion' of the Geographical Society. This was Sir Richard Burton, Orientalist and traveller, who, Hooker tells Darwin, 'has in a public despatch, filched away all poor Mann's credit for the ascent of the Cameroons, calls it his expedition, planned and carried out by him, and calls Mann his volunteer associate. I never read anything so gross in my life. Poor Mann had set his heart on the thing for 2 years, had failed the first time, and was actually leaving Fernando Po for the ascent, when Burton arrived at F. Po as Consul, did leave and had ascended the Mt. several weeks before Burton, following him, was at its foot; having prepared the way and provided guides and everything. I am quite disgusted, but hardly know how to act. I dislike and despise the Geogr. Soc. way of going on so much, that I do not like to bring the matter forward there, and as to having a quarrel with Burton, we all know what it is to touch pitch.'

> I rather like [he writes on June 2] to keep the Geog. Soc. as a sort of seton upon science: it draws all odium for scientific lion-hunting, toadying and tuft-hunting away from the Linnean, Royal and Geological—only that the latter are too fond of following in wake! For my part I eschew them all now, and intend to keep them and their society at arm's length.

And somewhat later, rejoicing that he was not on the Committee of the Geological, he remarks to Huxley: 'I am quite accustomed to seeing things done "more Geologico"—in fact the Geolog. Soc. and its attributes have been worth their price to me in the valuable introduction it has proved to Helter Skelter science and business.'

Through the earlier years of this decade Hooker was specially concerned with the reorganisation of the Linnean Society. His object was to see the Linnean take the same position with regard to Natural History as the Royal Society with Physics. He had been elected a Fellow in 1842, and was chosen a member of the Council in 1853, serving in this capacity for twenty-four years, during fifteen of these as Vice-President. Once on the Council, he endeavoured to carry out much-needed reforms. The famous Linnean collection had fallen into a bad state; Hooker's offer to help rearrange it the year before, when he and Thomson were sometimes meeting at the Linnean, had not been taken up: doubtless owing to Robert Brown's opposition to any change. The printed reports of proceedings presented their subjects in confused order, so that specialists had difficulty in finding what they wanted. It was most desirable to separate the reports, according to their kind and weight, into Proceedings and Transactions (a reform in which the Linnean was anticipated by the Royal Society, thanks to the efforts of 'the small band of us yclept the Philosophical club'), and to divide botany from zoology. Experience in other countries had shown this to be absolutely essential, for the sake of the botanical and zoological public alike, who were now forced to buy reports in which they had no interest; and for the sake of simplifying the already complex bibliography. Moreover, 'though you and I,' he assures Huxley, 'as joint editors may work well on a mixed Journal, the chances are that others would not,' among 'the hundreds of details that belong to both, i.e. to neither.'

References to the subject appear in the letters from November 1853. The Linnean had just elected a new president in Thomas Bell,[1] who held that office for the next eight years. Great things were hoped from his known administrative ability and his keen desire to resuscitate the Society. Hooker could recall one meeting in the old rooms in Soho Square when only five members were present to support the President and Secretary. The list of contributions from British botanists during the last ten years compared unfavourably with those made to other journals. The Secretary was chronically hard up for papers; not unnaturally, since 'for such advantages can the Botanists be expected to sail in such a coal barge, where zoology is little better than rats and cockroaches?' The meetings therefore offered small attraction. 'If something is not done the Society will certainly fall to pieces.' But 'I see no prospect of anything being done till you come up, and Lindley gets on the Council!' (To Bentham, November 1853.)

However, one after another the essential reforms were carried, despite temporary half-measures interposed by the President in order to meet Brown's uncompromising opposition to every point of principle and detail, whereupon Hooker exclaims, 'Save me from a vacillating man of all others,' but confesses afterwards, 'He is so good-natured and anxious that everything should go square that it is impossible to quarrel with him.' At the crucial moment, however, the President backed up the reformers, pacified Brown, and finally, with a rich man's liberality, guaranteed that the free distribution of the new Journal to all Fellows should have a fair trial,

[1] Thomas Bell (1792–1860) was distinguished as a dental surgeon and a zoologist. At Guy's Hospital he was for long the only good surgeon who applied scientific surgery to diseases of the teeth. He was most widely known for his popular Histories of British Quadrupeds, of British Reptiles, and British Stalk-eyed Crustaceae, as well as his edition of White's Selborne, a place where he spent his old age, having bought White's house, The Wakes. As Secretary of the Royal Society (1848–53), and as President of the Linnean Society (1853–61) he did excellent administrative work.

while to meet the ensuing expenses of reform, whether in publications or keep of library, MSS., and collections, £1000 was promptly raised among the Fellows, which 'showed the vitality there was in the old trunk.'

The position of the Society was still further improved in 1856. A great stir had been made 'to get Govt. to give us Burlington House as a site for the five chartered Societies who promote abstract Science.' Now the Treasury granted the Linnean apartments in Burlington House, whither the Royal and the Chemical went also, while the Geological and Astronomical refused to move from Somerset House.

Now that the Linnean was placed in juxtaposition with the Royal and on an equal footing as regards position and all other outward matters, it only needed a little active aid from its members to raise it to its former position, and Hooker was indefatigable in stirring up his fellow botanists to contribute papers. As he wrote to Harvey (November 1856):

> I have always considered that the service it rendered to science between 1790 and 1830, by purchasing the Linnæan collections at its own cost (for £3000), and by publishing gratis to its fellows 20 quarto illustrated volumes of important matter that could never else have seen light, were claims enough upon every man of science to support it.

But the resuscitation of the Linnean Society was only a step towards a larger scientific object. This was to induce Naturalists to concentrate their publications into well-established periodicals and if possible to check the indiscriminate scattering of their papers in numerous journals, many of which were virtually locked to science. It was a most serious evil, and he adds roundly, 'The number of badly edited and badly supported journals is quite incredible, and the present practice of cramming Zoological and Botanical researches into one periodical increases the evil many-fold.' Not that the reformers had any intention of interfering with the provincial societies or Natural History journals, albeit true of some that vehement exertions whip them into a spirited beginning, only to fall away soon and remain burthens upon science. Their immediate purpose was to establish the Linnean on a sound basis, and cultivate a catholic spirit amongst naturalists. 'The crying evil,' in Hooker's words, 'is that Naturalists are profoundly indifferent to one another's wants, and so long as each is regardless of whether it is reasonable to suppose that his fellow Naturalists will get access to his publications, science must drift into confusion.' Let the Linnean then provide the means of rapidly publishing abstract researches with the certainty that they would soon be in the reach of all European and American Naturalists. Then the time would come when all the best papers on such subjects would as certainly be sent to the Linnean as the French ones to the Paris Academy. In the same way, if circumstances compelled the dropping of the Kew Journal of Botany, the best of its material would be absorbed in the Linnean, with its wider circulation, to the advantage of science.

Another valuable piece of centralisation planned was a *compte rendu* from Burlington House, with a classified index of all important papers contributed to the various societies in the United Kingdom. In all these ways the minor societies might be brought together, while the highest flight of hope saw the Royal and Linnean publications issued together.

During the years of reconstruction, Hooker was unflagging in his support of the Linnean Journal, calling on his fellow workers to help, and receiving many promises. Even so it was difficult to keep all up to concert pitch, as appears from an urgent appeal to Henslow, apparently written in 1859.

> I now therefore beg and entreat you not to leave us in the lurch any longer; it is of greatest importance that authors of repute should contribute to the first volume of the Journal, and of all those who promised me two years ago to contribute, and who spurred me on to get up the Journal, scarcely one has kept his word. The responsibility of the thing very much lies upon my shoulders, and I am now calling upon those who induced me to take it, to keep their words: but some of the best are dead! and as to

others, these are promises which they do not see the moral obligation of keeping, or at any rate act as if they did not. None can so well help me out of the difficulty as you, for you could without trouble give us both Zoological and Botanical scraps; and it is scraps we want as much as papers.

Another undated appeal (probably in 1861) reiterates his own responsibility for the progress of the Journal.

Dear Huxley,—I find that we are really hard up for zoological matter for our Linnean Journal, which is now arrived at its critical period; so my dear fellow do not desert us and give us a yarn on the Crab's inwards without fail—it is almost a sin to press you to write, but I must be whipper in. We have plenty of good botanical matter and Lindley has rallied round us, but if zoological matter is not forthcoming, *the present plan* of the Linnean Journal will fall through and my shoulders will have to ache for it, as the onus of the undertaking rests so much with me.

I like your Museum thing[1] extremely, it is the only really sound elementary introduction to understanding Geological evidence that I have seen. I shall bring it with me on Tuesday.

Ever yours,

J. D. Hooker.

Thus the Linnean Journal came to fulfil its function as a record of the natural history sciences for workers in science, so far as focussed by the Society. As he wrote later, 'It is a gallant Society that struggles on amongst proverbially poor naturalists, spending its whole income on publications and Library and *giving* all its publications to its members.'[2] The Journal was the more needed on the botanical side, as the Kew Journal of Botany had for some time been going downhill. The best botanists had become chary of contributing, for Sir William Hooker, though unremittingly busy in his old age, had grown careless and uncritical in his editing, and his son had no

[1] 'Preliminary Essay upon the Systematic Arrangement of the Fishes of the Devonian Epoch,' Mem. Geol. Surv. of U.K., 1861.

[2] To Mr. Bolus, Feb. 4, 1873, who sought election to the Linnean (see ii. 4).

time to revise his editorial work. Indeed, he saw clearly that the Kew Journal could not advantageously continue, and with the help of old and trusted friends like Bentham and Harvey and Asa Gray, persuaded his father to give it up.

But the Linnean Journal was restricted to working men of science. To reach a wider public, to spread the general comprehension of scientific ideas, seemed very important to the advanced wing. To this end a scheme was organised, mainly through Huxley, whose energy was in touch with the literary as well as the scientific world in London. From 1858 onwards a fortnightly scientific column was arranged for in the *Saturday Review*,[1] to which Hooker was too busy to contribute, replying to Huxley's invitation as follows:

Kew: Wednesday, 1858.

I have long been under an engagement of honor to Lindley's *Gardeners' Chronicle*, a paper that has acted most liberally by me, and for which I have not written *a line* for 9 months, and have no present prospect of doing anything for, though I really ought and should. Now I cannot *bring myself to the scratch* to do articles (and however simple I am well paid even for notices of Botanical Events and translations of short foreign announcements); how can I expect to screw myself up to write pregnant columns (for they must be bellyfulls) for the *Sat. Review*?

Besides all this, as my non-original-work-duties increase here, I proportionately crave to be at original work. I want to get up good papers on obscure and difficult Natural Orders, and such work is quite inconsistent with reviewing.

I quite feel the want of such a class of articles as you propose and feel my own selfishness in withdrawing; but I doubt if the good effects would be at all commensurate with the time and labor that we should expend, and I am quite sure that both you and I would be much happier without such trammels. Further I am confident that the articles would in our cases be contributed at the expense of original work, and we should thus 'seek in certain ill, uncertain good.'

[1] It is amusing to find the *Saturday*, for all its excellence on the literary side, condemned as 'dreadfully sententious and priggish' and amateurish in its politics, whence its sobriquet of *Pall Mall Gazette*.

In 1860 a wider opening offered. Three years before that, the *Natural History Review* had been established in Dublin, its moving spirit and chief owner being Dr. Wright, whilst amongst others interested in it was Harvey, to whom Hooker wrote in candid condemnation of the first number and in particular of a careless survey of Hooker's views on Natural Orders.

> I beg that you will read what I have said, and tell me if you are not wholly mistaken in your suppositions. If that is the way you are to review Botanists' labours for *Dublin Review* I think we had better keep up the Kew Journal in self defence.

Indifferent success attended the Journal in its Dublin home. After nearly three years Dr. Wright proposed to transfer it to London, and to associate Huxley in the editorship, with practical control of the scientific side in his hands. Though the latter saw in the new scheme nothing but extra work for himself, it promised much for the interests of science, 'considering the state of the times and the low condition of natural history publications (always excepting *Quarterly Mic. Journ.*).' For three years he continued at this post, till overwhelmed by ever increasing work; then, paid editors being appointed, he handed over to them the responsibility of the 'commissariat' of the *Review*, which ran for two years more.

To limit the amount of this extra work, however, he had to get co-editors. Writing to Hooker a full account of what had been done, he remarks:

> Now up to this point you have been in a horrid state of disgust, because you thought I was going to ask you next. But I am not, for rejoiced as I should be to have you, I know you have heaps of better work to do, and hate journalism. But can you tell me of any plastic young botanist who would come in all for glory and no pay, though I think pay may be got if the concern is properly worked. How about Oliver? And though you can't and won't be an editor yourself, won't you help us and pat us on the back?

To the new *Natural History Review* Hooker, however, both contributed and offered criticism.

To T. H. Huxley

January 4, 1861.

My only fault with the 'Review' is its brevity as I told Currie to-day—I am extremely pleased with it and shall have some mild review for next number I hope if you have space. I still think there will occur a few cases where you must translate the German title—at least the German Botanists do often invent titles that are unintelligible except the book be read! It is the most useful Review I ever saw. Your article is very exhausting of all you propose, clear as to argument and extremely well put; the first three pages are also very happy, especially the prop. relative to man's duty. It will be a balsam to many short-witted and honest but timid enquirers.

Another point in which the organising spirit made itself felt was that of charitable funds for science. For such there was only the Civil List to fall back upon, and the demands made on it were ill regulated. The Treasury would be puzzled by receiving four applications at once for Natural History pensions —all the claimants being described as 'distinguished men.' Under such conditions it was useless to bring forward another who had not claims for Government aid.

Now a very deserving case occurred in the end of 1858, of a microscopist who had done excellent work, but had not achieved public distinction. To Hooker this hardly seemed a case for a Government pension, if it had been possible to obtain one. It was, however, a case for personal help from scientific men. A strong appeal was made on general grounds for £500 to buy an annuity, with the result that the amount was more than subscribed twice over. Instead then of sinking the whole sum in an annuity much larger than was proposed, a wider scheme was put forward—namely, to invest the capital, pay the annuity originally proposed to the beneficiary during his life, and in the end secure the capital as nucleus of a general scientific charitable fund, to be increased by voluntary subscriptions. Subscribers were given an option as to the destination of their own gift. With hardly an exception all agreed on the larger plan.

The following passages illustrate his point of view.

To the Rev. M. J. Berkeley

January 9, 1859.

I am quite sick and ashamed too of this constantly begging Govt. for pensions for persons whose claims can in no way be called national. Science suffers by the refusals we get, and really national claims suffer too. We should do much better to have a private fund for such unfortunate men as A., B., etc. whose most meritorious labors are neither sufficient to raise themselves to scientific hero-worship nor are directly beneficial to the Arts or otherwise. I do not think it fair to apply to the nation except in cases of great eminence or services of great practical value. It is the duty of Govt. to encourage and stimulate the first and to reward the second, but if the Govt. pensions such men as A. and B., they must also pension no end of literary characters with equivalent claims and less chance of private help. Few people look at this in a sensible manner, they regard pensions as State Vails to be scrambled for in the most undignified manner.

To W. H. Harvey

I see too, what I specially dislike, a sectarian view of the case arising—it is the Microscope versus all science; or Nat. Hist. versus all other branches. I strongly object on all grounds of policy and fairness too, to the establishment of a 'Naturalists'' fund, except indeed the Physicists prefer to have a separate one—when I shall gladly join the Naturalists; though even then I should feel myself in honour bound to join a Physical Science one too. Any attempt to segregate Nat. Hist. will do it great harm: it cannot stand alone, it owes the Microscope to Phys. Science, and all Physiolog. Botany too. Their narrow-minded views are the bane of science.

As to the particular encouragements to Science that consisted in the bestowal of medals for distinguished work accomplished, he came to find the whole thing unsatisfactory, after it had fallen to him both to receive and to allot these. The great difficulty lay in holding the balance between individual distinction and the claim of each branch of science for recognition in its turn, between rewarding the man who had arrived and encouraging the man who was working his way up.

Official recognition of this kind was very different from a worker's acknowledgment of his predecessors' labours; that was a proper recognition to receive, and indeed mere honesty to give. Personally, he was quite unconcerned if he found, on occasion, that certain continental botanists ignored the prior work of himself or his English friends, though he condemned such lack of frankness. 'I always feel,' he tells Asa Gray (March 29, 1857), 'that we must so often unintentionally ignore one another's observations, that we can ill afford to make the *least* of those we do know of.' The only thing that struck fire from him was neglect of his father's merits or the discourtesy of failing to acknowledge his abundant generosity.

The first of the letters that follow on the award of a Royal medal is in reply to a letter from Huxley, which is given in the 'Life of T. H. Huxley,' vol. i, chap. 8, under date of November 6, together with a response as generous as Hooker's from Edward Forbes. Huxley, who was on the Royal Society Council, explained to each of them, his close friends, why he could not vote for one to the exclusion of the other, and therefore voted for both!

November 7, 1854.

MY DEAR HUXLEY,—I am very much obliged for your kind note although quite uncalled for either as apologetic or explanatory, for I fully appreciated and approved your *springs of action.* I quite enjoyed having a competition and should have been very sorry for the sake of science and my own that no one else had been proposed. Of course I do not in any way look upon my claims and Forbes's as coming into competition, but do upon the claims of Botany and my etceteras and Palaeontology and Forbes's etceteras as having come into direct competition. There has been but one honour given to Botany by the R.S., that is the Copley medal to Brown, whereas Zoologists, Palaeontologists and Geologists *galore* have been honoured over and over again. I have always thought and still think that both

Lindley and Bentham in this country deserve a medal, infinitely before myself in Botany—men who are famous abroad but thought comparatively little of in this country from various motives. I should have been better pleased still if you or some other naturalist had proposed Forbes, for Grove[1] has no more *real* appreciation of Forbes's or of my claims than Graham[2] or De la Rue[3] have, and acted simply out of a vague sense of Geology being something more physical than Botany. In an abstract point of view I think Forbes's claims far superior to mine: but the R.S. should not look solely to abstract claims, but seek to distribute their rewards judiciously over all classes of science and the different branches of the classes, e.g. taking a hypothetical case—a man who (like you) works out a point of abstract science during the difficulties and discouragements of a voyage, has in my opinion an equal claim *at least* with a man who works the same in his easy chair; even though the latter works it better.

Bell told me of all the proceedings after I left Council on Thursday and spoke with undisguised satisfaction and pleasure of the parts you had taken.

Ever, dear Huxley, yours,
J. D. HOOKER.

Anything in the nature of sectionalism in making these awards was very repugnant to him; and he was doubtless

[1] Sir William Robert Grove (1811–96), a man of science and judge, was educated at Brazenose College, Oxford, subsequently receiving the D.C.L. in 1875, and the Cambridge LL.D. in 1879. Ill-health, which checked his early career at the bar, gave him time to follow his scientific bent. He became a member (1835) and subsequently Vice-President of the Royal Institution, and Professor of Experimental Philosophy in the London Institution. His invention of the gas voltaic battery in 1839 brought him election to the Royal Society the next year and a Royal Medal in 1847. His most important work on the Correlation of Physical Forces (1846) anticipated Helmholtz's essay on the same subject. Later, his scientific eminence brought him much legal work in patent cases. He was raised to the bench in 1871, retiring in 1887.

[2] Thomas Graham (1805–69), chemist; M.A. Glasgow 1824; Professor of Chemistry, Glasgow, 1830, at Univ. Coll., London, 1837–58; Master of the Mint, Keith prizeman and Gold Medallist of the Royal Society, first president of the Chemical and Cavendish Societies; F.R.S. 1836, and twice vice-president; Bakerian Lecturer 1850 and 1854; D.C.L. Oxford 1853.

[3] Warren De la Rue (1815–89) was one of those successful men of business with whom science came first. He was the author of various successful inventions, both for commercial purposes and for scientific research, and was especially distinguished for his work in celestial photography.

prompted by memories of this kind when, after privately naming certain botanists as worthy of a medal, he wrote to Henfrey in 1859:

> I may tell you that I am opposed to the whole system of medalising, as being quite beneath the dignity of real science and of the Royal Society; but if it is to go on, I shall hope to see it well carried out.

Beyond the question of scientific recognition of science work, lay the other matter of public recognition by knighthoods and the like. This concerned him later; but to summarise his opinion, services, not scientific eminence as such, should be 'rewarded' by distinctions.

Several letters illustrate his eagerness that due honour be paid to his father; the first is one to Bentham on his receipt of the Royal Medal (November 20, 1859).

> The first matter is the R.S. medal; I, and all other Botanists, are equally indignant with yourself, at my Father's merits being overlooked in the distribution of [the] Copley medal, the only one they could offer him—this is wholly Brown's fault, and will I fear never now be mended, greatly as it has been desired and tried for. The Copley is the only medal that could be offered him, and that medal is *theoretically* all but exclusively confined to great discoveries, or great generalizations of proved value to future investigators. I have long fought for its being given to general scientific merit of half a century or upwards—hitherto in vain. Happily the 2 Royal medals are in so different a category that they do not clash with the Copley, and they are further confined to our countrymen; but for this, your and my and Lindley's having a Royal medal would have been more than invidious. With regard to the claims of your line of research, it is true that *in Botany* they have (thanks to Brown) been altogether put aside, but those of a parallel character and value have always been acknowledged in Zoology and every branch of Physics; and 'better late than never,' is all I can say to the R.S. in your case—no medal was ever more richly deserved and it was I am told given unanimously.

To the Rev. M. J. Berkeley

1858.

I do not know whether I ever told you that there has been for years a hitch about electing my Father into the Academy at Paris, a matter now regularly jobbed. They have long felt that they ought to do so, but time has crept on and they have only cared to toady their own people. As it is, Wallich's place is not yet filled up!! because one party want my Father, another me, and a third (God help the mark) Parlatore!!![1] I have written privately to Decaisne (who is most honorable) to tell him that I must not be thought of by any one, for that it would be both an injustice and personal grievance to put me before my Father. I could not of course allude to the matter myself to any one but Decaisne (whom I knew from Brown and personal knowledge that I could trust), but it may be possible for you if you have occasion to write to Montagne to hint to him how astonished people are that my Father's claims are overlooked so long by the French Botanists. They are very welcome to stultify themselves by putting Parlatore before Bentham, Thomson, yourself, Harvey and half a dozen other men I could mention without including myself, but I cannot stomach this treatment of my Father. Please keep this matter private, and

Believe me,
Ever affectionately yours,
Jos. D. Hooker.

To Dr. Anderson

July 2, 1860.

Excuse my mentioning that any allusion to my Father in acknowledging your obligation to the Kew Herbarium (in Aden Florula) would gratify him very much. It is sometimes forgotten that he is its author and owner, and I know he has on such occasions felt hurt at the omission.

[1] Filippo Parlatore (1816–77) was born at Palermo; Director of the Royal Museum of Natural History at Florence and Professor of Botany. He is best known in England for his monograph on conifers and his unfinished *Flora Italiana*. He was President of the Royal Tuscan Horticultural Society and of the Botanical Congress in Florence, 1874.

Similarly to Harvey, July 1859, on the publication of his 'Thesaurus':

> I do not know on what principle you put Herb. Hook. to *MacKaya bella*, not to any other species, implying that that alone was in Hb. Hook., indeed I think that Hb. Hook. should be put to all those plants that were sent originally to it, and of which Herb. T.C.D.[1] received duplicates, especially seeing how indefatigable my Father has been in getting up correspondents for your books. . . . I would not mention this were it not that such trifles are made bones of contention and that my Father has himself diverted the current of Cape contributions to T.C.D. to a considerable extent.

[1] Trinity College, Dublin.

CHAPTER XXII

MISCELLANEOUS, 1850–1860

SEVERAL letters bear on his methods of work and illustrate his tendency to bring anomalies under established principles instead of inventing new principles to suit the exception; his passion to verify things for himself; his critical frankness in dealing with ill-founded ideas combined with readiness to accept well-founded criticism. Others are of personal interest.

Kew: Wednesday, Sept. 20, 1854.

DEAR BENTHAM,—I have just been examining a monstrous *Stachys sylvatica* with a long 4-lobed ovary consisting of 2 fore and aft carpels, i.e. one carpel with its back to axis and 4 parietal ovules in pairs at the sutures, thus (*diagram*).

I think this reduces your Labiatae to the ordinary type of carpellary structure. Was it not you? who once quoted Labiatae to me as opposed to Brown's marginal carpellary theory of origin of ovules?

I am a far better Tory than you are and like laws. I on principle object to nature having one law for carpellary produced ovules and another for free central ones. I would rather go the whole hog and call all placentation axial and all ovules produced on the axis, or adnate portions of it, or branched adnate portions of it, running along edges of carpellary leaves, than to hold to one law for the majority of plants and take another for the exceptions. In Botany there are no end to the 'morphological differentiations' (as Von Baer calls them in Zoology) which result in the most complete congenital obliteration of all traces of original design in the construction of compound organs. I had a talk with Lindley the other day about axial placentation, and he immediately knocked me down with Schleiden's argument derived from the ovule of Taxus being absolutely solitary and terminating a branch—this vexed my soul; for I confess to the most perfect distrust of Schleiden, which leads me to forget his writings, and I did, when reminded of it, remember his dwelling on that very point. After two days I modestly ventured to examine Taxus myself and behold, I found *two* ovules in every one of the first 3 buds I opened, and neither terminal, and when only one occurred it was lateral. Each had a rudimentary scale like ovarium. So much for that argument. On the other hand I can quite understand such a congenital arrest of organs in *Taxus* as should result in an apparent terminal ovule, without making a special law in the Vegetable Kingdom to account for it. I have also a monstrous *Primula* with parietal placenta and ovules; the Pink or Carnation is another common case in point and so on, all new facts tend to reduce the exceptions to the carpellary theory and none cut the other way.

I have commenced the V.D.L. Flora, and find it my fate to destroy species as I go on, and the more carefully I examine the more to fell; on the other hand I am extremely gratified with the multitude of good, new and undescribed species in the Australian Flora.

Passages may be quoted from two letters to Henslow which are too long to give in full. Henslow, struck by an anomalous structure in Nelumbium and several curious points new to him, and unaware of the light thrown upon these points by many observers, had founded an explanation of them on the structure as it was before him, and had assigned not only Nelumbium, but Nymphaea, to the Monocotyledons. Hooker had lately examined the germination of all the genera, and his lively criticism was directed, not against the facts observed, anomalous though they were, but against the reasoning, where there was so much evidence, direct and indirect, to be reckoned with on the other side.

3 Montague Villas, Richmond: January 24, 1855.

DEAR HENSLOW,—Thomson and I are aghast, and horrified, and thunderstruck, and doubled up at your conclusions about Nelumbiaceae. Here have we just printed off the result of the most long and patient study, of all the characters of all the genera, from the embryo, germination, rhizome, etc., etc., and come to a definite conclusion, that all these are in all respects dicots; and here you come in, and examining dried seeds of *Nelumbium* alone, knock all our results on the head, ruthlessly, remorselessly, wickedly and wantonly, perhaps with malice prepense! Only fancy, I have just printed 8 *pages* of arguments to prove that all are Dicots, root, stock (root-stock), and branch, leaf, flower and fruit! This is a blow to Flora Indica. Alas for Flora Indica, we shall go into mourning.

Joking apart, do you know that the point you have settled (?) is the most difficult and most disputed in all Systematic Botany, that it has occupied the attention of observers from Malpighi to Trécul, Hook. fil. & Thomson; that D. C., Richard, Planchon, Gertner, Asa Gray, Lindley, Henfrey, several Jussieus, and others have made a special study of it, and that within this very few months Trécul has published long essays on the subject? Like every other subject of the kind it cannot be settled by an examination of one organ or series of organs, but requires a very careful consideration of an immense number of facts in the comparative anatomy of plants. . . . Whether right or wrong in your supposition, you have, I assure you, good 2 months' reading and study before you would be justified in publishing on the subject; except indeed you have discovered some very novel fact. Thomson's and my belief is, that the resemblances to Monocots are pure analogies and nothing more; you must remember too that upon whatever individual point you may be inclined to ground your arguments in favour of Monocots, you have an enormous mass of evidence in favour of Dicots to subvert, besides the direct affinities with *Papaveraceae, Berberidaceae,* and *Ranunculaceae,* which I do not see how you are to get over. This one fact should engender caution, that Nymphs. have direct relations with these Orders, and none with any Orders of Monocots whatever. . . . Even Trécul, who considers the rhizome of Nymphaea as exogenous, agrees that the embryo is strictly dicotyledonous! I have examined all the genera in germination, Euryale, Victoria, Nymphaea, and Nelumbium, and these are all germinal, exorhizal, and dicot. in the process, besides the reticulated leaves and a host of other characters that you must find some explanation of, under your hypothesis.

. . . You may console yourself with the fact that there is no snare so great as an anomaly of this kind, in the way of a correct appreciation of the affinities of families. Of all branches of Botany the Systematic requires the most extensive knowledge of structure, and the most careful consideration of the relative (far more than the positive) characters afforded by the organs. Just look at Lindley's heterodoxies with all his knowledge, all arising from seeing only one side of the question. The older I grow and the more I study the affinities of plants, the more ignorant I feel, for it is a most comprehensive study. This is my homily on Nymphaeaceae.

Richmond: Saturday, 1855.

DEAR HENSLOW,—Many thanks for your exposition of Nelumbium. I think you have got hold of as pretty a paradox as ever graced the pages of Schleiden; however I will not prejudice your observation till I examine again. My great objection was however not against your making Nelumbium Monocots, which I always thought beyond assault, and which has never been assailed but by yourself, but Nymphaea, the structure of whose embryo and plumule is so totally different from your analysis of Nelumbium, that if your theory holds good then Trécul's paradox will be exactly reversed by you and Nelumb. will go to Monocots, and Nymph. remain in Dicots!!! I think however that your genius and originality have here led you deep into the slough of Paradox and that your emersion when it comes, will be with a rapidity directly proportioned to the buoyancy of your good understanding and the density of the said medium + the resilience resulting from the rapidity with which you descended. . . . I might have turned Buddhist, Romanist, Hindu or Mahomedan on half the evidence during the course of my travels.

A slightly condensed translation of Braun's[1] 'Rejuvenescence of Plants' appeared in 1854.

To T. H. Huxley

September 12, 1854.

I have been groaning over 'Rejuvenescence' que Diable! When is this German rubbish to end? Do read the first 20 pages and tell me your candid opinion as a scientific man: I confess to a want of poetic feeling or at least of that turn of it that appreciates aesthetics in its modern application to spiders and toadstools, or also (and really in this case to my sorrow) of power to grasp metaphysical subjects, and what some think high-class imagery too, and so I really would feel it a personal favour if you would tell me whether I ought to understand, or admire, or see any depth in, or at least see nothing that should convince me that there was no depth in, the first 20 pages of that blessed production, Braun's Rejuvenescence. Mind you, I am a personal friend of Braun's and like his real scientific work extremely, I cannot applaud it too much, but there appears to me a wide difference between exact studies upon the physiology and structure of cryptogamic plants, in which he excels, and upon the laws that regulate the development of organs, in which he is also good (though often fanciful), and these wild vagaries on the connection of life, soul, porridge, mouse-traps, and the divine essence. Braun's forte is mathematical precision and, like many other men of like mind, he cannot (at least so I think) distinguish between truth and nonsense when he takes up speculative subjects; after all perhaps I am fighting with a shadow and I have a notion that after the 20th time of reading Henfrey's execrable parody of the original, and after [Black?] (who is in Scotland) comes home, if I get him to enlighten me on the German, I shall find that Braun's mountain will sink into a mole-hill and that I shall find he is only clothing very old ideas in very cumbrous and far-fetched garments. I am far from condemning the Ray Club for

[1] Alexander Braun (1805–77) was born at Regensburg and educated privately till 1815, when he was sent to Carlsruhe. He contributed to botany while still a schoolboy. After study at Heidelberg (1824), Munich (1827) and Paris, he became Professor (1832) and Director of the Natural History Museum at Carlsruhe and later at Berlin. He wrote many papers; his most famous work is *Das Individuum der Pflanze, Species, Generations, &c.*, 1853.

translating these things, but I do condemn several of the translations as utterly unworthy of the Club and of England and as giving us the worst repute throughout Europe for our knowledge, or rather ignorance, of the spirit and language of Germany, and I protest boldly against such work as Oken, Braun, Schleiden, Meyer, and others, being given to the British public, without one word of explanation and without a sound preliminary essay on the subject, pointing out what can be understood from what cannot be, by 99/100 of the readers, let these be ever so clever or all (like me) ever so stupid! It would surely be much better to offer a little of the money spent on the laborious translation and printing of the worthless parts (the repetitions and verbiage and truisms and trash with which all these works abound) to a good preliminary essay and good notes. Good God! are these authors such Oracles that we must translate every syllable and render letter for letter, lest we lose a drop of their saliva, or a whiff of their flatulence? Darwin says he does not pretend to comprehend it! I have been reading Braun's Prize Essay on 'The Individual in Plants,' and like all other Prize Essays, you can see it is written for a Prize, only overdoes and mystifies what, in the only sense we can grasp it, is a very simple subject.

Braun reminds me of a kitten playing with its own tail. I could not help taking a dose of your Individuality Lecture after it as a curative.[1]

The following undated note, written while wife and family were away in the summer of 1856, is the echo of a controversy then proceeding in the *Annals and Magazine of Natural History*. Huxley, in his Royal Institution lecture 'On Natural History as Knowledge, Discipline, and Power,' delivered on February 15, 1856, had shown by various examples the inadequacy of Cuvier's doctrine, passed on by uncritical compilers, of a necessary physiological correlation of organs which acts as an infallible guide in the restoration of fossils. Given a tooth, then follows the shape of the jaw, the shoulder blade, the forearms, the claws; the diet and habit of the animal.

[1] 'Upon Animal Individuality.' A Friday evening discourse delivered at the Royal Institution, April 30, 1852. See T. H. Huxley: *Scientific Memoirs*, vol. i.

What then, says the critic, of the sloth? What structural distinction between herbivorous and carnivorous bears? The principle, 'valuable enough in physiology, is utterly insufficient as an instrument of morphological research.' Falconer attacked him in the June number. Huxley replied in July.

[June ?], 1856.

DEAR HUXLEY,—I have been dissipating the disconsolation of my solitude (rather fine that) by reading old Quarterlies as I nutrify and assimilate (better still) and find in xli. 313 a passage that will amuse you and rile Falconer—'Under the influence of this *delusion* "the necessary conditions of existence" the deservedly celebrated Cuvier is found asserting that any one who observes only the prints of a cloven hoof, etc., etc.—it is worth your reading.

Ever yours,
J. D. HOOKER.

In the letters next given, a masculine view of housewife philosophy blends with consideration for a 'kitchen revolution' which postponed a visit to the Huxleys. Mrs. Huxley, be it remembered, was for a long time something of an invalid.

Kew: Sunday [Nov. 1859].

DEAR HUXLEY,—My wife and I are going to arrange with Mrs. Huxley about our going to you on Wednesday week, anent which we abjure the dinner. It is all very well for us (you and I) to think and say what we please about it, but even the most modified dinners are sources of disquiets innumerable to ladies who are not well known to one another. I know from experience how it worritted my wife when she was in poor health, to have to provide for only one or two people whom she did not know; it generally knocked her up for the next day and she often knocked up before the evening was over. They will be anxious about matters that we care nothing about, let them go ever so far wrong; and about matters that *cannot* go wrong except by miracle, but then you see they do believe in more miracles than we do and that's the philosophy of it.

Now, as Hooker merely dated his letters 'Kew' or 'Kew Gardens,' Mrs. Huxley had no address at which to write to

Mrs. Hooker. Being constrained to send his wife's second letter, as he had sent her first, under cover to Hooker himself, the Professor, while roundly asserting that 'the first lieutenant scorns the idea of being "worritted" about anything,' took occasion to poke fun at his friend: 'The obstinate manner in which Mrs. Hooker and you go on refusing to give any address leads us to believe that you are dwelling peripatetically in a "Wan" with green door and brass knocker somewhere on Wormwood Scrubbs, and that "Kew" is only a blind.' (See 'Life of T. H. H.,' i. ch. 17, under the erroneous date of 1861.)

Kew Gardens: Saturday, November 19, 1859.

MY DEAR YOUNG FRIEND,—When you are wanted you will find out where I am. Very soon I shall have a half sheet of probabilities for you to calculate for me (in which you may find that $x = 0$).

I have elected to dwell in obscurity for past 3 months and should like to continue to do so for the future, and shall try to. I have neither house, wife, nor children,[1] and were I not as uxorious as a guinea-pig, and philoprogenitive to a fault, I should not sigh for change. I am living with my ancestors who take their turns of taking to bed—it being now the Mater who is prostrate, with a bad leg. As to going to town, I have not the smallest idea of doing so till my wife comes to wake me up, which will be when the house is ready for her and she for it, and Henslow ready to part with her,—he being absolutely lone now but for her.

I have avoided suicide by working extremely hard with my head, hands, and legs, have finished 2 papers for Linn. Trans., 2 for Linn. Journal, the Tasmanian Essay which has run to 130 pages, and the Flora of that ilk in 700. Except a week in Norfolk where I geologised 3 days with Lyell and Gunn, I have been nowhere but for an occasional Sabbath (I forget how to spell it, but know when it comes) to Hitcham.[2]

[1] Mrs. Hooker and the children were staying with Professor Henslow at Hitcham, while the house into which they were moving was being painted.

[2] A little later he tells Huxley how, besides his own ordinary duties and works, he had in one week 'revised proofs for five different authors' works, contributed stuff for two lectures [by non-botanical friends] and precious stuff too! and read three authors' MSS., and reported on a long fossil paper.'

Amid 'all this mental rumpus' without apparent end which made him

I read the history of the unctuous meeting of Philos. at Aberdeen and have read the severe remarks of barbarians on the toadying and tuft-hunting and buttering. Judging from titles of papers only, I should say there was never so much good matter in science brought to a head at once. Whilst you were sporting your science I was for 6 hours a day engaged in the philosophical pursuit of distributing 86,000 duplicate named Indian plants. I liked it passably well! I could think all the time and to some supposed purpose too. A good daily allowance of purely (or almost purely) manual work upon scientific materials is a most wholesome thing. I have thought my best thoughts when collecting and arranging, and now that I do not intend to collect or arrange any more, I find myself a fool for having snubbed these mechanical exercises that have secured the opportunities of opening up so many trains of ideas, that would otherwise never have fructified.

Huxley had asked for specimens of some insect pests from the hot-houses of Kew.

I send a brood or two of common mealbug, a piece of old cactus with Cochineal Cocci, and a few leaflets of a fern with '*Scale insect*' on it.

Fortunately we cannot supply you abundantly by this post, as my Father and I have had such rows with the foreman and gardeners about the prevalence of these beasts, that they are nowhere very abundant in our houses just at present. Asking us for Cocci is like asking a decent Boarding School Lady for a few crabs and other *Pediculi* from her pupils! However for Science's sake we will forgive you.

Unnecessary questions are a trial. He writes to Professor Henslow:

January 20, 1855.

Many thanks for your letter; I have been bothered out of my life with enquiries about *Gynerium argenteum*, and of all the γυνῆs she is the most troublesome. If

altogether dizzy with his own and his neighbours' affairs, there was a grain of comfort: 'I have but one grim abiding source of satisfaction—I don't lecture and I never will.'

Sir J. K. would only read the *Gardeners' Chronicle*, he will find out all about the plant and that the *male* is *not* now to be had at Kew—any more than apple flowers are at Xmas. I like your account of Sir J. K., he promises well, but these people are always promising well, and they make me as snappish as a turtle by asking questions that are answered a hundred times over in the weekly periodicals. Some other people bother me in like manner about Rhododendrons; and I am tempted to say 'read my book and you will find out all about them'; it is hard to have to write books and read them to the public afterwards!

A similar case occurs years afterwards.

To T. H. Huxley

December 2, 1869.

A. is a good soul, but is cursed with a Microscope.

I proposed a tax on microscopes some years ago, exempting Professors only. Recommend to him a mild course of study—to be followed by a reperusal of your lecture, after which you may tell him safely that he may write again!

The following touches on the sense of home. In 1854 Bentham had just decided to give his valuable herbarium to the nation and leave his beautiful but remote home in Hereford for Kew. With characteristic self-depreciation he had even contemplated giving up botany altogether, but the Hookers urged him to join them at Kew, where he could have the run of their own herbarium and library, and help to bring out the Colonial Floras projected by Sir William. Hooker had suggested this already, writing in 1853:

Do you know we often speculate on your coming to live in Kew, with plenty of botanical society for yourself and of friends for Mrs. Bentham; how glad we should be of you. You are suffering from a common calamity in the country: the migration of neighbours, and one you cannot guard against and which will grow with your years. If I saw any prospect of an advantageous settlement of your collection at Kew I would urge your cutting Pontrilas and having a small establishment here. I think you could live here comfortably for £600 including as much fly-hire as you pleased.

Then, were your Herb. at the K. of Hanover's and your Library with yourself, you might get on very comfortably. This will be my resting place no doubt, and I do not think we should quarrel, and I am sure our better halves would hail the event. If you should think of such a change (and it strikes me that feeling as you must, the comparative solitude of your present position, you may do so) I need not say how happy I should be that you put it into execution.[1]

To George Bentham

February 16, 1854.

MY DEAR BENTHAM,—I am heartily glad that your mind is made up now, as I cannot but in my humble judgment think that it is so for the wisest and best in every point of view. I have turned the matter over in every possible way, as I have been going through the daily dull routine of distributing tickets and specimens for 'Herb. New Zealand' and 'Herb. Ind.' or 'Hook. fil. and Thom.' I do not wonder at your regret in leaving Pontrilas, seeing that I have always felt leaving a home, however bad, and even for a better. In your case, so far as the change is concerned of *house*, yours will not be for the better, as you certainly will not get so good, large and airy a one here, and I fear nothing so much as your feeling the change. Still as I have always become attached to a home however bad, I quite expect that you will warm to a small abode here. It is very odd, but I left my detestable cabin on board the *Erebus* with real regret, and no less my wretched tent in the Himalaya: not from a maudlin romantic regard, but because I felt I had been happy and comfortable (after a sort) under their respective shelters and fulfilled so much of my destiny under them as was appointed to me without wishing or caring for better.

Whenever it was possible, during this period, a summer trip to Switzerland, then a more primitive playground than in these days, was planned. The Hookers enjoyed making up a small party of intimate friends, travelling in cheerful companionship and with the economy that attends on numbers. One such group which set out in 1852 became immortalised

[1] In 1855 Bentham moved to London, taking a flat in Victoria Road, whence he visited Kew daily.

in their inner correspondence as Brown, Jones, and Robinson after Doyle's delightful Tourists. Brown was Harvey; Jones, Hooker; and Robinson, Thomson, then established at Kew with the Hookers. In the autumn after their return the first letter to Harvey (November 4) opens:

> My dear Brown,—Your letter greeted us well and we were greatly delighted to receive it. Robinson says 'he would not like to insure your scrag in Tipperary'; Jones says he would, petikularly Mrs. Jones says so.

And a few days later:

> Mrs. Jones begs to report that all at Kew are flourishing; Mr. Robinson especially is in high feather, and evidently much the better for his Swiss trip. Has Mr. Brown heard that Auguste Balmat is expected in London next month? The Miss Martineaus informed Jones of the fact, hoping he might be able to assist in finding some employment for him during his stay in England—a difficult affair.
>
> A thick yellow fog necessitates the writing of these lines by candle light! Finally Mrs. Jones begs her kind regards, and will be very glad to see Mr. Brown at Kew again some day.

Afterwards the nicknames were regularly kept up in personal messages about 'Mrs. Jones' and 'the little Joneses,' or in planning future trips, as in 1858, when Mrs. Hooker, after drawing up a plan of campaign, adds:

> Now do, Mr. Brown, join your faithful friends the Joneses on this beautiful little tour, which looks so charmingly tempting on paper; it would add so much to our pleasure to have you with us. We don't mean to be away more than a month, and I shall set to work soon to lay it out in days, so as to get it all in comfortably—and I'll keep all the accounts, and you shall have no bother at all, but just enjoy yourself, and I am sure it will do you a great deal of good. Don't say no all in a hurry, but take time to consider. Joe sends his love.

It was a year when, owing to press of work, Hooker confessed he grudged the very time for a holiday, and suggested as a variant to stay 'two or three quiet weeks at some cheap, out of the way place (Tyrol or Pyrenees) and work up some of my florating materials, and afterwards go on to Sardinia or not.'

> My pleasure [he writes to Harvey, July 20, 1858] would be to go to only 2 or 3 places and spend a week at least at each—as one week at the Distel-Alp or elsewhere in Saas valley—one in valley of Ansasca, a day off, and one somewhere else, hard by, doing some work at each and enjoying some very moderate walks at each. I have no love of climbing any more, or of cleaving glaciers, but I should like wandering for an hour or two in a day about such places out of the way of tourists or tripping excursionists.

But alternative plans had to be made nearer home, for Mrs. Hooker could not go far away from Bath, where her aunt, Miss Jenyns, was lying seriously ill.

Thus a few days later:

> We proposed the Cornish tour because my wife would be as near Bath there as here. I am charmed with your Kilkee plans, not so Mrs. Jones who has an aversion to the sea, no taste for that *sea*nery and besides *Flea* rhymes with Kilkee. The great objection is however that it is as far from Bath as Switzerland. There is also the Hewmeedity of W Ireland, and 16 days' wind and rain out of a fortnight, plus colds and neuralgia, is no joke on a holiday tour.

'But whatever be decided,' he adds, 'I am like you, I bargain for the sea or the snow—all else is dull, flat, tame, stale and unprofitable.'

So again botanising is a leading attraction in the unfulfilled holiday plan for 1859, and he declares to Harvey:

> I would ten times rather go to Cadiz than to top of Mt. Rosa for a month; specially as there is something to be got and much to be seen in Spain, and especially if the trip brought in the contrasted regions of the Atlantic and Mediterranean coasts, followed by the crossing of the Pyrenean pass to the Biscayan coast, so as to secure comparative results beyond the mere numbers of species.

In 1855 the Great Exhibition in Paris, rival of its English prototype of four years before, drew everyone to France. 'When are you going to Paris?' he asks Henslow on June 1. 'The Benthams have taken lodgings there for 6 weeks. I am all in uncertainty whether I go at all or no. I am desperately busy.' After the fashion of such shows, it was not half completed by the end of the month; still 'I hear that it is really a very fine sight indeed already, and that the public are grumbling unreasonably and unnecessarily.'

On July 3 he writes to Bentham in Paris that he has 'partly plotted a trip to Germany with Nat. (Lindley) [1] about the middle of August,' adding:

> I really do not know what to say about going to Paris; I can't speak French you know, and am indomitably repugnant to exert myself in conversation. I am pretty ashamed of my ignorance, and hate myself quite sufficiently for my indolence and *mauvaise honte* not to wish to expose myself to my own reproaches. You that wrote a book on Logic may unravel this if you can. Then too I do not care to go without Fanny; altogether, in short, I am in a muddle. I did half promise to go with Henslow, but he is disgusted with his wax models having collapsed. I do not feel happy at the thought of going anywhere with this huge Indian collection on hand.

Eventually he joined Henslow at the end of September, on his way back from a visit to Germany, for the Queen was going to Paris for a week in mid August, and the place would be impossible for lesser folk.

From this trip he returned on October 3 'via Paris, from Vienna, Tyrol, Como, Mt. Rosa, Alps, Oberland, &c. (in inverse order).' The journey is described in the following letter by Lord Lindley:

The Lodge, East Carleton, near Norwich: June 19, 1912.

DEAR LADY HOOKER,—Many thanks for your kind letter and the Photograph of Sir Joseph which I am very pleased to have.

[1] Nathaniel Lindley, son of Dr. John Lindley, Ph.D., F.R.S., the Professor of Botany at University College, London; LL.D., D.C.L., Fellow Royal Society and British Academy; called to the Bar 1850; Master of the Rolls 1897–1900; Baron 1900.

I have no notes of my trip with Sir Joseph in 1855, but I have a lively recollection of its main incidents. The cholera was raging and we were fumigated on the frontier of Italy on the Stelvio Pass. Milan was stinking with Chloride of lime; Venice was deserted, and the Scientific meeting at Vienna which Sir Jos. was to attend was put off. We saw the caves at Laibach and went to Breslau to see Göppert's celebrated collection of Amber containing seeds, insects, &c., which Hooker was very desirous of seeing. We wound up our trip by staying a week in Paris to see the Great Exhibition there and got home penniless.

Our trip cost us £50 apiece; and we often saved hotel bills by travelling at night when passing through uninteresting country. I could talk French and German well then—I wish I could now! Hooker had introductions to Scientific men, but I cannot recall their names—Humboldt and Koch, I think, at Berlin; a Botanist at Vienna, Göppert at Breslau, and several in Paris. I think there was some one in Dresden and another in Munich; and we went and spent a night with a friend at a house on a lovely lake not far from Munich, but I forget the name of the Man and the place. Von Martius may have been the man, but I am not by any means sure.

I wish I could help you further. We met Henslow and, I think, a daughter of his when in Paris, and stayed at the same Hotel.

CHAPTER XXIII

LETTERS TO DARWIN, 1843–1859

In one of his letters Darwin makes special mention of preserving his friend's letters. The answers to scientific questions are detached and placed among the memoranda of that subject; the other parts are put among his general correspondence, so that it would only be a matter of half an hour to rearrange them in case of need. In spite of his care, however, a large number of the earlier letters from Hooker have disappeared wholly or in part. From the remainder I give a selection to illustrate their correspondence before the appearance of the 'Origin.'

Darwin's first letter to Hooker (December 1843) is printed in the 'Life of Charles Darwin,' ii. 21. He had then sent his Galapagos collections to Hooker through Henslow, who had had them in keeping (see 'More Letters of Charles Darwin,' i. 400); the next in sequence, which answers the following of Hooker's, is given in 'More Letters of Charles Darwin,' i. 39.

J. D. Hooker to C. Darwin

December 1843.

The Galapagos plants are far more extensive in number of species than I could have supposed, and are the foundation of an excellent Flora of that group: Mr. Henslow has sent with them those of Macrae which hardly differ from yours. I was quite prepared to see the extraordinary difference between the plants of the separate Islands from your Journal, a most strange fact, and one which quite overturns all our preconceived notions of species radiating from a centre and migrating to any extent from one focus of greater development.

I do not think there is in the North any instance of the floras of two such remote spots as Kerg. Land and Cape Horn being identical. Two Floras appear in the Northern Hemisphere, the American and the European. The former is confined to the American Arctic shores and islands, the latter to all Arctic Europe, Asia and Greenland: Western Arctic American to the W. of the great chain of the Rocky Mountains, and North of the Oregon River may also belong to the European Flora and is likely to, but I have not compared, having no materials in the *Erebus*. The abrupt line of demarkation is most remarkable in Baffin's Bay and Davis Straits, the most common European Heathers and some other plants being found abundantly along the Eastern shores and islands of those waters, but never on the Western. Of course a multitude of plants are common to both Hemispheres, which makes it in one sense the more remarkable that two or three of the types of Northern European Botany should not cross to the Westward of Longitude 60° W.

I have been progressing with the Antarctic plants, using yours, King's and my own at once, and each according to the Nat. Ords., beginning with Ranunculaceae, where the value of every scrap tells better than it is possible to suppose. The little *Cardamine* or Cress I prove, by comparison with about 50 states of it running through the whole continent of S. America, to be the same as the most common European weed, *C. hirsuta*. This is not wonderful, but it is, that Winter's Bark, *Drimys Winteri*, should extend through the whole continent of S. America and Mexico, from 25° N. to 56° S. It is true that the extreme states vary, and apparently specifically, but take the regular series of specimens, beginning with my own Cape Horn ones, your and King's Fuegian, Bertero's and Bridge's and Cuming's Chilian, the Brazilian ones of many collectors; Peruvian and Bolivian States from others; and finally, end the list with the Mexican, and no one (not even the most determined species-monger) can make them specifically distinct. It is further proved by the later Brazilian Botanical authors considering their species the Chilian, and contemporaneous Mexican writers, not aware of this last re-union, uniting theirs to the Brazilian. I do not suppose that there is another plant of so great a size having one third as great a range in Latitude.

The Govt. have not as yet granted anything towards my publication, but I hope they will ere long. Not being a good arranger of extended views, I rather fear the Geographical distribution, which I shall not attempt till I have worked out all the species, especially as I hope that more facts of as great importance as the range of the Winter's Bark may turn up. With many happy returns of this season,

Believe me, my dear Sir,
Your most truly and obliged,
JOS. D. HOOKER.

We have just had a pretty little Barberry of your Chiloean collection [*Berberis Darwinii*] engraved for the Icones Plantarum, as it will not come into the Antarctic Flora, save in a note.

Early April 1845.

I do not doubt the Flora of the Sandwich Islands being very peculiar, but the difficulty is to settle what amount of new species or of new genera produces peculiarity. One species will sometimes render a whole vegetation peculiar in the eyes of some. In some instances, which I mentioned to you before, and which Hinds[1] has wholly overlooked, the Flora of the Sandwich group is quite singular, in the preponderance chiefly of *Lobeliaceae* and *Scaevoleae* (if I remember); they are not however likely to strike a casual observer or to give a feature to the vegetation. Wilkes is probably indebted to his Botanist for the observation, which is just: no missionary book, nor does Cook (I think) nor any other unpractised observer, particularize the group as having any peculiarities of vegetation, but the contrary. I have not read Wilkes yet. Our ideas of peculiarity are most loose, we have no standard; in the first instance we must know the absolute numerical amount of peculiar species; this must ever be the primary point, the leading fact; all

[1] Richard Brinsley Hinds (*d.* before 1861) was surgeon to H.M.S. *Sulphur*, and made the first collection of Hongkong plants which reached England. He was author of *The Regions of Vegetation*, 1843, and edited the botany of H.M.S. *Sulphur's* voyage, and contributed several papers on shells to various publications.

other causes of peculiarity, as a preponderance of species, genus or higher group, or insulation of individuals, &c., &c., must be secondary considerations. Except Brown and Humboldt, no one has attempted this, all seem to dread the making Bot. Geog. too exact a science; they find it far easier to speculate than to employ the inductive process. The first steps to tracing the progress of the creation of vegetation is to know the proportion in which the groups appear in different localities, and more particularly the relation which exists between the floras of the localities, a relation which must be expressed in numbers to be at all tangible.

Edinburgh: July 1845.[1]

Bother variation, development and all such subjects! it is reasoning in a circle I believe after all. As a Botanist I must be content to take species as they *appear to be*, not as *they are*, and still less as they were or ought to be. You see I am annoyed at my own incapacity to fathom or follow the subject to any good purpose (open confession is *good* for the soul).

I think I can give you plenty of instances of peculiar genera with several good species in very small islands. [A list follows.]

I have always felt opposed to Bory's (who is a great Gascon! but not to be despised) views of the variableness of insular species. I certainly have no good evidence in favour of the loose statement I made and which corresponded with a vague idea I held, of insects being scarce on islands; yet 13 species is surely very few for Keeling if size is to be regarded; how often may you not find 13 on your own window? Kerguelen Land has only 3. New Zealand and V.D.L. are certainly poor—in Trinidad (of Brazils) I saw only 3, I think, a *Hemerobius* and the House flies and Cockroach, introduced from a wreck: Canaries and Madeira are poor, I think; Cape de Verds are too dependent on the W. coast of Africa to judge from. Nothing struck me as so marvellous as the appearance of 4 Insecta and many Arachnida you mention as on St. Paul's rocks. Still I agree with you on the main point that such few as there are would be enough for impregnation if they only went to work about it.

[1] For Darwin's answer, see *More Letters*, i. 51.

I cannot prove that there is much hybridising[1] in nature, but do not see why there should not be, as we do not doubt that species require the pollen of other individuals, exactly as in the higher animals you must not 'breed in' (I think the term is).

I cannot hook my Kerguelen trees or climate on to the vacillating temperature of S. America: many thanks for the information though. Do you connect the union of the Conchogeographic districts at the Galapagos with the currents?

Every young Irish Yew bears berries; there is a sort of Irish Yew in Ayrshire which I believe, like the Goddess Diana of the Ephesians, dropped down from Heaven, and picked itself up in a garden; when I hear whether it bears berries I will tell you if she be equally chaste. If the Yew had been Italian and bows made it would have been dedicated to Diana.

And now to bother you for the last time. The re-appearance of plants in certain situations is a curious phenomenon of which instances are multiplying daily in this neighbourhood: there are doubtless series of seeds in some grounds lying dormant but not dead: what a curious principle life must be and what an uncomfortable abode it must often have. Cutting open railways causes a change of vegetation in two ways, by turning up buried live seeds and by affording space and protection for the growth of transported seeds: so that it is often very difficult to determine to which cause the appearance or superabundance of a plant is attributable. The Dutch Clover case is constantly quoted, but the Stirling Castle one is more curious. The King's Park was dug up in about 1650? during the 1st rebellion; wherever the cuts were made for encampments, the Broom appeared, but in a year or two disappeared. In the rebellion of 1745, it was again encamped upon and again Broom came up and disappeared: it was afterwards ploughed and immediately became covered with Broom, which has all, for the third time, vanished.

To conclude (I have been reading Scotch Sermons!) how curious that water plants should be so widely diffused. Water must have been a mighty agent in dissemination; not only though are these diffused but are diffusable.

[1] The word is used in the sense of the later 'cross-fertilisation.'

Aponogeton, a Cape plant, not native of cold regions, bears a freezing every winter in our ponds: no one would have dreamt of it.

Edinburgh: July 1845.

I am exceedingly glad that l'Espèce [by Godron] has interested you, and will try and get you a copy from Montagne, through whom my father received this. I am not inclined to take much for granted from any one who treats the subject in his way and who does not know what it is to be a specific Naturalist himself. Those who have had most species pass under their hands, as Bentham, Brown, Linnaeus, Decaisne, and Miquel, all I believe argue for the validity of *species* in nature; they all direct attention to the cases where *salient* characters are unimportant, though taken advantage of by the narrow-minded *studiers* of overwrought local floras, and these facts, thus noticed as cautions to others, are taken up by such men as Gerard, who have no idea what thousands of good species there are in the world. Nature may have both made and muddled species; we shall never know what are species in some genera and what are not. Generally cultivation will prove the validity of a species; Gerard says that 'varieties of apples, &c. are more distinct than many species,' but how soon all revert to crabs; again, the wheat is always adduced as a permanent variety of some unknown plant and it ought on that account to rank as a species, but I do not think so because it will never run wild; it is to me very marvellous that the wheat seed is destroyed by being left in the ground of our country and that we see so little next year on a field that has supported millions of ears during the present.

Gerard evidently is no Botanist, he talks of having found both *Prunus spinosa* and *Rubus rusticans* without spines. Now spines are only abortive branches, and their absence or presence is never, of itself, a botanical character; as a spine is not an organ *per se*: and again, no *Rubus* ever had or ever will have *spines*; the *prickles* of *Rubus* are mere *appendages of the cuticle* and have no organic connection like spines with the pith and wood of the plant: species vary in the *prickliness*, just as they do in *hairiness*, according to the amount of spines or hair produced; but they vary in *spininess* according to the number of branches that are checked in growth which is much affected by want of moisture. You are right then to query that bit about the plants developing spines in bad soil; for they only lose the power of nourishing the new leaf buds sufficiently and do not develop a new organ. (Hence hairiness is of more importance than spininess in distrib.). The *Persicaria* becoming hairy when removed from moist places is natural: hairs are believed to be provided as hygrometric appendages, to modify respiration and transpiration, water plants don't want them. It is facts such as the Irish Yew presents that afford fair ground for argument on such a topic. Noting instances by tens or hundreds of variation in individual species is nothing new; few have an idea of the labour required to establish or destroy a species of a mundane genus. You have a *Senebiera* from Tres Montes, its capsules are much larger than the common *S. pinnatifida,* but that is so universally diffused a plant and so variable in the size of its *leaves* that at first sight no one would be inclined to grant specific dignity to the Tres Montes plant from the capsules. It struck me to put this subject to a Geographical test, the result is, that the *S. pinnatifida* is probably a native of the *Plate alone,* whence it has spread by ships all over East and West America, all West Europe near the coast, in fact both shores of the Atlantic, from Britain to the Cape and from Patagonia to Canada, wherever ships touch and cultivation ensues, and on W. from Valparaiso to California, wherever ships go, but through many hundreds of specimens there is no variation whatever in the size of the pods, and I therefore conclude that the Tres Montes plant is the W. coast representative of the E. coast plant. Now though De Candolle had hinted that *S. pinn.* was an American plant, he did not define its limits and retained two or three identical plants as different species which came from out of the way localities: to define its limits I had not only to consult all floras where it was described, but all where it was not, for such a mundane plant creeps into every flora. My troubles did not end here, for I had no Valparaiso *Senebiera,* and Bertero has an undescribed one from that port, which is alluded to as *S. diffusa,* Bert. MSS. I naturally concluded yours was this, but thought I would write to Brit. Mus. to confirm it, for fear of accident, but Bertero's was genuine *pinnatifida*, he gave it a new name *taking for granted* it was a new species. So as *S. pinnat.* does not at Valparaiso vary into big pods I am more persuaded that yours is a representative species of W. coast of N. America. That neutral territory of representative species you ask about is just what I want to work out, but it needs great materials.

Ever yours most truly,

J. D. HOOKER.

The following comes between Darwin's letters given in M.L. i. 411 and 414, of which the latter is dated April 10, 1846.

One of the great objects I had in view in my notion above alluded to [of the distrib. of Galapagos plants] was to group the plants according to their derivation, and I have a class in reserve for '*apparently peculiar species, possibly the altered forms of introduced plants.*' It is quite true that in most islands there is a lot of very dubious species, by no means to be confounded with their countrymen, and not polymorphous in the said island, but wofully near certain continental congeners. Thus I would divide the Galapagos plants into 4 groups: 1. Ubiquitous, e.g. *Avicennia.* 2. Of nearest continent, as *Baccharis.* 3. Possibly altered state [illegible]. 4. Original creations, as *Pleuropetalum* or *Scalesia.* The third group may not be a large one in the Galapagos (according to my notions) but its acknowledged existence is a matter of some importance. In the cases of Madeira, the Canaries and Azores, said group 3 must be very considerable. Such however is the difference of opinion amongst Botanists as to what should or should not be a species, that the question in any shape will be a troublesome one, though not on that account to be dismissed unconsidered.

I stumbled on a splendid fact the other day, that the *Lycopodium cernuum* is only found in the immediate neighbourhood of the hot springs in the Azores. When alluding to its distribution at p. 114 of my Flora I dared not mention that it was not known to be an inhabitant of Madeira or the Canaries, as I thought it *must* turn up there; now however I do not expect it and feel sure that the presence of this torrid plant in the Azores is due to the hot springs. What I am most pleased at is the apparent proof of the universal suspension of the sporules of this genus in the air and the consequent strengthening of my hypothesis, that the genus should be decimated *sparing only every tenth*! Of course it is a strong fact for migration, and for the existence of the impalpable spawn of Fungi, &c., in all air.

I have been more coolly analysing the bearings of the Forbes Botanical question[1] lately, and with the distressing result, that I fear I must haul out of all participation with him. You will think me unstable as water, and I must blame myself for speaking too much without thinking. It is not from a reconsideration of *his* facts and arguments that my faith is weakened, but from an independent examination of the Flora of the N. Atlantic Isles and W. U. Kingdom, which shows that there are plants in those regions which have been more put to in getting there than the Asturias ones need have been. Such are the American plants, *Eriocaulon septangulare* in the Hebrides and W. Ireland, American *Neottia* in S. Ireland, and *Trichomanes brevisetum* in W. Ireland and Madeira, all of them American plants not found further E. on continents of Europe or Africa. Also the *Gymnogramma Totta*, a fern of the Cape only in Madeira and Azores, and *Myrsine africana*, which positively skips from the Cape across all intermediate Africa on one side to Abyssinia and on the other to the Azores! I hope to be allowed a conversation with Forbes on the subject, for really with his Sargassum weed, &c., he is going too far.

It is very easy to explain on what sort of ground Botanists make one class of plants higher, and as easy to prove them futile by their results. I do not however think your objection valid, urged on the grounds of Owen's observations on organs which are developed in the animal kingdom,[2] but which organs are valueless for systematic purposes, if present even, in the vegetable. It is upon the modifications of the

[1] Viz., that several Spanish plants in Ireland could not have been transported by any known agencies; hence they supported the argument for a Miocene continental extension between Ireland and Spain, and from Spain to the N. Atlantic Islands.

[2] A. St. Hilaire used a multiplicity of parts—e.g. several circles of stamens, as evidence of the highness of the Ranunculaceae: Owen conversely used the same argument to show the lowness of some animals, urging that the fewer the number of any organ by which the same end is gained, the higher the animal. The subject of 'high' and 'low' is touched upon further, pp. 460, 463.

sexual organs and their accessories that all the Nat. Orders are defined. The organs of locomotion afford the Botanist no characters, those of digestion next to none: and the mode after which the various component parts of a compound body (a plant) are arranged is valuable only for the 3 highest groups, Monocot, Dicot, and Acot, and not absolute even amongst these. Generally speaking, in Botany highness and lowness are synonymous with complexity and simplicity of structure. I can hardly conceive either simplicity or complexity of one particular organ indicating the rank of a being in the scale of creation.

November 1851.

Coprosma is almost peculiar to N. Zealand, and for the life of me I do not know how to draw the line between there being only one species or 28!—it covers the country in every form of herb, bush and tree, from sea to mountain top,—but it is no worse than Rubus, Willow or Rosa are in Gt. Britain, and on the whole I ignore Bory's theory.[1] Generally speaking, the N. Zealand species are as well or better marked than the European, or the Australian, where *Eucalyptus* and various other genera are not to be surpassed in Protean dispositions. For the rest, recent discoveries rather tend to ally the N. Zeald. Flora with the Australian —though there is enough affinity with extratropical S. America to be very remarkable and far more than can be accounted for by any known laws of migration. I am becoming slowly more convinced of the probability of the Southern Flora being a fragmentary one—all that remains of a great Southern continent. A *second* species of the otherwise strictly great S. American genus *Calceolaria* has turned up in N. Zealand, and of the two only genera of N. Zeald. Leguminosae, one, a *tree* (*Edwardsia*), is common to Chili and N. Zealand and to no other countries—the other is confined to N.Z. and allied to nothing. Several of the truly wild grasses are European I think, and yet not found in Australia!

Hitcham: June 1854.

Will you oblige me with your ideas of what constitutes highness and lowness in the Animal Kingdom? e.g. in

[1] See p. 439.

plants I should say that a high development in the scale is indicated by special adaptations of organs to the discharge of functions, great deviations in those organs from the type upon which they are constructed. Thus Ranunculaceae are low in the scale because the floral organs are apt to run into one another and revert to the type (a leaf) on which they are constructed—because calyx and corolla are so often alike—stamens often reverting and the follicles present little deviation from a leaf folded on itself. Hence Monopetalous flowers are higher than polypetalous, inferior ovaries a higher type than superior, Dicotyledons than Monocot, Exogens than Endogens, &c., &c.

Darwin's answer is given in 'More Letters,' i. 76: the distinction he draws lies in the amount of morphological differentiation and the division of physiological labour. (See below, p. 463, letter of December 26, 1858.)

Darwin had been making out various Grasses from book descriptions, and sent one that baffled him for identification.

Richmond, Sunday.

My dear Darwin,—Your grass appears to me to be *Festuca pratensis,* and agrees as *ill* with the descriptions as most plants appear to do. How on earth you have made out 30 grasses rightly is a mystery to me. You must have a marvellous tact for appreciating diagnoses. I am sure that I could not have done it. I very much rejoice at your feats, as it will afford us many subjects of interest in common when we meet again. I think that some structural points would interest you—as that of the inflorescence of Grasses. Amongst facts of interest which will one day be licked into shape pro or con species and migration, is that of the South Coast of Australia. I have just made a résumé of the Australian Leguminosae, about 900 species. Of these some 450 inhabit the *South West* Corner, Swan River, &c., and about 300 the South East (New South Wales, &c.), but there are not 10 *species* common to both! Now what can migration be about, trans-water or trans-land?—and what a busy time of it Dame Nature has had in making so many species, whether by creation or variation.

I am busy at Indian Compositae. There are two very common English Thistles, a small one, *Carduus acanthoides,* and a big one, *C. nutans*. I never heard of their being supposed to be varieties by any one, and they differ in many points; but the Himal. specimens are all of an intermediate form—its small states identical with *acanthoides*, its large with small *nutans*. These facts shake species to their foundation—but according to my view of species, as contrasted with other systematists, there are sore few of them. In fact if there were a possibility of bringing your and my opinions *to book*, it might prove that we were not so far divided. The more I study the more vague my conception of a species grows, and I have given up caring whether they are all pups of one generic type or not—that the main forms remain so long distinct; that we may through their characters trace their distribution, is certainly all we can expect to prove in our day; and the laws of that distribution more than we shall establish in our life-time.

I have a glorious fact for you. A tropical species of *Cyperus* (*polystachys*) and a tropical Fern, *Pteris longifolia*, grow in the *hot soil* of the Volcano of Ischia and nowhere else in Europe or the Mediterranean: see Hooker's Journ. Bot. for Nov. 1854, p. 351 (it is on Athenaeum table). Now I can wriggle out of the Fern case by allowing ubiquitous meteoric dispersion of Fern spores, but the Cyperus is a disgusting and detestable fact that disgusts my soul within me. I must however have a bite at you if I can, and so will ask why if the Cyperus and Pteris got there no other migrants did?

March 2, 1855.

I am going on with the Tasmanian Flora and find the subject very interesting. Some of the scarcest and most local Alpine plants reappear on the isolated summits of the Australian Alps, and thence too I have the English *Sagina procumbens*, which, as far as I know, has not been found in the South Hemisphere, except in the Falklands (this wants study though). I am also preparing as I go on for a general work on Geogr. distrib. of the whole Australian Flora—this is ambitious, but it is really the most extraordinary thing in the whole world. The Flora of Swan River, i.e. of extratropical S.W. Australia, will I believ turn out to be the most peculiar on the Globe and specifically quite distinct from that of N.S. Wales—also generically to a much greater degree than any two similarly situated areas.

[For Darwin's answer see C.D. ii. 44, which leads to the following]:

To Charles Darwin

[March 1855.]

[Wollaston][1] adduced one fact as opposed to Forbes' Atlantis theory, which is *Ophrys*, an abundant S. Europe genus of many common species, but unknown in Madeira. Now this has such minute seeds and such millions of them, that if the Madeira plants were transported aerially, one cannot conceive the absence of Ophrys. To me such cases as Ophrys are extremely important, as indicating a sequence in the creation of groups, for if Ophrys was as abundant and wide-spread when Atlantis existed as now, it must have been there too then and we take for granted would be now; on the other hand, assuming the wind as the agent, if Ophrys had existed in Europe as long as the other species that are common to Europe and Madeira, its seeds must have got wafted across.

The fact of apterous coleoptera strikes me too as extremely curious and reminds me of an old remark I made that not only the few beetles of Kerguelen Land were apterous but the only lepidopterous insect in the island was so too!

Your final cause for so many insects being apterous is very pretty and no doubt good, but how does it square with the fact, that so large a proportion of Desert (Sahara, Pampas, Australian) Coleoptera are apterous—that in fact where wings would be most wanted and where it is to be assumed that great areas must be traversed for either animal or vegetable food, that there the insects have smallest powers of locomotion—that where the deer, birds, and carnivora have the longest legs the insects have the shortest. Had the Madeira coleoptera unusually strong powers of flight, would we not have said that this was to enable them to make for shore again after being blown out to sea?

I have just (thanks to Bentham's kind aid) concluded a good and complete catalogue of the Australian Leguminosae, and shall probably work it yet. There is but one European species, the common *Lotus corniculatus*; it abounds in

[1] Thomas Vernon Wollaston (1822–78), entomologist and conchologist; M.A. Cambridge 1849; F.R.S. 1847; made collections and published works relating chiefly to the coleoptera of Madeira, in addition to other writings.

marshes of N.S. Wales and Tasmania, but is not found wild elsewhere out of Europe that I know of in the Southern Hemisphere—these are the extraordinary facts that will not be accounted for. Out of full 800 species I do not think that there are a dozen common to South-East and South-West Australia ; whole well marked genera, containing many sections and species, are absolutely confined to S.W. Australia. There is nothing like this in any other part of the world : it is utterly astounding, and though I thought myself well up in the Australian Flora, I was not prepared for this to such an extent. Also taken as a whole the Flora of Tasmania does not present so many species *hardly distinct* from S.E. Australia as it ought. The Tasmanian species are either very distinct, or quite the same, and what is most curious, this applies as well to the alpine plants, though the climate of the Australian Alps must be a good deal different from that of the Tasmanian ones.

There is another point to be worked in your apterous insect case—viz., the proportion of apterous European species in Madeira great or small. If over-sea migration were the means of peopling Madeira with insects, then the European species should be winged ones. There is still another point. Do you suppose that the majority are apterous because the winged ones have been blown out to sea and perished miserably ? Really these questions are like Cerberus and his heads—the more arguments one disposes of the more rise up in your way.

Kew : November 9, 1856.

I have finished the reading of your MS. [on Geog. Distrib.] and have been very much delighted and instructed. Your case is a most strong one and gives me a much higher idea of *change* than I had previously entertained ; and though, as you know, never very stubborn about unalterability of specific type, I never felt so shaky about species before. The first half you will be able to put more clearly when you polish up. I have in several cases made pencil alterations in details as to words, &c., to enable myself to follow better—some of it is rather stiff reading. I have a page or two of notes for discussion, many of which were answered as I got further with the MS., more or less fully.

Your doctrine of the cooling of the tropics is a startling

one, when carried to the length of supporting plants of cold temperate regions, and I must confess that, much as I should like it, I can hardly stomach keeping the tropical genera alive in so very cool a greenhouse. Still I must confess that all your arguments *pro* may be much stronger put than you have.

I am more reconciled to Iceberg transport than I was also, the more especially as I will give you any length of time to keep vitality in ice, and, more than that, will let you transport roots that way also. Many of these subjects which I never myself studied for myself, I wanted put in the systematic form you have put them, for proper appreciation.

I think you might support your cause by making more use of Gulf streams and oblique lines of transport—you appear to dwell too much upon meridional lines of migration. This mode of travelling at once suggested the query, are the Arctic and Antarctic American genera more allied than the Tasmanian and Siberian—the former offering every possibility in continuous land—the latter none? It also makes you appear to shirk the question of transport from East to West or vice versâ. You offer no explanation of the vegetation (not littoral) of Abyssinia and India Peninsula being so similar; or of the Carnatic, Ava, and N.W. Australia being in so many points alike; of the curious parallels or representatives between Madagascar, Ceylon, and the Sunda Islands. In short meridional migration alone occupies you. Nor do I like putting Iceland, Faroe, and Spitzbergen out of the category of the glacially peopled countries, and leaving Shetlands, Orkneys, Scotland in it; this is however a trifle. Ch. Martins' [1] arguments seem to apply no more to these islands than to any other area continental or insular. If they presented any anomalies as the presence of Lapland plants or Greenland ones, I might then believe them to be peopled by accidental migration—but if Icebergs are to be so powerful why did they bring no Greenland, American, or other plants to these islands which are so well situated for the purpose?

Thanks for A. Gray's letter. I do rub my hands and chuckle (like Lyell) at the happy idea of my being caught in a

[1] Charles François Martins (1806–89), born at Paris; geologist and botanist. He was Correspondant de l'Institut, Hon. Professeur à la Faculté des Sciences at Montpellier, where he was Director of the Botanical Gardens. He wrote on the Creation of the World and on Topography.

paradox. I know the human soul loves paradox, even to miracle, and that this love of it is only one of the curses of Science, but Lord bless you, my dear Darwin, it is the greatest paradox in the world to think of Conifers as anything but very high in the Vegetable Kingdom.[1]

April 11, 1857.

If you knew how grateful the turning from the drudgery of my 'professional Botany' to your 'philosophical Botany' was, you would not fear bothering me with questions. The truth in its primitive nakedness is, that I really look for and count upon such questions, as the best means of keeping alive a due interest in these subjects. I indulge vague hopes of treating them some day, but days and years fly over my head and all I do is done in correspondence to you, but for which I should soon lose sight of the whole matter.

Harvey's observations on *Fucus* varying much and yet in some way under *most* different conditions goes with me for a good deal and I would endorse it. . . .

There are I think heaps of such cases, they have so often struck me, that one of my sketched out methods of treating the Indian plants common to W. Europe and India is by dividing them into:

1. Identical unvarying species.
2. Identical variable species.
   (*a*) Variations equal and similar in both countries.
   (*b*) Variations unequal, or dissimilar, or both.

In answer to Darwin's letter of June 25, 1857 (C.D. ii. 102) about the curious character of the seedling leaves in young Furze, after quoting some parallel cases, he proceeds:

A great stumblingblock in development to me has been the very great differences between the cotyledonary leaves of plants, even of the same Nat. Order. Leguminosae for instance: this has always prevented me from understanding the embryonic development in plants being so good an evidence of affinity as in animals. Comparative development would appear to begin with the post-Cotyledonary leaves, and the Cotyledonary may be regarded as placenta? amnios? &c., which vary in allied animals. Is this not a

[1] See also the letter to Asa Gray, p. 480.

shadow of a generalisation? I have often recommended germination and first formed leaves as the most interesting enquiry a young Botanist could take up, and particularly urged it upon G. Henslow.

Towards the end of the year, when about to visit Down, he sought some Darwinian information from his old friend Berkeley.

Have you ever made any observations on *inducing varieties* by playing tricks with plants? as by high manuring wild species; plucking all their flowers off for several years; pruning; &c. Darwin wants to know who has done such things.

Writing on January 12, 1858, Darwin refers to his own former belief, and Asa Gray's strongly expressed opinion, that Papilionaceous flowers were fatal to his notion of there being no eternal hermaphrodites among plants. He now brings forward evidence to show that in this class of plants cross-fertilisation takes place through the visits of bees, and that since the latter were introduced into New Zealand, clover had begun to seed, which did not happen before. Several questions arise for Hooker to answer.

January 15, 1858.

The Leguminous affair is extremely curious, I am quite gone over to your side in the matter of eternal hybrids and hermaphs. *Carmichaelia* and *Clianthus* have closed flowers, and hence probably require artificial hybridization, but *Edwardsia* has exserted genitalia and should not be a parallel case. With regard to the Wellington Clover case, it really looks too good—my impression is that Wellington was hardly a colony before 1842, and that there could not be sufficient clover cultivation there before that to warrant any conclusions, but I may be wrong. At any rate I should like some definite details of the state and extent of clover crops before 1842, say in 1839–1840. I will show your letter to Sinclair who will be here to-morrow.

None of the New Zealand Legumes have flowers quite as small as clover, though those of *Carmichaelia* and of *Notospartium* are very small. Is it not dangerous to assume that Humble bees would not visit small flowers in New Zealand, because they do not in England? In England I fancy the more numerous and active hive bee forestalls the Humble bees in the matter of small flowers—if indeed the Humble bees do not visit the latter. They surely visit Heather flowers in Scotland?

It would indeed be curious if a relation could be traced between no bees and no small flowered Leguminosae, but you must remember the strange absence of small Leguminosae in Fuegia, Falklands, and the Pacific Islands generally. The question hence becomes a very involved one and forms part of a larger one, viz., is there any relation between the Geog. distrib. of bees and of Leguminosae?

Bentham's late researches into the British Flora have so greatly modified his views of the limits of species, that in my eyes they invalidate the results of local Floras very materially. He has completed the MS. of his British Flora, having studied every species from all parts of the world, and most of them alive in Britain, France, and other parts of Europe. Well—he has turned out as great a lumper as I am! and *worse*.

Then did you see a paper of Decaisne's on *Pyrus*, translated in *Gard. Chron.* about 3 weeks ago—in which he adopts Thomson's and my views of species and says that if he had to monograph Plantaginaceae again he would reduce whole *sections* to one species and of course as many species, i.e. marked forms, would then rank as varieties. Now it was Decaisne (a most admirable Botanist) who on receiving the Flora Indica, wrote me most kindly and earnestly begging me to reconsider my mode of viewing species, and hinting that I was going to the devil. All this does not directly affect your results, but it shows that you should draw them from materials of all kinds—local and general, and from systematists. . . .

[The following is in answer to Darwin's letters of February 9 (M.L. i. 107) and February 23 (C.D. ii. 110 and M.L. i. 107), suggesting that the small genera vary less than the large.]

February 24, 1858.

I will answer your query about big genera, deliberately, in the affirmative and give answers. I have been thinking a great deal on amount of variability in great and small genera, and find it exceedingly difficult to explain logically the practical reasons there are against Botanists making varieties of *well marked* species, i.e. of small genera. Many of the small genera still kept up would never have been made at all, had the whole of the Natural Order as now known been known when those genera were made. *E.G.*, in Europe we have, say 3 very different members of a large *unknown* Asiatic group of plants, certainly 100 species: of these 3 as many genera are made in Europe: but after getting all the 100 Asiatic species, though these show that the said 3 genera are naught, we do not therefore cancel them, but in 9 cases out of 10 we group the Asiatic species as best we can under the 3 European genera. A thousand unphilosophical reasons occur, of considerable (*present* practical) weight to keep up the said old genera.

We must never forget that Systematists have two very different ends to meet: 1. To provide a ready nomenclature without which the science cannot advance and which we change as little as possible—and further use every means to avoid even a necessary change—so important is it for all to get up the nomenclature, and so bulky and complicated is this nomenclature. 2. To arrange the members of the Vegetable Kingdom scientifically, which is only done for the sake of scientific followers. Now we repeatedly find that to express our views scientifically we must break up the whole nomenclature, and rather than do this excessively, we confine ourselves to stating our views without acting upon them. In no respect do we sacrifice more to the utilitarian purpose of nomenclature, than in keeping up small *bad genera.*

Practically no one (except a few of us) hesitates to remove a very distinct species of an old genus, especially if its characters are constant and it is an *invariable plant,* and to make of it a new genus, just because it is more unlike its 20 neighbours than they are unlike one another. The probabilities in this case are that the 20 are varieties of 8 or 10, and being variable have varieties made of themselves, whilst the one constant plant goes to a new genus, and is a small genus with no varieties.

Again, practically very few do up an old genus of one

or a few well marked unvarying species, especially if its generic name is a very familiar one, hence *Amygdalus, Prunus, Cerasus,* are kept up, though certainly not good genera in a scientific view of Rosaceae. Few plants are more variable than Hawthorn—it is a small genus dismembered from Pyrus, but no British author makes varieties of it. Genera in short are almost purely artificial as established in Botany : some are *objective* like *Salix* and *Rosa,* i.e. every ignoramus recognises them and they are called natural genera, good genera, &c., &c. Others are *subjective,* they require a special knowledge of the Order to which they belong to know them—ignorami do not recognise them : such are genera of Grasses, Cruciferae, Umbelliferae, &c. But between what the ignoramus does recognise and does not there is no limit ; and the first rate Botanist, working upon a partial knowledge of a group, is only in the position of an ignoramus after all. His two very distinct groups of an Order are to him two genera ; had he the whole species of the Order he would never have recognised the groups at all, *as groups.* This is a terrific screed.

[Darwin replied on February 28 (M.L. i. 105) and March 11 (C.D. ii. 102), and Hooker responded] :

March 14, 1858.

I quite see in what respects local Floras are much the best suited to your purpose ; or rather, *how they would be so,* if they were worked out upon the same principle as the general Floras, but the fact that they are not so, and that they are hotbeds of bad big genera, is a very serious objection to the use of them.

I shall be however most curious to see the results of Bentham's British Flora. He reduces the Rubi to 6 species, I think (and about 11 varieties, I suppose), which gives you a small very variable genus, whilst Babington has 28 species or so, besides varieties—so *Callitriche,* of which Babington has several species but which Bentham reduces to 1 with 2 ? varieties. You must however take care not to get *entêté* with your results. I shall certainly go over the Tasmanian Flora for your sake, and see whether or no I should not have noticed varieties to many small genera, to make their species consistently worked with the big. I am quite sure I should. The object of these books, you must remember, is not to tell everything about a plant, and perhaps least of all to tell the amount of their variation, but to lead others to :—1st, name; 2, affinities; 3, distribution; 4, uses—and so on. As a rule the amount of variation is a speciality affecting the species differently in different localities, and is therefore only recorded when the omission of its record might lead to the non-recognition of the plant by the character. All plants are variable: see how the descriptions teem with '*vel*,' '*aut*,' '*et*,' &c.

The long and short of the matter is, that Botanists do not attach that *definite* importance to varieties that you suppose; they do not treat large and small genera equally and similarly, and the sum of inequalities thus produced tends to make the species of small genera look more invariable than big.

Had I been doing the Flora Indica as I should have done with an eye to making it a descriptive book of variation, I should most certainly have added varieties to most of the small genera, thus—

Naravelia *a* and *b*,
Adonis *a*, *b*, *c*,
Callianthemum *a*, *b*,
Aquilegia *a–z*,
Ceratocephalus *a*, *b*, *c*, *d*,
Caltha *a*, *b*, *c*, *d*,
Isopyrum *a*, *b*, *c*,

to render them equivalent to the varieties in Clematis and other big genera, and confounded your statistics.

Just look and see how much more frequently we notice under the monotypic genera, its variations and variability, than we do in the polytypic (excuse the coined phrase).

So my dear Darwin do not be in a hurry with your conclusions. I am quite sure that had monotypic genera or oligotypic been at all materially less variable than polytypic it would not have escaped the sagacity of men like Linnaeus, Brown, D. C., or Bentham, and that it would force itself on the attention of any cautious observer.

March 18, 1858.

You have set me thinking much on varieties in great versus small genera. I am obstinately inclined to take general monographs for data in preference to local Floras, for the general works alone seem to me to give a fair chance of the species being uniformly treated, because local Floras consist of: 1. Local plants—these we agree are not so variable as mundane plants.

2. Mundane plants, of which only one form is found in the said local area, and which are hence not treated as variable in the Flora of that area.

3. Mundane plants, of which two or more varieties are found in the area.

Now as you increase your area the small local (i.e. invariable) genera do not reappear, but the small mundane genera do with an increased number of variations, and the large mundane genera with their variable species also relatively increase. Is this not so? Be that as it may, I have just got Weddell's monograph of Urticeae back from binder, and as I told you that I thought it would prove as unexceptional food for analysis as may be, I have roughly tabulated the results and enclose them. Weddell has reduced both the large and small genera enormously and consistently, and I attach the greater confidence to his work from the close accordance between the relative number of species to variation in large and small genera.

Again, if the species of small local genera are themselves local, it follows that we procure fewer specimens of the species of such genera than of large, and hence make fewer varieties. This any general Herbarium shows. A genus of one species presents only a single *specimen* much oftener than a genus of 10 species does only 10 specimens.

Again, suppose I am naming by comparison or otherwise a species of a large genus, I find it agrees a little with many species, exactly with none, but most nearly one—I hesitate and am in difficulty and my tendency is to make a var. of that it is nearest, all the more if the latter is itself variable; but in naming in the same way a species of a small genus I find no such difficulty—it is perhaps not exactly like the species to which I refer it, *still it is not the least like anything else.* So I made a var. of my first plant partly lest my successor should refer it to any of the other species which it resembled from missing it under the vars. of that I refer it to, but in the second case no such precaution is necessary.

In the early summer, Hooker had read in MS. Darwin's discussion of 'what to call varieties.' Cheered by his criticism, Darwin subsequently sent for further criticism what he had to say about genera, in the discussion of 'the "Principle of Divergence," which with "Natural Selection" is the keystone of my book.'

Kew: July 13, 1858.

I went deep into your MS. on variable species in big and small genera and tabulated Bentham after a fashion, but not very carefully. After very full deliberation I cordially concur in your view and accept it with all its consequences. Bentham's book confirms you, though with modifications. The larger genera I believe to be groups of more presently variable beings than the small and I think you have quite made good your point. Still I would not abandon the arguments against, for I still think that the *disposition* or rather the necessity of making more book varieties in large genera than in small is a very important fact.

I have also well considered Bentham's Exceptional Orders, and am inclined to attribute that also partly to his idiosyncrasy; upon thinking well over his method of writing I have often seen that he will make rather hastily a new species in a larger genus of which a vast number of good species have recently turned up. The mental process is: 'Such and such a country teems with Astragalus or Pedicularis (which he has himself first elaborated), here is a new province of that country just supplied us with a lot of specimens and the chances are that heaps of them are new, and that more specimens will rather tend to prove doubtful new species to be distinct than the contrary.' It is not easy to explain to you how fully I appreciate this tendency in another person—but I am convinced it is so, and that that is the key to the Benthamian Exceptional Orders. This does not, however, apply to Weddell's *Urticeae* which I must tabulate more carefully. This was the case when Bentham and I did the Afghanistan and Thibetan *Astragali* and *Pedicularias*—he pronounced many new which I thought varieties, always saying: 'Oh that country is the headquarters of Astragali, you must expect heaps of novelty.'

In some passages of your MS. you rather underrate I

think the influence of associations of this and other sorts on descriptive systematists.

On re-reading your MS., I find the same objections as before, viz., that you overrate the extent of my opposition to your method. My great desire was to put every possible objection as strongly as I could. I did not feel myself a dissenter or opponent to your views, so much as a non-consenter to them in the present state of my knowledge, nor till you had weighed my objections which I thought of greater weight than I do now.

July 15 [1858].

The E.I.C. Examinations are cutting my time to shreds which must account for some of the incoherence of the foregoing. I have had more time for thinking over the subject at odd half hours and have endeavoured to grapple with the whole question. That point of the hypothetical behaviour of large genera when on the decrease puzzles me.

As a corollary to your law, large Natural Orders should have fewer genera in proportion to species than small; i.e. fewer definable groups. Cruciferae, Compositae, Umbelliferae and Grasses bear you out in this—true no end of genera are made in them, but they are bad—other Natural Orders are opposed.

I think I have thought of a better reason than you give for whole Nat. Ords. being worse for your purpose than local Floras—viz., 1. That conditions do not go on varying with the area beyond a certain point; there are limits to the combinations of climate and soil. A genus inhabiting 1000 square miles will survive such and such conditions and under their influence form *x* species;—all these conditions may occur in 100 miles of the said area, and adding the other 900 miles adds no more conditions. 2. Many large genera are absolutely confined to the tropics or to temperate regions or to districts and do not stand in the same relation to one another as the mundane genera do. This I think you have expressed, but not more clearly than I have. This would lead me to suggest the propriety of working one or two of your Floras by purging them of stragglers and such plants generally as are typical of other climates and exceptional in this—of stragglers in short—e.g. *Panicum*. I think this process would intensify your results.

Darwin submitted a definition of the great groups into which flowering plants are divided. Hooker in reply defines these, and adds:

If you take reproductive organs as test of highness or lowness, then Coniferae are top of Vegetable Kingdom; if you take coverings of those and neglect the organs themselves, you may place them below Monocots, but in so doing you neglect the vascular system, germinative and embryological characters which are all as in Dicots, not as in Monocots.

P.S.—I am very busy with the Introductory Essay to the Tasmanian Flora, and am dealing with the Australian as a whole. The only thing that will strike you is that the vast majority of the trees are hermaphrodite; this arises from the preponderance of arborescent hermaphrodite Orders (Myrtaceae, Leguminosae) and absence of Amentaceous.[1]

The great preponderance of local distinct species in the Flora I must hook on to the destruction of seeds somehow, restricting the multiplication of forms. In the Swan River Flora, where an incredible number of species are crammed into a very small area, the climate and soil seem most unfavourable to the germination of seeds by nature, and further the most local and peculiar Order, Proteaceae, ripen very few seeds and are a long time about it.[2]

I however want you to *print* before I make up my mind to go into this subject. I also want you to print that I may take up your refrigeration doctrine, to which I think I should have come clumsily at last by myself as the only way of accounting for the spread of European species to Australia.

It is curious that so many more European species should be in Australia than in Fuegia and S. Chile, especially considering the enormous distance of Europe to Australia and no continuous mountains.

[1] This exception to the rule, proved in England, New Zealand, and the United States, that trees have their sexes separated more often than other plants, is noted in the first edition of the *Origin*, p. 100. In the sixth edition, the qualification is added, that 'if most of the Australian trees are dichogamous, the same result would follow as if they bore flowers with separated sexes.'

[2] For Darwin's caution on this point, see his reply to this letter given in M.L. i. 445.

Put end of string on globe on England and other end on V.D.L., and it will run through the most continuous masses of land on globe; it is the greatest stretch of all but dry land that you can find, and I can connect the Botany the whole way by mountains of (1) Borneo; (2) Java and Ceylon and Penins. Ind.; (3) Khasia; (4) Himal.; (5) Caucasus; (6) Alps; (7) Scandinavia. I can thus connect botanically England with V.D.L. better than I could Canada with Fuegia!

Kew: December 21, 1858.

I am and have been working hard at my Essay and make about as slow progress as you say you do. I am utterly staggered by some of the facts of distribution: here is wild rice and lots of other plants identical with the Indian, in N.W. Australia, several hundred miles from the coast, and there is a most typical American plant (not found in India) from the same locality. I have now got together about 500 tropical Indian species in Australia, many of them very peculiar, besides many generic types almost all Peninsular Indian, not Malayan or Javanese types, but plants of the sandstone ranges of Australia and India. Now though there are several wet-country Australian *types* (not species) in Malayan Islands and Peninsula, there are *none* in the Indian Peninsula, nor are there any of the hundreds of Australian sandstone and dry tropical types in the Indian Peninsula. Now I never can believe that 500 Indian plants got transported by existing causes to tropical Australia, and that the said causes did not return one tropical Australian *Acacia, Eucalyptus, Stylidium, Proteacea, Goodenia, Casuarina,* or *Restiacea,* &c. to the Indian Peninsula.

Weeds, herbs, shrubs, and trees of *many* Indian families have gone S.E. to Australia and nothing has come back. N.B. Eucalypti, Casuarina, and Acacias grow magnificently all over the Peninsula where planted and ripen loads of seed.

You kindly promised me the loan of your Chapter on transmigration of forms across tropics and I should be glad of it. I am grievously troubled to know at what date to assume this transmigration; am I safe in assuming that the Antarctic types entered Australia at same Epoch, and what was general character of Australian Flora at that

Epoch ? Jukes,[1] I find, speculates in his sketch on Australia being two groups of islands ; was your review on Waterhouse anterior to this ? [2]

Highlands of Abyssinia will not help you to connect the Cape and Australian temperate Floras ; they want all the types common to both and, worse than that, India notably wants them. Proteaceae, Thymeleaceae, Haemodoraceae, Acacia, Rutaceae of closely allied genera (and in some cases species) are jammed up in S.W. Australia and C.B.I. [Central British India] ; add to this *Epacrideae* (which are mere § of *Ericeae*), and the absence or rarity of Rosaceae, &c., &c., &c., and you have an amount of similarity in the Floras, and dissimilarity to that of Abyssinia and India in the same features that does demand an explanation in any theoretical history of Southern vegetation.

I still hold to a large Southern Continent characterised by these and the Antarctic types. Perhaps during the Cretaceous and Oolitic periods some of these types existed in the N. Hemisphere also ;—hence the *Araucaria* cones in Oolite, *Banksia* wood of the sands at Chobham (what age are they ?) and cretaceous fossils supposed to be *Proteaceae* in Belgium, &c. ???

Are the coal and sandstone fossils of Australia Palaeozoic ? and is there in Australia a gap in the Geolog. series between these and modern tertiary beds ?

I also still regard plant types as older things than animal types. I have a fossil *Araucaria* cone from the Oolite identical to all appearance with *A. excelsa* of Norfolk Island, and the Chobham fossil Banksia wood is identical with Tasmanian. I do not suppose specifically in either case, but that such highly organised types should be so similar, indicates a great age for them as types.

[For Darwin's answer, dated Dec. 24, see C.D. ii. 142.]

[1] Joseph Beete Jukes (1811–69), an admirable field geologist and writer, a pupil of Sedgwick, did pioneer geological work in Newfoundland, 1839–40, and spent four years on H.M.S. *Fly* as naturalist to the expedition which surveyed N.E. Australia. Returning to England in 1846, he joined the Geological Survey, and in 1850 became Director of the Irish Survey. His book referred to in the text is *A Sketch of the Physical Structure of Australia*, 1850.

[2] In this unsigned review of *A Natural History of the Mammalia*, by G. R. Waterhouse, vol. i., in the *Annals and Magazine of Nat. Hist.*, vol. xix., 1847, pp. 53–56, Darwin had speculated on the S.E. and S.W. corners of Australia having existed as two large islands, and only recently been joined. (M.L. i. 448.)

Kew: December 26 [?], 1858. *

I wish we could have a little work together. When shall we ever get to a reasonable agreement? I am horrified to find that you think Australian forms lower than Old World ones; because under *every method of determining high and low in Botany the Australian vegetation is the highest in the world.*[1]

1. The proportion of Phanerog. to Cryptog. is infinitely greater in Australia than elsewhere (this as being a mere condition of climate I do not give much for).

2. Monocot. to Dicot. are in same proportion as elsewhere.

3. Petaloid (higher Monocot.) are in greater ratio to Glumaceous in Australia than in Europe.

4. The four Orders of Dicots, considered by different systematists as highest, are Compositae, Myrtaceae, Leguminosae and the Ranunculaceous, including Dilleniaceae &c. Now, I believe (I have not tabulated yet) that all these are in greater proportion and more varied in Australia than in any other country.

5. Then, granting with the heretical J. H.! that Conifers are highest Phaenogs., and they are as numerous and *most* varied.

6. There are very few Monochlamideous or Achlamideous Dicots in Australia.

Now I have been using your line of argument to my own purposes in this fashion: 'Granting with Darwin, that the principle of selection tends to extermination of low forms and multiplication of high, it is easy to account for the general high development and peculiarity of Australian forms of plants, these being the remnants of an extensive Flora of great antiquity and which covered a very extensive and now developed Southern continent, &c., &c., &c.' How often do I say all our arguments are two-edged swords.

Again, some Australian plants are rapidly running wild in India, as *Casuarina,* and I believe several Acacias in the Nilgherries and some other Leguminosae.

We cannot argue anything by contrasting the multiplica-

[1] Replying on the 30th, Darwin explains his meaning to have been the competitive superiority of the Old World plants when they met the Australian (M.L. i. 114). See also the letter to A. Gray of January 2, 1858, p. 480.

tion of European forms in Australia and New Zealand with the absence of the converse in England; our spring frosts account for the difference. In South Europe I believe various Australian forms are rapidly becoming naturalised. Consider too the current of export of European agricultural notions and plants to Australia and consequent alteration of conditions and that nothing of that kind comes back to Europe.

Your letter has interested me more than any you ever wrote me (because we are both *ripening I hope*), but it staggers me too. It opens a much wider question upon which I have often pondered in vain and have hoped làtterly to have made more of: it is this—are we right in assuming that the development of plants has been parallel to that of animals? I sent out a feeler in the concluding notices of my review of A. De Candolle where I indicate my view that Geology gives no evidence of a progression in plants. I do not say that this is proof of there *never* having been progression—that is quite a different matter—but the fact that there is less structural difference between the recognisable representatives of Coniferae, Cycadeae, Lycopodiaceae, &c. and Dicots of chalk and those of present day, than between the animals of those periods and their living representatives, appears to me a very remarkable fact. . . .

CHAPTER XXIV

ON SPECIES

ILLUSTRATIONS of the way in which Hooker's own conceptions of species and their problems took shape may be drawn from his correspondence during this period. He viewed the question from two sides, for he was the shining exception who gave point to Darwin's complaint:[1]

> How few generalisers there are among systematists. I really suspect there is something absolutely opposed to each other and hostile in the two frames of mind required for systematising and reasoning on a large collection of facts. (C.D. ii. 39.)

His mind was scientific in both the wider and the narrower sense. It combined observation with generalisation, the need for orderly detail with the equally impelling need for principles to give these ordered details an intelligible interpretation.

The primary object of science is order, and order is expressed by classification. A perfect classification seeks its basis in all the criteria gradually brought to light by research. Collection, labelling, grouping by external likeness is not enough. Each advance in scientific order expresses more truly the inner workings of nature, and to improve classification, therefore, is of more vital importance than to add to the store of accumulated material. Thus a group given individual importance on inadequate grounds became an offence against science, and obscured yet further the dark question of the origin of species. To reason on the ill-defined was hopeless. And so ill-defined

[1] See further, vol. ii. pp. 18, 26-31; Essay on the Distribution of Arctic Plants

were species that it could be cynically said by one of the older school that a species was anything that had received a specific name. Hence, in the botanist's phrase, it was better to reduce one bad species than to make a score of new ones.

The sense of this was strong upon Hooker even in his student days, the days of botanical tramps through the British Isles ; as he writes to Harvey in 1845 about a much disputed variety of heath found in Ireland and in Spain :

> *Erica McKayi* I never thought distinct from *tetralix* and have many dried intermediate states. Many a battle I had with Balfour in Connemara on the subject ; he would never own it a variety, even when I showed him living specimens. I did not and do not give in to Bentham's verdict, as he knows well, who retains the species in consideration of the glabrous ovarium.

This view of species was only accentuated with time. The more material he worked over, the greater the amount of variability found. Conversely, to establish the limits of a species properly, required the examination of a vast amount of material. As he begins the Indian Flora with T. Thomson, where his aim is 'to introduce some order into the confused mass of bad genera,' he tells Bentham (October 15, 1852) :

> Except for an enormous mass of species and specimens it would be impossible to come to a right conclusion as to their limits, yet the species are *very distinct indeed* when species, however close they run to one another ; it is very pretty to see different species running into analogous varieties and yet holding their characters.

So he gets ready for Col. Munro, the authority on grasses, 'a huge collection of duplicates, which will be absolutely essential in working up such genera as *Arundinella*.' (1853.)

It is the same with the Laurels :

> Nees has certainly overdone the species greatly, but that is not to be wondered at, or visited severely, as it is impossible to do them satisfactorily without flowers, fruits; and leaves, and a host of specimens. (To Bentham, September 3, 1854.)

'Many specimens,' he exclaims to Bentham, when he finds two of his *new* New Zealand species are *old* Tasmanian ones (July 30, 1856), 'always break down characters,' and he avows, 'it is a bad sign of genus when it is extremely difficult to refer new species to any of the others.' (February 5, 1852.)

Long before he impressed the fact on Darwin (p. 457) he was well aware that those who deal with an incomplete flora or a small number of specimens are apt to define isolated varieties as so many new species. Accordingly, to arrive at trustworthy fact, these irregular results of the 'personal equation' among describers must be regularised, at whatever cost of labour in examining new or re-examining old material, and so he groans at discovering in the work of a voluminous botanist 'an unfathomable gulf between him and right understanding.'

A few examples may be given of his dealing with the excessive multiplication of species and the consequent overlapping and confusion.

On September 24, 1851, just when the last boxes of his Indian collections have arrived, he tells Bentham:

> Klotzsch [then in Berlin] offers to make a frightful mess of the Rhododendrons, cutting the genus into 20 and placing varieties of one species into two or more genera, and allied species into each throughout; it is dreadful; he wants me to be partner in his crimes.

Three months later he describes himself as 'swimming in synonymy,' and on March 20, 1852, writes to Harvey:

> What a glorious Grass-man Munro is; he reduces my father's Herb. to about 1600 species! I quite expected they would come down to 2000.

Six days later:

> Munro has named nearly all my *Paniceae* and finds 5 new species! I think I should have sent them to Steudel, who (Munro tells me) is going to make a monograph of *Panicum* alone, containing 500 species! Munro and I made 86 as I think in Herb. Hook.

De Vries has just finished a monograph of *Angiopteris*, making 60 species out of what Daddy, I, and Jock Smith call 1. What with De Vries, Klotzsch, and Steudel we shall have Phaenogamic Botany messed like *Algae*, except we show a bold front.

Again, November 4, 1852 :

We have pitched into Clematis. Steudel has 40 Indian species, Wallich 18, and we 12 ! And yet we have all Wallich's ! Royle's ! ! Edgeworth's ! ! ! [1] etc., etc. species. The fact is that there are only 15 species in India and that's a plenty ! The Fl. Indica will cut up ridiculously small.

And there is a world unsaid in the brief ejaculation (May 18, 1858) : 'So Sonder [2] makes 106 Oxalises—humph.'

To Col. Munro

September 9, 1853.

I have rough polished Berberideae and had such a job to get through the *B. vulgaris* and *aristata* groups, which by the way I cannot distinguish *specifically* from one another. I quite expect great opposition in the first group and I may state once for all, that I take no person's opinion on them as worth a snap, who has not studied the varieties of *B. vulgaris* itself ; and no one who *has not* can have any idea of what they are ! I have also carefully studied all the garden species of the N. Hemisphere.

Madden [3] came here two days ago and spent the morning.

[1] Michael Pakenham Edgeworth (1812–81) was an Indian civilian who had studied botany under Graham. He contributed papers on the botany of India and Aden, and on the Indian Caryophyllaceae to the *Flora of Brit. Ind.*

[2] Otto Wilhelm Sonder (1812–81). He was the author, or part author, of several works, *Plantae Preisscanae*, 1844–7 ; *Revision der Heliopticleen*, 1846 ; *Flora Hamburgensis*, 1851 ; *Die Algen des tropischen Australiens*, 1871 ; *Algae Ost. Afrikanae*, 1879 ; and *Algae Australianae hactenus cognitae*, and he assisted Harvey with his *Flora Capensis*.

[3] Edward Madden (*d.* 1856). He was Lieut.-Col. Bengal Artillery, and President of the Botanical Society of Edinburgh, F.R.S. Edinburgh. He collected in Simla and Kumaon. He published *Brief Observations on some of the Pines and other Coniferous Trees of the Northern Himalaya*, in the *Journ. Agric. Soc.* of India, 1845, and a Supplement to it in 1850, and *Nepal Plants* in 1856. The genus *Maddenia Rosaceae* was called after him by Hooker and Thomson.

I showed him the Berberis which confounded him; his only objection (or crotchet) was the simple racemed form of *aristata*, as different from the panicled form of the same plant. These two he had studied living and found them always distinct though growing side by side; I showed him loads of specimens he could not decide between! but the fact of his having found them distinct *side by side* outweighed all others. Now what are we to give for such facts? They are most important, but are we to admit every collector's (however good a botanist) testimony on such a point, as of specific importance? Thomson thinks *lycium* the only good one on the same grounds, whereas Madden vows he found these passing into one another every way. I took *asiatica* for the best marked of them all, and that again Madden denies *in toto*. I wish you would kindly tell me what your own 'particular variety' was amongst them.

In September 1853 he tells Munro:

I am travailing through an Essay on 'Species, their distribution and variation,' for the New Zealand Flora Introduction ['which I hope;' he afterwards tells Bentham, 'will be read, though I cannot flatter myself it will be of any great use'], chiefly intended to open students' eyes to the great leading facts of the case and to inculcate caution, or they will have their Flora in a pretty mess, for it is a frightfully variable one.

But his apparently destructive tendencies were really constructive. He tells Harvey (January 1852):

I am combining very many species with Tasmanian and South American plants—many are identical without trace of change, which led me to claim some variation for others which belong to very widely different genera. . . . The upshot will be the total bouleversement of our previous ideas of the extent &c. of the Flora and a very close alliance indeed with Australia. I am really extremely anxious to get the thing well done, but greatly doubt people's being satisfied with my destructive propensities, which however are far more really *constructive* than those who have few materials to work from and judge by can form any idea of.

His care in working out species detail is illustrated by a friendly scolding of Harvey in 1859 for rejecting on inadequate grounds the identification of an African cress, *Cardamine africana*, with the widespread *C. hirsuta*, whose range is described in the letter to Darwin of December 1843 above.

Criticism should at least be as well equipped as the opinion criticised.

Kew: February 19, 1859.

MY DEAR HARVEY,—I am really sorry for your disappointment with the lithographer, it is very disheartening, though by the way it is I fear only a righteous and well merited retribution for your most unjust, ungenerous, ungracious, and unphilosophical attack on my Cardaminologia. Thwaites makes the Ceylon C. = *hirsuta, suâ sponte, without any hint from me*. He sent it thus named years ago before the Enumeration[1] was conceived of; though I altogether agree with him.

Who are you? that you, without seeing my materials, say that *africana* and *hirsuta cannot* be the same. You might at least go over my evidence before you condemn. I just wish that you had spent as many hours over the wretched weed as I have. I assure you that when I did the N.Z. Flora I spent *several mornings* at that plant alone, and had spent a long time at it when I did the Antarctic, and have since on doing the Indian plants; always with the same result. I do not demand infallibility, but I have a right in common with every man of science that my conclusion be not put aside without my evidence being examined. You who know *Plocamium coccineum* and *Ceramium rubrum* might be careful I think of *Cardamine hirsuta* in another man's books! Scolding apart, my belief is that C.h. is one of those plants of which you may make 20 species or one, if you make 2 you must make many more, and seeing that *C. africana* is in my apprehension joined to *hirsuta* by intermediate forms of habit, of foliage, of inflorescence, and of pod, it ranks according to my philosophy as a variety and not as a species. As soon as geological or other causes have destroyed said intermediates then I will make it a species. If each Botanist is to insist on keeping two dissimilar things

[1] *Enumeratio Plantarum Zeylaniae*, Thwaites, published 1859-64.

species because *he* has not uniting forms, though others say they have, then there is an end of the matter.

Ever yours affectionately,
Jos. D. Hooker.

Finally, in March, after showing how they are at cross purposes in the matter, he concludes :

The principles we should go on are to unite what nature unites *wherever* she may have done so, and not to *assume* that she ought to have done so elsewhere. However, as I am sunk in the sink of creation of species by variation you may do what you like with the *Cardamine.*

So in February he tells Bentham :

I have made sweeping reforms in the New Zealand Flora, upon which I am quite hot and am egregiously pleased and interested ; somehow I have taken greatly to working out species and genera and examine a great deal more than I used to.

It was the same with the Introduction to the 'Flora Indica' by himself and Thomson.

So complete a bouleversement of all former nomenclature perhaps never occurred to any considerable Flora since Linnaeus' Vegetable Kingdom. It has, however, been impossible to avoid doing battle with all our predecessors' species, whose utter disregard of one another and of any other part of the world's Flora but India has produced inextricable confusion in many cases. (To, Munro July 1853.)

The said Introduction [he tells Bentham in 1853] is to be a tremendous long essay on all things botanical in general and Indian in particular ; we have taken up the subject of Indian Bot. Geography in a comprehensive manner, and have gone at great length into geographical divisions and the collections and some works of our predecessors. Also we have several pages on the study of systematic Botany in general, and the use of Herbaria ; the prevalence of bad species ; narrow prevalent ideas of variability and too much stress laid on habit. In all this we do not expect you to agree, but for my part I am convinced that time will prove our estimates of species very false indeed. I do not know a greater snare than that of habit; we take an ideal of a herb, tree or shrub, and carry it with us through all countries. Take the common oak, what is its habit apart from the English park variety? Compare it with the Scotch oak in the Highlands or the long gaunt things that flourish at the Cape of Good Hope. We have been doing up our Indian Coniferae and find *Juniperus excelsa* quite identical in all botanical characters with *Sabina, chinensis, Dahurica, virginiana, occidentalis* and several others, as was indeed pointed out by my Father, Fl. Bor. Am., and again by Spach who goes much further. Now supposing these to be all the same, will any one tell me what is the habit of the species? Suppose them different if you please and I answer that in the Himalayas the one species assumes the habit of all the others.

Take the ordinary Scotch Fir in Switzerland; what is its habit? certainly not that of the Scotch plant; nor of the German; it is a curious fact that I rarely could recognise by the eye our common English trees in Switzerland, so altered is the habit. I wish you could have gone with us to Dropmore 4 months ago, to have seen the cedars of all sizes, hues, habits, and shapes: all of Lebanon and amongst them all the *Deodar*, looking *anything but a very distinct variety*. Lindley was quite taken aback and has been mum ever since about *Deodar* and *Lebanon* being different species. To-day *Ephedra* has brought the same thing under my notice and I would far rather take C. A. Meyer's only (and microscopic) character from the micropyle to distinguish *helvetica* from *vulgaris* than any amount of difference of habit. I am quite disquieted with the fictitious nature of characters as now given in books. There are in said book of Meyer's 4 species without a single character important or *unimportant* between them. To take Endlicher's Coniferae; is it not pure fraud to go on enumerating species with specific characters that are mere play upon words? and this without a syllable of remark or excuse. What single character is there for any *Taxus* but *baccata*?—the keeled scales of the bud is all he gives and it breaks down in *T. baccata*!

The deeper I go the more convinced I am that Brown

is right, and that there are not 50,000 species of flowering plants known. Wallich has 8 names for *Pteris aquilina*, and I do think he has two names for ¾ of the species in the early part of his catalogue, besides Don's, Royle's, Edgeworth's, Roxburgh's, and often De Candolle's. This however is an old story. I admire your great caution and desire to curb my rabid radicalism: but the tide will turn one day and the reducing species will go on apace, and then the reaction will be terrific. After all there is something to be said for me. I am a *rara avis*, a man who makes his bread by specific Botany, and I feel the obstacles to my progress as obstacles to my way to the butcher's and baker's. What is all very pretty play to amateur Botanists is death to me.

The following letters to Asa Gray deal with the Introduction to the New Zealand Flora.

Kew: Wednesday, January 26, 1854.

MY DEAR GRAY,—I was extremely pleased by your letter last night, and quite as much with the mere fact of my treating of the subject having been thought worthy your attention, as with the many too flattering things you say of it. Such Essays attract so little attention in this country, that one feels, at least I did, that I was writing for the dead more than for the living, though amongst other men Agassiz had a prominent seat in judgment before me. After all I regard the whole Essay more as a résumé of general impressions than a specimen of *close reasoning*, for of the latter, in truth, the subject *does not admit*. There is not a single argument that will not cut both ways, and may not be turned *pro* and *con* species, specific centres, &c., &c. Your turning my arguments against myself on the point, that two originally created distinct species so similar as to be almost undistinguishable, may exist in two widely sundered localities, is an awful staggerer, and I have always felt it to be the most impracticable objection of any to the possibility of determining what is and what is not a species. I have touched on that very point at ch. 2, § 2, towards end. 'These considerations; etc.,' but perhaps too gingerly, also in the Fl. Antarct. I think, see *Empetrum*. I combat this theory more upon principle than upon facts;—once admit it and the flood gates are opened to species-mongers, and it is cast in your teeth every moment, as an argument for making every slight difference, if only accompanied with geographical segregation, of specific value.

Nevertheless I am quite aware that such species must exist; I do not deny, nor would I blink, the evidence in favour of it, nor that it is the gravest of all objections to the pronouncement upon species in our present state of knowledge. I therefore admit its application to practice only in exceptional cases. The long and short of it is, that if you admit two centres you may as well admit all Agassiz, you cannot draw the line, and Geographical distribution is hence a vain study, the connection of life with the revolutions of our globe and with all the physics of nature is naught, and nothing can come of its pursuit but the temporary gratification of taste and ingenuity.

I am amused by fancying you 'fall into the snare you lay for another'—the following, which shews how all these arguments cut two ways. You say *generic resemblance* is a strong point, and not enough dwelt upon. I grant it fully. I suppose I thought it too hackneyed, though it is far from being so in a philosophical point of view. But you go on with consummate sangfroid to tell me of Dorking fowls and Manx cats, starting off at a tangent without rhyme or reason! This I grant too, but let me ask you what would be done by Gould or Agassiz with a Dorking fowl, if it were shot and skinned in the Andamans and brought from thence as its only habitat? Not only would a new genus be made of it, but its toes would lead to a deal of pen, ink and paper, analogies, affinities, relations, &c., &c., &c. Ditto with the Manx cat, an osteological specific character would be found for it as easily as Cuvier found one for the Falkland Islands rabbit, which had not been 30 years out from Europe! Oh dear, oh dear, my mind is not fully, faithfully, implicitly given to species as created entities *ab origine*, but it is to the imperative necessity of sticking to one side or the other and, without being bound by it, referring, arranging, and reasoning by it. I take that side which, though apparently the most narrow and prejudiced, is the only one which really keeps the mind open to investigate, which co-ordinates all the elements of geography, system and physiology, and which keeps the observer's attention alive to the importance of studying collateral phenomena.

I have long been aware of Agassiz' heresies. His opinions are too extreme for respect and hence are mere prejudices. They are further contradicted by facts. Lyell and I have talked him over by the hour. Lyell and Agassiz are great personal friends. I always think Agassiz an extraordinarily clever fellow and a treasure too as a scientific man, but there are many people whom personally we like and men of science too, but whose views on individual points are best left alone. Giving too much attention, even to oppose, the startling views of such people rather encourages them, and there is an inherent love of getting fame *at any price*, i.e. *getting notoriety*, amongst these French, Swiss, and Italians that leads them to commit themselves on such questions. The long and short of it is, that we have too many *clever* people in the world, too few sound ones. When you Yankees take up the higher branches of Botany more generally you will turn out far more and better work than we do, for you are a far better educated, sounder, more practical people, and I look to you for the great discoveries, come when they may.

Is your N. American Larch different from ours? Is there more than one Yew in the world? How many Junipers have you? Coniferae are I am sure much more variable and widely distributed than is supposed, and whilst all our commonest wild and cultivated Junipers, Yews and Scotch Pines are telling us by every specimen that their habits vary with every local circumstance, we are still quoting *habit* as a specific character for Coniferae. I showed Bentham two yews in a hedge at Pontrilas [Bentham's house in Wales] side by side, of which he owned that specimens from each would make two species, and their habit was so different, that were they growing side by side in a garden, the *habit* would have confirmed the difference. Take *Juniperus communis*, I found it in the Rhone valley growing like *recurva* of India, with a straight trunk and conical coma. As to our Deodar avenue of Kew, it is the seediest, most ragged affair you ever saw, many of the trees far more like young cedars. These were all seed raised; had we planted cuttings as nurserymen do, of the most weeping glaucous, long leaved stirps, what a different thing we should have had. I do think habit a perfect snare with many people; we stereotype an ideal habit and refer everything to it. Of the many people ready to swear and declare that they can never mistake an Oak, Beech, &c., &c., by habit, how many can prove their words?

You say that we are not to pronounce species the same because they are united apparently by certain forms of each —I grant this fully, but how are we to act upon it and deny local Botanists specific value to their small fish? This is no good argument; a better one is, that we do not know which is the originally created state that you call the type, or that I call the connecting form. *E.G.*, You may say Cedar and Deodar are distinct though apparently united by a few exceptional forms of each. I say no, the exceptional intermediate forms present no new character different from either. The original type *cedar* was intermediate in character, but is extinct, one extreme form is retained, driven to the top of Mount Libanus, and hence called *Libani.* Another extreme form is retained in the humid Himalaya. We cultivate the Libanus stirps which retain to a certain degree its rigid character, but often lose it. We also cultivate the Deodar stirps, and because beautiful we propagate by cuttings from the states most typical of Deodar, i.e. most extremely unlike Cedar, and propagate the error by artificial means.

Kew: March 24, 1854.

DEAR GRAY,—Very many thanks for your capital long letter, which begins by agreeing with me that, 'the subject does not admit of close reasoning'; and goes on with as pretty a specimen of admirable close, clear, and accurate reasoning as I ever wish to peruse. I only wish you had taken up the subject instead of me, for you throw out your grapnels with a judgment and precision that put my loose ratiocination (is that the word?) to shame. You must (probably do) know that I am one of those cross grained fellows who, after building up a tall tottering castle, get sick of it and can't bear a kind friend coming to prop it up; neither do I like an enemy to knock it down; so there is no pleasing me but by praising my castle in the abstract, whether it stands or falls.

I entirely agree with all you say about representative species, and groan over the hitch in deciding what we are to agree to call a species in such cases. I also fully agree that the fundamentality of the argument derived from generic resemblance is not fully appreciated by myself; one is apt to overlook its real whole weight, from being accustomed to bear it, like atmospheric pressure. It is *per se* unanswerable, and hence put aside for less valuable facts that afford scope for reasoning and debate. I am hence the more glad that I wound up my chapter with the quotation from you; for which I do not deserve the credit which I hope others will attach to its introduction. I put it in as much for the sake of strengthening my argument by quoting one known to be so able to judge as you are, as for what it said. I believed in *you* in short, quite as much as in what you wrote.

To Asa Gray

March 29, 1857.

My Father has just asked me to review Berkeley's Introduction to Cryptogamic Botany for him a little in detail. It is no joke to read it to begin with. It is a wonderful book, chock full of observations, full of reflections, full of able thought, accurate analysis, as carefully and honestly done as a book can be, and a result of a mastery of the subject which I believe no other man living possesses. Unfortunately it is abominably written and arranged, and the really admirable correlations of facts and phenomena in the different organs and orders of plants dealt with in the most higgledy piggledy fashion. It is like a country parson's sermon all over, without a beginning, middle, or end, the leading ideas are here, there, and everywhere, bound together by the jolliest rigmarole of conjunctions, prepositions and adverbs. These parsons are so in the habit of dealing with the abstractions of doctrines as if there was no difficulty about them whatever, so confident, from the practice of having the talk all to themselves for an hour at least every week with no one to gainsay a syllable they utter, be it ever so loose or bad, that they gallop over the course when their field is Botany or Geology as if we were in the pews and they in the pulpit.

Witness the self-confident style of Whewell[1] and Baden Powell,[2] Sedgwick[3] and Buckland. Berkeley has avoided this latter snare but has got thoroughly imbued with the idea that it matters little how his matter is served up. The book, however, pleases me amazingly; there is a lofty tone throughout it, an aiming at the highest principles and an earnest desire to make his readers think for themselves as much as he does for them.

Bentham's résumé of our views will appear in the Journal Linnean. The Germans have got to dreaming on the subject as usual, and A. Braun is groping amongst the blacks for the characters of the whites. There is a story somewhere of an Englishman, Frenchman, and German being each called on to describe a camel. The Englishman immediately embarked for Egypt, the Frenchman went to the Jardin des Plantes, and the German shut himself up in his study and thought it out! How can Braun, who has no practical knowledge of large masses of species, know where the generic idea and name is to be fixed, how far, in short, systematic language is to be carried into the subdivisions of plants? Seemann has got some twaddle about whether genera are objective or subjective, a point easily disposed of, *Rosa*

[1] William Whewell (1794–1866) was the famous Master of Trinity, Cambridge, from 1841, of whom Sydney Smith said that science was his forte and Omniscience his foible. He had held the chairs of Mineralogy and Moral Philosophy, and his memoirs of the Tides had won a gold medal from the Royal Society in 1837. His universality of learning was shown in his *History of the Inductive Sciences* (1837), and his *Philosophy of the Inductive Sciences* (1840), which, his friends recognised, produced a greater effect on study than any specialisation of his. His other celebrated work, *Of the Plurality of Worlds*, appeared in 1853.

[2] Baden Powell (1796–1860) was Savilian Professor of Geometry at Oxford from 1827. He wrote especially on radiant heat, optics, and the general history and study of science. A liberal churchman, he took his part in theological controversy, and was a contributor to *Essays and Reviews*. Among his best known books were those on the Unity of Worlds, Natural Theology, and the Order of Nature.

[3] Adam Sedgwick (1785–1873) was one of those men whose influence was due as much to his warm affections as to his powers of preaching, teaching, and research. As Woodwardian Professor of Geology from 1818, he reorganised geological teaching at Cambridge; was President of the Geological Society 1831, and received the Wollaston Medal in 1851 and the Copley in 1863, and refusing other preferment, became Canon of Norwich. His research into British geology resulted in the establishment of the Cambrian system; but though a pioneer in his own department, he was unreceptive of new and progressive ideas, such as Lyell's uniformitarianism, and vehemently opposed Darwin.

being clearly an objective genus, as is *Salix* and a heap of others, whereas almost every genus of Umbellifers is a subjective idea, and a confoundedly bad one too.

Mutual criticism took the liveliest form between these best of friends. The allurement of paradox has already been referred to, p. 450 *sq*.

To Asa Gray

1857.

Many thanks for your letter and the swishing review of Berkeley. It serves him right, but he certainly will not like it. He has made no remarks on my review in the Journal of Botany; I suppose that like another friend of mine (the last letters of whose name are Asa Gray) he thinks I am wrong when I find faults!

I am charmed with your criticisms on my ideas of Physiology, &c., &c. Your ideas remind me of a firework called the serpent which makes fiery circles,—ascends, makes more circles,—descends, then flares up and goes out. Mine you may compare to a similar work called a whirligig cracker, which does the same in a less methodical form. They both end as your ideas may end—in a blaze, a bang and a stink. We neither understand one another nor our subject in one another's eyes, and the stink of each alone remains to each. I shall be very glad to take any amount of vital force when I find any one else doing so. With me it stands in the same relation to other forces that magnetism does to heat, electricity, sound, sight; each of which is a *tertium quid* investigated by the following up the laws of the others. With you Physiology = Biology, with us they have a totally different meaning. I mention this to show you how far we are at cross purposes in diction. Development = growth, I agree and generally use the latter term, but it is raw and undignified.

Heaven defend me from my friends! I put Bentham up to Ranunculanths! I who cannot tolerate English names in any shape! They are Henslow's children, and bad, though the best; being infinitely better than *ads*, *worts*, and *aceae*. I think Bentham right to adopt them, because they are now solemnly sanctioned by her Majesty's Government, no less, for the delectation of National Schools;

and as the Henslow diagrams will be the great engine of instruction for schools, ladies, parsons and the like, it would *meo sensu* be most unwise of B. to have ignored them or adopted any new-fangled ones. I hate and despise the whole English system both for ordinal and generic names. You know how difficult it is to get any really good books put into Govt. circulation, and it would be a most serious drawback to the good Bentham's would do were he not to make his uniform with the system in vogue. These things are trifles to us, but terminology is a serious affair to the classes the book is intended for; so whatever you do, do not put Bentham off using *anths.* I advised saying Ranunculaceae—Ranunculus family, and in brackets (Ranunculanths) after.

To Asa Gray

January 2, 1858.

Yours of the 19th has just arrived and gratified me very much. I am, I need not tell you, in the habit of saying at *least* as much as I think, when I have fault to make or find, for I hate to let it be supposed that I have held back any growl, or grudge, or stone of offence in hat or pocket.

I am glad that you have taken up the Balanophoreae matter and that of high and low specialization. I hope you note that I do not commit myself to the theory of perfection being expressed by consolidation, but state all hypothetically. I wish I could see my way clearly through the maze of high and low amongst Dicotyledonous Exogens. Formerly I felt inclined to exalt Tiliaceae, Malvaceae and Euphorbiaceae, and to assume as the highest type of flower that which has (1) complete series of whorls; (2) those whorls all distinct from one another; (3) each whorl being of numerous members; (4) each member being highly specialized; (5) each carpel to contain many perfect ovules and albuminous dicot. seeds;—thus in short returning to DC. Still the question remains, is a large imperfect group to be placed at the top of the vegetable ladder because one or a few of its members presents these attributes in greater degree than any other vegetable does?—this cannot be conceded, and so the whole fabric falls to the ground. Destroy all Euphorbs, except the monandrous genus *Euphorbia,* and all clue to its affinities and rank are lost. We must therefore turn to higher considerations than mere organic complexity and perfection of whorls and make these secondary—when the physiology of the reproductive organs at once suggests itself and Gymnosperms jump up from the bottom of the scale to the top! for they *superadd* to the perfect Phanerogamic reproductive apparatus an exaggeration of that of the highest Cryptogam, and this without showing the slightest trace of low development in trunk, embryo, pollen or ovule, and without displaying any of the peculiarities which keep Cryptogams below Phaenogams, except always the want of a stigma, which does not imply however any modification of pollen or pollen-tube!!!

I am atrociously busy, as, if you knew anything about me, you would know by this long letter.

The Floras of New Zealand and India are based on the acceptance of the reigning belief in the fixity of species. The change takes place between 1855 and 1859, when the Australian Flora was published, more especially, as has been pointed out, after the full argument of the Origin was first put together in 1858 and resolved the chief difficulties which his own work had left unanswered.

Thus he avowedly adopts a new principle in his Introductio n to the Tasmanian Flora, which he explains in the following letters to Harvey, whom he is consulting as to affinities between the Cape Flora and the Australian, to Asa Gray and Bentham.

Kew: Sunday, January 1, 1859.

DEAR HARVEY,—I am labouring right hard at the Introd. Essay on Australian Flora,[1] whose only hope of utility is the quantity of curious stuff it may contain; for as to elaborating from it a theory of the origin, etc., of Australian Botany, it is hopeless, I fear. What I shall try to do is, to harmonise the facts with the newest doctrines, not because they are the truest, but because they do give you room to reason and reflect at present, and hopes for the future, whereas the old stick-in-the-mud doctrines of absolute creations, multiple creations, and dispersion by actual causes under existing circumstances, are all used up, they are so many stops to further enquiry; if they are admitted as

[1] First volume published 1859.

truths, why there is an end of the whole matter, and it is no use hoping ever to get to any rational explanation of origin or dispersion of species—so I hate them.

January 6, 1859.

I am determined to start in my investigations on a different principle and to try and square all my facts with (or arrange them by) the most modern doctrines without therefore adhering to or accepting those doctrines. The old theory of absolute creations, of single individuals or pairs is used up! Grant them, and what's the use of arguing any more? Grant too that all migration has been effected under existing relations of sea and land, and there is an end of that matter, we may whistle for another force to effect migration, other than the known agency of animals, winds, and waters. If we are to assume nothing but these, we are stumped! If the course of migration does not agree with that of birds, winds, currents, &c., so much the worse for the facts of migration! No religious creed could be more exigent, exclusive, and repressive. I should be wrong to say I *disbelieve* these doctrines simply because they do not explain my facts, so long as they do not contradict them. I should be as wrong to say that I *believe* them so long as I think that other doctrines may explain the facts as well or better than these. I now then start on the assumptions: (1) That all vegetable forms are in a state of unstable equilibrium. (2) That the rate of change and extent of change vary at different times and places, depending on physical conditions, i.e. on extent of surface to change over and of conditions of surface to promote and perpetuate change. (3) That the majority of main types of existing forms have survived all Geological changes from the Palaeozoic era downwards to our time. (4) That during this interval many of these type forms have migrated from one hemisphere to another, some of them remaining specifically unchanged, others generically, others subordinately. (5) That during their migration they have expanded and contracted, i.e. sometimes thrown off constellations of varieties that (by selection) have become new species, at others few, at others none. (6) That during some epoch there has been any amount of change of land and water.

This does not touch the aboriginal condition of all

types, i.e. of species, my object being to account for existing distribution.

These hypotheses square with all my facts, for from them you would expect to find :—

I. That, as regards extent of variation, all existing plants are made up of two classes or assemblages, (1) A large number of species so distinct from one another that no one doubts their constancy or disputes their limits, and which we cannot connect with others or with one another except by intercalation of an immense series of intermediate forms that do not now exist. (2) Of a vast assemblage that range themselves in clusters of variable forms so slightly distinguished that no two Botanists agree as to their limits, and any one admits that one, or a few, small characters alone distinguishes each from its allies.

II. That, as regards rate of variation, some forms have remained specifically unchanged from the Oolite downwards, others only generically, whilst others are more changed still.

III. That Australian forms are found only in the old rocks of Britain.

IV. That the Floras of sinking (Volcanic) islands contain a larger proportion of distinct types than those of continents.

V. That some of those types are not at all represented on the continents, others only on the nearest continents.

VI. That the further the island is from the continent the greater is the peculiarity of its Flora.

VII. That the number and variety of ordinal types is as great in the S. temperate Zone, where there is so little land, as in the North.

The numbers and proportions of orders (and numbers of genera too ?) remaining the same in both. This I can understand if you will allow me in the South as large and varied an available surface as Europe, Asia, and America present ; for if you were to destroy 2/3 of Europe, N. Asia, and America you would not reduce materially the number of genera, nor of orders at all, but a vast number of species would be destroyed.

To Harvey

March 1859.

I am delighted to hear of the progress in Thesaurus and Flora [of the Cape]. You are a brave man indeed. I am groaning and growling and making an awful ado about my Introd. Essay to V. D. L. Flora, which is a heretical, hypothetical, clumsy, laboured, cumbrous rigmarole of what I believe to be the correct ideas not yet fully developed, owing to backward state of science.

To George Bentham

Kew: July 17, 1859.

The Introd. Essay goes on *very slowly indeed*, many thanks for your valuable hints, I have modified some of my expressions (which conveyed more than I intended) accordingly. On two points you and Gray are rather hard. You expect me to *prove* or *make out* my case, and Gray calls me hasty, precipitate, etc. Now my case is no more capable of proof than the opposite doctrine of separate creations, and I do not pretend to be able. I think I show better cause for its probability than creationists can for theirs, but this a matter of opinion: at any rate the doctrine is conceivable and there is an immense deal in all the steps that lead to it; whereas all the avenues to further research are blocked by the opposite. On this point and on Gray's objection I have said a few words in the concluding paragraphs which you have not yet seen. Thwaites has written to me on the subject evidently on Thomson's suggestion, for Thwaites was once a devoted variationist and I suppose is so still, though he writes cautioning me not to commit myself. One of your arguments against is favourable to me, if logically pursued, viz. that as to the age of man being illimitable, and yet never exceeding a certain amount, viz. 1–200 years. Were then Methusaleh and his contemporaries different species? Then again as regards Camelopard and shorter legged animals of its tribe—their difference in that respect is not so grêat as between a Skye terrier and Greyhound. After all the case is quite analogous to the Science of Geology; Lyell's views of uniformity of action and immense periods were laughed at by those born and bred to the doctrine of successive cataclysms in a world only 6000 years old, and I cannot help feeling that the difficulty in this case of species is to conceive time enough; that however is not an impossibility, but that of special creations of highly organised beings is an impossible conception. I am much influenced too by the progress of Physical science and 'Natura nihil facit per saltus.'

To Bentham

August 8, 1859.

Very many thanks for your last letter, and the notes on the Essay. I have revised the paragraphs on anomalies, but not altered much, as I think that such as they are, the peculiarities of the Flora are much more objective than of any other Flora, and more pervade the whole vegetation. . . . I was afraid of overdoing the peculiarities, and have failed to do them justice. I agree with you that my allusion to them is not sufficiently discriminative. Take *Eucalyptus* altogether as a genus and it is really a remarkable vegetable, considering the number of forms its Bark assumes; that alone would make it notable.

CHAPTER XXV

THE MAKING OF THE 'ORIGIN': SCIENCE AND FRIENDSHIP

MODERN Science dates from before or after the 'Origin of Species.' The publication of the book was, so to say, the Hegira of Science. By it the science of living things was revolutionised and every other branch of natural science was stirred. After the vested interests of current opinion rose up in a great turmoil, Philosophy took a new element into her reckoning. The Natural Sciences claimed their rights as knowledge, discipline, and power.

But the making of the 'Origin' is not only a history of science—it is the history of a great friendship. In its fabric the two strands are indissolubly interwoven. As Darwin exclaimed to his friend, 'Talk of fame, honour, pleasure, wealth —all are dirt compared with affection, and this is a doctrine [in] which I know from your letter that you will agree from the bottom of your heart,' so the achievement is ennobled by the warm human affection that so long sustained the worker and aided the work. For twenty years the materials for the task were being amassed; for fifteen of these years Hooker was Darwin's confidant and helper. Without Hooker's aid Darwin's great work would hardly have been carried out on the botanical side.'[1] In his quiet isolation at Down, cut off from the ordinary converse of the world by the perpetual uncertainties of ill-health, Darwin found refreshment and delight in pouring out to his friend his schemes of research and his wonderful experi-

[1] Sir F. Darwin and Professor Seward, in M.L. i. p. 39.

ments on the living action of plants, sure of sympathy, yet begging Hooker, if he could spare time to read these letters, at least to waste none of his too busy hours in answering them, saying:

> It is a pleasure to me to write to you, as I have no one to talk to about such matter as we write on. But I seriously beg you not to write to me, unless so inclined; for busy as you are and seeing many people, the case is very different between us (June 19, 1860). It is the greatest temptation to me to write *ad infinitum* to you (July 19, 1856).

As to direct botanical aid, he wrote with enthusiastic appreciation and careful criticism of Hooker's publications, which bore so closely on his own work. But this was the smallest part of their scientific interchange. Though he repeatedly insists 'Do not answer questions merely out of good nature' ['of which towards me you have a most abundant stock' (April 8, 1857), 'as wonderful as mesmerism' (1846)], it was the unstinted privilege of the elder friend to ask, as it was the privilege of the younger to answer from the fulness of his botanical knowledge, a host of questions bearing on the relations and distribution of individual plants and groups of plants, wherein lie answers to some of the riddles of life.

The beginnings of this friendship have been told by Hooker himself in the 'Life of Darwin,' ii. 19.

> My first meeting with Mr. Darwin [he tells us] was in 1839, in Trafalgar Square. I was walking with an officer who had been his shipmate for a short time in the *Beagle* seven years before, but who had not, I believe, since met him. I was introduced; the interview was of course brief, and the memory that I carried away and still retain was that of a rather tall and rather broad-shouldered man, with a slight stoop, an agreeable and animated expression when talking, beetle brows, and a hollow but mellow voice; and that his greeting of his old acquaintance was sailor-like—that is, delightfully frank and cordial.

It has already been told how the proofs of the 'Voyage of the *Beagle*' reached him through the Lyells in the spring of that year, while he was hurrying on the last of his medical studies in order to take his degree before sailing with Ross, and how, there being no other time available, he slept with them under his pillow, and read them before getting up in the morning.

> They impressed me profoundly, I might say despairingly, with the variety of acquirements, mental and physical, required in a naturalist who should follow in Darwin's footsteps, whilst they stimulated me to enthusiasm in the desire to travel and observe.

In the letters from the Antarctic there are several references to Darwin, who saw various of these letters through the Lyells. The correspondence between them, as has been told on p. 169, began in December 1843, when Darwin wrote to congratulate him on his return (C.D. ii. 21) and urged the importance of correlating the Fuegian Flora with that of the Cordillera and of Europe, at the same time offering his own collections of plants from the Galapagos Islands, from Patagonia and Fuegia for examination.

> This led to me sending him an outline of the conclusions I had formed regarding the distribution of plants in the southern regions, and the necessity of assuming the destruction of considerable areas of land to account for the relations of the flora of the so-called Antarctic Islands. I do not suppose that any of these ideas were new to him, but they led to an animated and lengthy correspondence full of instruction.

Only the first two or three letters open with the formal 'My dear Sir' of the period; by February 1844 Darwin inaugurates 'Dear Hooker' to his 'co-circum-wanderer and fellow labourer,' while from the day of his impending departure to India the 'very truly' or 'very sincerely' of either signature, gradually merging in 'Ever yours,' is lost in 'Your affectionate friend' or 'Yours affectionately' maintained by both to the end.

Acquaintance ripened swiftly into friendship. 'Farewell!' Darwin concludes a letter in 1845. 'What a good thing is community of tastes! I feel as if I had known you for fifty years. Adios!' And 'forty years on' the sympathetic bond between them was as strong as ever. In 1881 Darwin writes:

> Your letter has cheered me, and the world does not look a quarter so black as it did when I wrote before. Your friendly words are worth their weight in gold.

One of the starting points of Darwin's 'presumptuous work' had been the striking impression made on him by the distribution of the Galapagos organisms; hence his eager desire to know whether the botany of this isolated group was as suggestive as the zoology.

The correspondence began in December; by January the momentous confession was made:

> At last gleams of light have come, and I am almost convinced (quite contrary to the opinion I started with) that species are not (it is like confessing a murder) immutable.

He had instantly recognised Hooker's capacity. 'I am pleased to think,' he writes on Hooker's rejection at Edinburgh in 1845, 'that after having read a few of your letters, I never once doubted the position you will ultimately hold among European Botanists.' And in the next letter, 'It is absurdly unjust to speak of you as a mere systematist.' More than this, he recognised that Hooker also believed, as he put it in the Preface to his Flora Antarctica, that 'Geographical Distribution will be the key which will unlock the mystery of the species.'

But true views of geographical distribution were impossible without full and accurate Floras. Here no doubt was a redoubled motive for the ardour with which Hooker flung himself into his unending labours, the extent of which called forth the first of many anxious warnings from Darwin as early as 1845, to beware of overwork, doctor though he be, and a novel prescription, 'You ought to have a wife to stop your working too much, as Mrs. Lyell peremptorily stops Lyell.' The perfecting of his great Floras involved the re-examination of his vast materials and the more or less incomplete work of his predecessors, so as to sweep away the existing synonymy and overlapping, and to readjust the systematic details by making clearer the true affinities and world-range of disputable genera and species. Complete and accurate classification according to nature was the first step towards finding the key to it all.

Thus Darwin, in the act of asking his aid, stimulated his native bent. He was encouraged in his inclination to deal with the wider bearings of his observations, which, in Darwin's eyes, made his Flora and his letters so different from the works of so many other systematists, remarkable for their lack of instructive general results. And though special researches such as these appeared to distract him from his main work on the Southern Floras, yet they shaped his own views and added to his reputation.

> I am almost sorry for your eternal additional labours on the Galapagos Flora [writes Darwin in September 1846; but adds emphatically], as yet your work assuredly has not been thrown away, as many have referred to your curious geographical results on this archipelago.

Similarly, of a preliminary sketch of his Tasmanian results, in 1844:

> I trust that your sketch will not have cost you ultimately loss of time, as, judging by myself, preliminary sketches and resketches do much good. . . . Seriously, I almost grieved, when I saw the length of your letter, that you should have given up so much time to me. Sir William will think me a bad friend to you, but anyhow, I trust, the sketch part of your geographical results will not turn out lost time.

These generalisations gave special value to his work and led Darwin to repudiate his description of himself as not possessing a philosophic mind, 'one of the greatest falsehoods ever told by implication; read your own Galapagos paper and be ashamed of yourself' (the whole passage is given in C.D. ii. 37). In short (March 31, 1845):

> Nothing would do you so much good as a little vanity, and then you would not talk of collecting facts for others, when, say just what you please, I am sure no one could put them to better use than yourself.

It was a unique relationship of minds. Each had had the same kind of experience in world-travel, and had observed nature, animate and inanimate, with a special interest in the same question—namely, how the different forms of life had reached their present habitats. In this, indeed, the younger man had taken the elder for his model. Before their friendship and alliance began, Darwin, the born scientific enquirer with philosophic breadth of mind albeit small technical training, had advanced far along his memorable line of research. He took everything for his province that bore on heredity and variation, fertility and decline in living forms, the competition they had to meet, their range and movement, the relation of them to their fossil predecessors in the same area, the geological changes which had determined the ancient courses of migration. Hooker, master of a whole branch of science, with technical training in it from his childhood up, and equally awake to the part played by geologic change in the problems of distribution he longed to solve, eagerly placed his vast knowledge, his sound criticism, his special observation during his later travels, at the disposal of the inspiring friend and fellow-worker who had gone so much further on the same quest as himself and had pushed it into wider fields than his own.

Each was deeply conscious of his debt to the other. Of the discussions they used to have, Hooker records ('Life and Letters of Charles Darwin,' ii. 27) : 'I at any rate always left with the feeling that I had imparted nothing and carried away more than I could stagger under.' Darwin from the earliest time feels the immense value of his help, in books lent, summaries of results, in published works, letters, conversations. 'For my own part,' he writes after a visit of Hooker's to Down, 'I learn more in these discussions than in ten times over the number of hours reading.' And again, after reading the Antarctic Flora, he speaks of having 'extracted more facts and views from you than from any other person,' while 'my pen runs away with me when writing to you' (March 19, 1845).

The thanks of a later period are foreshadowed by the thanks of the first twelvemonth, as :

> Really I do not know how to thank you half for all you have done for and sent to me. I might with truth do so for every single paragraph in your letter and every one volume. . . . Your remarks are exactly the thing, which ever since being in Tierra del Fuego, I have felt a keen curiosity about, and have often complained to Henslow how rarely I could find any such general remarks in Botanical works.

And in 1845 the prospective break in their personal intercourse, if Hooker were elected to the chair at Edinburgh,

> is a heavy disappointment to me; and in a mere selfish point of view, as aiding me in my work, your loss is indeed irreparable. . . . I assure you deliberately that I consider all the assistance which you have given me is more than I have received from anyone else, and is beyond valuing in my eyes.

More than this: they can express themselves with animation to each other, without risk of being misunderstood.

Hooker contributes much from his own knowledge. Distribution is his favourite subject, and he supplies statistics in the form desired to show range and migration, struggle and survival, from the Floras of the Southern Hemisphere or India or the Polar regions, all of which have fallen within his direct research. Moreover, he is particularly able to tell much about variation, for, as the preceding chapters show, he had long been struck by the incertitude of botanists on this head, and comparing detailed results all over the vast fields he had covered, had found many species as defined by local observers to be but varieties of a common species with every intermediate gradation. He can put Darwin in the way of answering the question whether large genera with wide ranging species, as should be the case with strong and increasing kinds, produce more varieties than smaller groups. At the same time he adds a warning as to the different impression of distinctness made on botanists by a given degree of difference occurring within the large or small group, so that what here would be ranked as a variety, would there be ranked as a species, to the confusion of any statistics that merely compare the relative numbers in existing lists. This is one of the cases where Hooker, after raising all the possible objections which must be overcome, is himself converted to Darwin's view by the facts which he has elicited for him.

He vehemently repudiates the notion (suggested by a geological article) of coal having been formed in shallow seas, and about this Darwin long continues to poke fun at himself and the botanists, to whom he finds it is the proverbial red rag. They differ as to continental extensions. While both condemn Forbes' unrestrained speculations in this direction, Hooker is too liberal for Darwin, who, though on occasion claiming and accepting great geological changes in land and sea, stands out against volcanic islands in the ocean being thus linked to continents, or the invocation of vast upheavals and depressions without other and independent evidence, as a simple way of accounting for a single phenomenon in distribution. Later, however, we see him constrained to accept Hooker's claim for a continental extension to New Zealand, as one of the cases that 'required it in an eminent degree,' but through a vanished Antarctic land, not directly to Australia.

Meantime he debates with his friend every other possible form of transport. Seeds may be carried by winds, ocean currents, berg transport, in mud clinging to a bird's foot, in the crops of birds, even the most unexpected birds, as when to his triumph a petrel is found helping in the transport of certain nuts. He confounds the popular belief that seeds of every kind must inevitably be destroyed by immersion in sea-water, through a series of experiments on temperate and tropical seeds, the latter supplied often from Kew, where also some of the experiments are repeated. He makes a salt-water tank, and tests the power of seeds to sink or swim, discovers how many will germinate happily after this treatment. He tells how his children at Down anxiously watch the trials to see whether he will 'beat Dr. Hooker.' Then as the experiments proceed and a seed to be experimented on happens to be delayed, he chaffs his friend merrily: 'I believe you are afraid to send me a ripe Edwardsia pod for fear I should float it from New Zealand to Chile!'[1] And so he quickly routs Hooker's cautious scepticism. The latter, confident that nothing will happen, has planted some seeds that the Gulf Stream has carried across the Atlantic to the coast of Norway. They germinate perfectly, and in answer to his confession of defeat (the letter is not extant), Darwin writes (June 1, 1856):

> I read your note as far as 'unutterable mortification' and was in despair, for I came instantly to the conclusion that probably Government had determined to give up Kew Gardens! and you may imagine how I laughed when I came to the real cause of mortification. It is the funniest thing in the world that you do not rejoice; for you have (as I never have) put in print that you do not believe in multiple creation, and therefore you surely should rejoice at every conceivable means of dispersal. Well, I and my wife have enjoyed a jolly laugh, and all the more from fully believing for a second that some great calamity had befallen you.

To quote a few more of the points with which the letters teem: Does the evidence show that in plants as in animals variability increases in parts which are abnormally developed? Do experiments in the Kew greenhouses show that cross fertilisation improves the fertility of the plant? Do statistics indicate that trees, where the abundance of adjacent blossom would tend to self-fertilisation, counteract this tendency by being more often dioecious than other plants? What of hybridism in botany; or of the part played by insects in fertilisation? On what definition does a botanist rank a class of plants as high or low in the scale, and how is competitive highness measured, i.e. that superiority in development which enables, say, the recent forms of Europe and Asia to oust Australian forms when they meet, especially as some particular adaptations in a 'high' class represent a retrogression according to the usual standard, which measures 'highness'

[1] The plant is only found in these two countries. It was shown that leguminous seeds as a rule were destroyed by immersion, thus suggesting a reason for the peculiarities in the distribution of the Leguminosae.

by increasing complexity of structure? How far do physical conditions alone effect similar changes in different plants? How far do the curious facts of distribution among Arctic plants indicate an extended glacial climate? Does the evidence from the migration and variation of temperate and subarctic plants indicate that this cold spell was world-wide, and was a factor in producing 'representative species' now isolated from each other?

Without further quotation of detail here is enough to illustrate the range of Hooker's abounding help in matters of fact or of theory. Unfailing also is his information about books to be consulted or papers in scientific journals dealing with special points. Many were not procurable even from the Linnean Library, where Hooker arranged that Darwin could take out what volumes he wanted. Many he lent to his friend from his own botanical library to be studied and lightly marked on the margins for the purposes of his analysis, sometimes to be borrowed afresh that the marked passages might be consulted anew when some better scheme of analysis had presented itself or some flaw had been detected in the previous scheme. 'I never cease begging favours of you,' writes Darwin in August 1855, when asking for the loan of the copy he had before of Asa Gray's Manual.

The parcels generally go from Kew to the Nag's Head in the Borough, the headquarters of the Down carrier, whether botanical parcels or a 'magnificent and awful box of books,' though in the case of a rare orchid in flower, Parslow, the immemorial butler, would travel to Kew and carry it back in his own safe hands.

Once, when Hooker had a fair copy of one of Darwin's MSS. to read, a misfortune happened which recalls, though it happily did not equal, the catastrophe to the sole MS. of Carlyle's 'French Revolution' in J. S. Mill's house. The bundle 'by some screaming accident' had got transferred to the drawer where Mrs. Hooker kept paper for the children to draw upon—and they 'of course had a drawing fit ever since.' Nearly a quarter of the MS. had vanished when Hooker prepared to read it at the end of a busy week.

I feel brutified, if not brutalised [he confides in Huxley that evening], for poor D. is so bad that he could hardly get steam up to finish what he did. How I wish he could stamp and fume at me—instead of taking it so good-humouredly as he will.

Nor did Hooker merely leave to his friend the tabulation of these important statistics of variation and distribution from the sources thus supplied. He often undertook it himself as a side-work in the flora on which he was at work, whether of New Zealand or India or Australia or the Arctic regions, for no other worker and no published book could provide the answer.

By a happy compensation these free gifts of time and labour for friendship's sake brought their own reward. With Hooker, as with others, such as Asa Gray, whose opinion Darwin had asked on similar points, the consequent research independently enriched his own books, widened the scope of his results, and pointed the way to a revivifying theory. Writing to Hooker in January 1857, Darwin says:

You know how I work subjects, namely if I stumble on any general remark, and if I find it confirmed in any other very distinct class, then I try to find out whether it is true, if it has any bearing on my work.

From this sprang many of his special researches. It was an additional merit in his procedure that he not only saw the crucial points that needed investigation, but inspired his most open-minded friends to independent research on the same lines, leading them to generalise on their results, instead of resting content with mere statements of fact. Thus, when Hooker writes (in December 1857):

I have begun my Introd. Essay to Tasmanian Flora. I think I shall confine it to a clear exposition of all the main features of the Flora of Australia and leave all conclusion drawing to others:

I am very sorry [he replies] to hear you do not intend to give generalisations in your Tasmanian Introduction but I do not believe you will be able to resist; what is in the spirit must come out.

Happily this resolve was broken by the impulse of Darwin's compulsory publication.

However, Hooker's long established conviction that species are more variable and less easily defined than most naturalists believed, did not bring him at once into the Transmutationist camp. He accepted the considerable variability of species and their spread by migration each from some one original starting place, a point less difficult perhaps to define than the perplexing modes of migration: he accepted even their relationship to allied species, their fossil predecessors in the same area, but to accept so much was not to accept their transmutation from other species. He went to India 'possessed, but not converted' by Darwin's theories, and was somewhat disappointed not to find them cleared up by the discovery of transitional forms in Sikkim, the meeting ground of tropical and arctic flora. The actual process of transition had not been observed; the partial light thrown on the question in fragmentary discussions was not enough, and until 1858–9, after the consolidation of Darwin's arguments in the famous Abstract, Hooker, as has been already noticed, worked avowedly on the accepted lines of the fixity of species, for which he had so far found no convincing substitute.

His critical attitude so long maintained may be regarded less as opposition to the tendencies of Darwin's speculations, than as the caution of a judicial mind, that required wholly convincing proof for itself before accepting the theory and all its consequences, and was equally desirous that the proof be wholly convincing for the credit of the friend who advanced it. Darwin never tires of telling how he values his criticisms. They led not to destruction, but to reconstruction. 'You never make an objection without doing much good,' he exclaims (November 18, 1856). After a long talk together, 'fighting a battle with you clears my mind wonderfully' (October 19, 1856), or, touching Hooker's help over the question of large genera varying largely, already mentioned, 'Again I thank you for your valuable assistance. . . . Adios, you terrible worrier of poor theorists!'

But as long as the full argument of the 'Origin' had not

been presented in consecutive form, there was the constant probability that criticism on a single point could not know that it was already outflanked by a previous argument, developed elsewhere by the author, but not impressed on the critic in this particular connection. Thus replying to Hooker, who finds the changes effected by external conditions inconstant and unequal to modifying species, Darwin urges (November 11, 1856) that the external conditions by themselves do very little in producing new species, except as causing mere variability upon which selection can work. He feels strongly that to make this clear, he ought to have sent Hooker a preliminary note on variation and its causes.

In this connection it may be noted that even after the publication of the 'Origin' Hooker continued to lay more stress on external conditions than did Darwin, who explains (May 29, 1860) that he sees in almost every organism (though far more clearly in animals than in plants) *adaptation*, and this, except in rare instances, must, he thinks, be due to selection.[1]

Again (March 16, 1858) Darwin finds the reason for various difficulties raised by Hooker in the fact that probably he has not yet sufficiently explained his notions, and begs his friend to await the MS. dealing with these points. So when he does send

[1] Thus, in March 1862, Hooker wrote to Bates: 'I am sure that with you as with me, the more you think the less occasion you will see for anything but time and natural selection to effect change; and that this view is the simplest and clearest in the present state of science is one advantage, at any rate. Indeed, I think that it is, in the present state of the inquiry, the legitimate position to take up; it is time enough to bother our heads with the secondary cause when there is some evidence of it or some demand for it—at present I do not see one or the other, and so feel inclined to renounce any other for the present.' Hereupon Darwin finds it 'curiously satisfactory' to see him and Bates 'believing more fully in Natural Selection than I think I even do myself'; but he startled Darwin in November with the frank confession that every single difference which we see might have occurred without any selection, having got right round the subject and viewed it from an entirely opposite and new side. 'I do and have always fully agreed,' is Darwin's answer, but under certain provisos, which in fact do not seem to occur. See M.L. i. 212, 199, and 223.

[Henry Walter Bates (1825–92), the 'Naturalist on the Amazons.' His boyish zeal for entomology took fuller shape under the inspiration of A. R. Wallace, with whom he set out in 1848 for these unharvested regions. Here he spent eleven years. His wide researches into the insect fauna and the problems of mimicry led him towards the theory of natural selection; and he became at once a staunch supporter of Darwin when he returned in 1859. From 1864 until his death he was Assistant Secretary to the Royal Geographical Society, and he was elected F.R.S. in 1881.]

fairly complete sections of his MS. to his chief critic, his words, 'Believe me I value to the full every word of criticism from you, and the advantage which I have derived from you cannot be told,' are a measure of the delight and relief at that critic's appreciation of the finished argument. The process bears out the phrase of June 2, 1857:

> Although we are very apt, I have observed, at the first approach of a subject, to take different views, we generally come to a near approach after a talk.

Indeed, in writing on the subject, Darwin confesses, 'I try to give the strongest cases opposed to me. I have been working your books as richest (and vilest) against mine' (July 12, 1856). But in the end, when the first paper expounding his views had been read at the Linnean, he concludes:

> You cannot imagine how pleased I am that the notion of Natural Selection has acted as a purgative on your bowels of immutability. Whenever Naturalists can look on species changing as certain, what a magnificent field will be open,—on all the lines of variation—on the genealogy of all living beings—on their lines of migration, &c., &c.

At the end as at the beginning he was keenly aware of all the help Hooker had lent, help which, as has been said, Hooker himself rated at nothing. Darwin, however, exclaims:

> You speak of my having 'so few aids'; why should you? [you] yourself for years and years have aided me in innumerable ways, lending me books, giving me endless facts, giving me your valuable opinion and advice on all sorts of subjects, and more than all, your kindest sympathy.

Again, when the Abstract had been set going after Wallace's paper had come like a bolt from the blue,[1] he cries, 'in how

[1] It will be remembered how Wallace, on realising the vast work already done by Darwin to establish the theory on an incomparably broader basis than the observations which had suggested the same theory to himself, generously waived all claim to priority. When in May 1864, in his paper on the Evolution of Man, in the *Anthropological Review*, he repeated his disclaimer, Hooker writes to Darwin (May 14): 'I am struck with his negation of all credit or share in the Natural Selection theory—which makes one think him a very high-minded man.'

many ways have you aided me.' Yet again, when this delicate situation had been arranged, he adds, 'You must let me once again tell you how deeply I feel your generous kindness and Lyell's on this occasion; but in truth it shames me that you should have lost time on a mere point of priority.' Still, perhaps the greatest service of all was 'making me make this abstract; for though I thought I had got all clear, it has clarified my brains much, by making me weigh relative importance of the several elements,' and 'I shall, when it is done, be able to finish my work with greater ease and leisure.'

Perhaps the most remarkable tribute paid by Darwin to his friend is that which is given in the 'Life and Letters,' ii. 138. The date is October 1858, while he was hard at work on the Abstract. Hooker the critic had seemed strangely unmoved by the arguments advanced, but a rather despondent note praying him not to pronounce too strongly against Natural Selection till he had read the Abstract, brought an enthusiastic reply, declaring that Darwin's speculations had been a 'jampot' to him. To this Darwin rejoins:

> I wrote the sentence without reflection. But the truth is I have so accustomed myself, partly from being quizzed by my non-naturalist relations, to expect opposition and even contempt, that I forgot for the moment that you are the one living soul from whom I have constantly received sympathy. Believe that I never forget even for a minute how much assistance I have received from you.

But Darwin, with his usual generosity of spirit, watching the increasing parallelism of their views, feared lest he had checked Hooker's original thoughts by discussing his own views with him so fully and freely. Hooker would have been the last to admit anything of the sort. He, as has been said, while gradually loosening the foundations of his former opinions, was slow to reach conviction as to the new, and only under stress of the completed argument of the 'Origin.' His original interest in their common problems connected with Geographical distribution and the unsatisfactory views current about species, was ever intensified by their constant discussions, while the special investigations, the result of which often helped to push him along the Darwinian path, were frequently prompted or stimulated by Darwin's enquiries. His own ideas involved mutability of species. Yet so long as he remained unpersuaded of a true cause for mutability, he could hardly have carried these ideas to their full completion.

Darwin's feeling, well expressed in the letter of December 25, 1859, which is given in the 'Life and Letters,' ii. 252, appears further from an as yet unpublished passage in his letter of November 14, 1858, the remainder of which is given in C.D. ii. 139 and M.L. i. 455.

> I have for some time thought that I have done you an ill-service, in return for the immense good which I have reaped from you, in discussing all my notions with you; and now there is no doubt of it, as you would have arrived at the mixture [?] independently. My only comfort is, that without you were prepared to give up species, you must have been greatly bothered in your conclusions, for the ranges of identical and representative species are so mixed up in this case, as hardly to be separated. And I can most truly say that I never thought that I might be interfering with your independent work.

And again, on January 28, 1859:

> I never did pick anybody's pocket, but whilst writing my present chapter [Geographical Distribution] I keep on feeling (even while differing most from you) just as if I were stealing from you, so much do I owe to your writings and conversation: so much more than mere acknowledgments show.

Hooker, however, took the opposite view in the missing letter to which Darwin replies on April 2:

> Do not fear about interfering with me in your publications. I have little doubt your views will be, and have arisen, independent of mine.

> [And on Ap. 7,] The Fl. Austr. and Origin contain much of the same, but yet somehow everything is taken up from

such different points of view, that I do not think we shall injure the originality of our respective books.

[In short,] You may say what you like, but you will never convince me that I do not owe you *ten* times as much as you can owe me (Dec. 30, 1858, M.L. i. 114).

But Hooker would never admit this, and five years later, when Lyell, in his forthcoming 'Antiquity of Man,' proceeded to give him large credit for his services to the Darwinian theory, his native impulse was to send Darwin a flat disclaimer (March 15, 1863):

He has written to me also about the date of publication of the Australian Essay, as preceding your 'Origin'—in this matter he has got into a fix by giving said Essay a prominence which in the history of the discussion it (and its author) do not deserve. I have such an extreme aversion to intrude myself personally into such matters, and such an abomination of reclamations, that I cannot set him right, even did the plan of his book now admit of his giving the Essay less prominence. As it is, I am ashamed of seeing it paraded with an italicised heading, just as you and the 'Origin' are, and an importance given to its priority of publication which it never dreamt of claiming. Had I really believed that your 'Origin' would have been out so soon after it I really think I should have delayed the Fl. Tasmanica rather than antedate you; but though I knew you were actually printing the 'Origin,' I knew how long it had been delayed, I knew how uncertain your health was, and I was working myself to death to get the Tasmanian Flora and its (for me) gigantic expenses off my hands. As it is Lyell seems to think me entitled to a goodly share of the credit of *establishing*, though not *originating*.

1. Because of your over-generous acknowledgment of assistance from me in the 'Origin.'

2. Because it was my making him eat the leek of variation, that so stupefied his senses that he was enabled to swallow Origin and apply Selection (as gastric juices).

3. Because I forced the card of non-reversion of varieties.

4. Because I first applied many of your results to the class and district of one Flora and country, in a way intelligible to him.

5. Because he understood my arrangement of the subject better than yours—at least so he said, some 18 months ago.

All this is no reason for putting me *in the same category with you as propounder* of the doctrine, which his work seems to me too much to do. However, I have not alluded to this subject to him, nor should I, if he had been as careful never to mention my name, as Huxley would seem to be, not that he really is so in the least I am sure.

To this Darwin replied (March 17):

What a candid honest fellow you are, too candid and too honest. I do not believe one man in ten thousand would have thought and said what you say about your own work in your letter. I told Lyell that nothing pleased me more in his work than the conspicuous position in which he very properly placed you.

CHAPTER XXVI

PUBLICATION OF THE 'ORIGIN' AND THE 'INTRODUCTION TO THE TASMANIAN FLORA'

DARWIN was well content that his ideas, given to the world in November 1859, had already won the support of Lyell and Hooker, the first geologist and the first botanist of the age. The publication, nearly a month earlier, of the Introductory Essay to the Flora of Tasmania, though of course unable to refer to the store of material and argument in the printed page of the 'Origin,' was scientifically the strongest possible buttress of Darwin. It took the crucial case of the Australian Flora which presented so many exceptions to the rule of Distribution elsewhere. In a country of relatively uniform physical features, the botanist expects to find a large number of individuals of comparatively few kinds. Here the case was reversed. The number of genera and species was very great. More than that, the crowded forms of the S.W. were singularly different from those of the S.E. Though so near, they had not intermingled, while in Tasmania, joined to the S.E. region at no very remote geological date, appeared a larger proportion of extra-Australian plants, notably those of Antarctic and European types.

Beginning with a reference to his large materials, and the fact that in the five years of his work he had personally examined 7000 out of the 8000 species discussed, he avowed his revision of the views expressed in the New Zealand Flora, set forth not as his own views, but as the current working hypothesis, namely the immutability of species as created.

Now the aspect of the problem had been changed by Darwin and Wallace ; writers must be freer to adopt such a theory as may best harmonise with the facts adduced by their own experience. For they had greatly influenced the theoretical questions as to the origin and ultimate permanence of species, though he still held, as then, that consideration of existing species alone was insufficient to decide as to ancestry or originally created types. The answer was to be drawn from the patient study of variation with its causes and checks of the distribution over the globe of living and fossil forms, leading to survival and extinction.

In the New Zealand Flora his experience had already led him to insist on the variability of plants, far greater than was generally recognised, and he had indicated that it is to the extinction of intermediate species and genera that we are indebted for our means of resolving plants into definable genera and species, a position generally accepted by believers in the permanency of species. He was now moved to show how far we may extend this view to the limitation of species themselves by the elimination of their varieties through natural causes.

Still, though it is only an arbitrary line, a question of degree, that separates genera and species and varieties, he continues to use the term species as the coin of science, which for practical purposes of description passes current among believers in mutability and permanence alike.

The moment had come to write those general essays on variation and distribution in plants which Darwin had often urged him to write, reviewing in the light of all the new evidence those questions which, on the botanical side, he had made his special study for so many years. The conclusions which emerge as to the extent of variability and the balance between the forces of nature which make for change and for permanence immediately arrested the attention of his fellow workers, who were often met by statements that variation on a large scale did not exist, or that if it did exist, all specific distinctions as we know them would have been obliterated.

Thus he shows that :

This element of mutability pervades the whole vegetable kingdom; no class nor order nor genus of more than a few species claims absolute exemption, whilst the grand total of unstable forms generally assumed to be species probably exceeds that of the stable.

He adds a doctrine of 'centrifugal variation':

The tendency of varieties, both in nature and under cultivation, when further varying, is rather to depart more and more widely from the original type, than to revert to it.

In the New Zealand Flora he had quoted the current opinion of the tendency to reversion in cultivated stocks as supporting the theory of permanency in species. This, on further evidence, he now doubts. The reversion is one of habit, not of specific character. He agrees with Vilmorin, the famous horticulturist, that when once the constitution of a plant is so broken that variation is induced, it is easy to multiply the varieties in succeeding generations.

On the other hand, if nature has provided for the possibility of indefinite variation, she regulates it as to extent and duration, by methods such as cross fertilisation, indicated by Darwin. Thus 'it is doubtful whether the natural operations of a plant tend most to induce or to oppose variation'; hence both views on species find support in nature, and the question cannot be decided by investigating variation alone. It is these checks on indefinite variation aided by the extinction of unprofitable varieties, that give a temporary appearance of fixity to existing species. In support he brings forward the *modus operandi* of Natural Selection.

The facts of distribution when analysed point in the same direction towards connected change. Species are replaced in distant areas by allied forms; the same varieties do not appear to repeat themselves at different periods when the sum of conditions cannot have been identical. The three great classes of plants are distributed with tolerable equality over the surface of the globe; so are some of the larger orders. If, then, the existing species have originated in variation, the means of distribution have overcome impediments and the power to vary is shared equally by the different classes.

A résumé of the effects of physical conditions on plants leads to discussion of the problems suggested by the traces of world-wide migration of polar and cold temperate forms left on the mountains, even in the tropics, and by the outlying Oceanic islands; present geological conditions are insufficient to account for these.

At the same time, the earliest known fossil plants are so high in development already that subsequent evolution of species cannot be said to support the doctrine of 'progressive development'—the doctrine, namely, that the course of development is an advance from 'lower' to 'higher.'

> Only be it said by way of caution [he characteristically adds], we have no accurate idea of what systematic progression is in Botany, or the relation, progressive or retrogressive, between the simpler and more complex co-ordinates in a group.

From the sum of these theories, as arranged in accordance with ascertained facts, he sets forth in § 35 his working 'assumptions' of genealogical continuity since the earliest known period; the rise of differences through individual variation; their definition through the extinction of intermediates; their stability due to cross-fertilisation; the temporary stability of physical conditions, and the successful germination of those seeds only which are adapted to these conditions.

All these points are fundamentals in Darwin's theory. That Botany, where no Lamarckian 'effort' could be predicated, pronounced so plainly for the natural working of his generalisations, was of the first importance.

As to the choice between the opposed principles as working hypotheses, neither can offer absolute certainty as to the origins of things; but while the one forbids the progress of enquiry, the other opens the field to fruitful inference.

As he puts it, in §§ 38–40, the arguments for the immutability of species have neither gained nor lost by further investigation and observation. The facts are unassailable that we have no direct knowledge of the origin of any wild species; that many are separated by numerous structural peculiarities from all other plants; that some of them invariably propagate their like; and that a few have retained their characters unchanged under very different conditions and through geological epochs.

If we conclude from such arguments that species are immutable, all further enquiry is a waste of time, until the origin of life itself is brought to light.

The most important of these facts is that of genetic resemblance. To the tyro in Natural History all similar plants may have had one parent, but all dissimilar plants must have had dissimilar parents. Daily experience demonstrates the first position, but it takes years of observation to prove that the second is not always true.

And the systematic study of the classification of species, which are fixed ideas, draws off the mind of the botanist from the history of the ideas themselves, i.e. the species, with which he works.[1]

> If it be urged that the origin of species by variation of pre-existing species be a hasty inference from a few facts in the life of a few variable plants, it appears to me that the opposite theory, which demands an independent creative act for each species, is an equally hasty inference from a few negative facts in the life of certain species.

Worse still, the doctrine of immutability leads to the denial of a rational relationship between the phenomena involved and of any vital *rationale* of classification. All is swallowed up in the gigantic conception of a power intermittently exercised in the development, out of inorganic elements, of organisms the most bulky and complex as well as the most minute and simple. Such a conception is unrealisable: the boldest speculator cannot conceive of its occurrence in any field of his own careful observation; the most cautious advocate hesitates to assert

[1] Darwin (M.L. i. 175) found the same difficulty in convincing naturalists; they had 'a bigoted idea of the term species.' His ideas were more easily understood as a rule by intelligent people who were not professed naturalists. Among scientific men, they were accepted most commonly by geologists, next by botanists, and least by zoologists (to de Quatrefages: M.L. i. 187).

this of the simplest organism, because it would commit him to the doctrine of spontaneous generation of organisms of every degree of complexity.

> If the barren facts under such a theory may receive a rational explanation under another theory, the naturalist should use this as the means of penetrating the mystery of the origin of species, holding himself ready to lay it down when it shall prove as useless for the further advance of science as the long serviceable theory of special creations, founded on genetic resemblance, now appears to be.

Only the application of these principles could explain rationally the apparent anomalies of the Australian Flora, its ancient types reinforced by European migrants whose course could be traced along the intermediate highlands, and its two southern corners, only recently joined by the rise of the barren land between, possessing each the remains of separate floras developed on different portions of a large but now vanished Antarctic continent.

The Tasmanian Introduction was for the scientific world only. Hooker was right in his estimate of its popularity, though wrong about the 'Origin,' which had an unimaginable success, the first edition being sold out at once on the day of publication, November 24. Thus he writes to Darwin in April (?) 1859 :

> From what Boott said I thought Lyell had exceeded so much my estimate of the public's interest in such works, that I could not help saying so to Boott. How glad I shall be if it proves the contrary for Science's sake. As to my Essay, if Reeve does not print it separately [this was done] only 150 copies will be printed and 75 sold, as of the Flora Tasmanica ; if he does, I shall buy 100 for distribution, and the sale of the remainder will, judging from the New Zealand Essay, be 2 copies ! In point of sale or awakening interest our books cannot interfere—the number who read both will be inconceivably smaller.

The publication of the 'Origin' elicited the following : it will be noted how Hooker continued to lay more stress on factors other than Natural Selection.

Athenæum: November 21, 1859.

DEAR DARWIN,—I am a sinner not to have written to you ere this, if only to thank you for your glorious book. What a mass of close reasoning on curious facts and fresh phenomena; it is capitally written and will be very successful. I say this on the strength of two or three plunges into as many chapters, for I have not yet attempted to read it. Lyell, with whom we are staying, is perfectly enchanted and is absolutely gloating over it. I must accept your compliment to me and acknowledgment of supposed assistance from me as the warm tribute of affection from an honest (though deluded) man, and furthermore accept it as very pleasing to my vanity—but; my dear fellow, neither my name, nor my judgment, nor my assistance deserved any such compliments, and if I am dishonest enough to be pleased with what I don't deserve, it must just pass. How different the *book* reads from the MS. I see I shall have much to talk over with you. Those lazy printers have not finished my luckless Essay,[1] which beside your book will look like a ragged handkerchief beside a Royal Standard.

Kew: (? before December 14, 1859).

DEAR DARWIN,—You have, I know, been drenched with letters since the publication of your book and I have hence forborne to add my mite. I hope that now you are well through Edition II., and I have heard that you were flourishing in London. I have not yet got half through the book, not from want of will, but of time—for it is the very hardest book to read to full profit that I ever tried; it is so cramfull of matter and reasoning. I am all the more glad that you have published in this form, for the 3 vols., unprefaced by this, would have choked any Naturalist of the XIX century and certainly have softened my brain in the operation of assimilating their contents. I am perfectly tired of marvelling at the wonderful amount of facts you have brought to bear, and your skill in marshalling them and throwing them on the enemy. It is also extremely clear as far as I have gone, but very hard to fully appreciate. Somehow it reads very different from the MS., and I often fancy that I must have been very stupid not to have more fully followed it in MS. Lyell told me of his criticisms. I

[1] The reprint.

did not fully appreciate them all, and there are many little matters I hope one day to talk over with you. I saw a highly flattering notice in the 'English Churchman'—short and not at all entering into discussion, but praising you and your book and talking patronisingly of the Doctrine!

Bentham and Henslow will still shake their heads, I fancy.

Ever yours affectionately,
JOS. D. HOOKER.

P.S.—I expect to think that I would rather be author of your book than of any other on Nat. Hist. Science.

Kew: January, about 20th, 1860.

DEAR DARWIN,—I have had another talk with Bentham, who is greatly agitated by your book—evidently the stern keen intellect is aroused and he finds it is too late to halt between two opinions; how it will go we shall see. I am intensely interested in what he shall come to and never broach the subject to him.

I finished Geolog. Evidence Chapters yesterday: they are very fine and very striking, but I cannot see they are such forcible objections as you still hold them to be. I would say that you still in your secret soul *underrate* the imperfection of Geol. Record, though no language can be stronger or arguments fairer and sounder against it. Of course I am influenced by Botany and the conviction that we have in a fossilized condition $\frac{1}{\infty}$ of the plants that have existed, and that not $\frac{1}{100000}$ of those we have are recognisable specifically. I never saw so clearly just the fact that it is not intermediates between existing species we want but between these and the unknown *tertium quid*.

You certainly make a hobby of Nat. Selection and probably ride it too hard—that is a necessity of your case. If improvement of the creation by variation doctrine is conceivable, it will be by unburdening your theory of Natural Selection, which at first sight seems overstrained; i.e. to account for *too much*. I think too that some of your difficulties which you override by Nat. Selection may give way before other explanations,—but oh Lord! how little we do know and have known, to be so advanced in knowledge by one theory. If we thought ourselves to be knowing dogs before you revealed Nat. Selection, what d—d ignorant ones we must surely be now we do know that law.[1]

The reviews of the 'Origin' were for the most part consistent in passing over the strongest lines of the argument, and either fixing solely on the confessed difficulties or making simple appeals to prejudice. Reasoned opposition was worthy of respect, and could be met with argument; but such effusions as Dr. Haughton's[2] address to the Geological Society of Dublin on Darwin and Wallace's papers evoked the exclamation to Harvey (May 27, 1860), 'What a conceited puppy H. must be and how deplorably ignorant of the first principles of Natural Science, to see nothing in the papers, let them be ever so wrong.' And later, 'it will do Haughton a lot of mischief.'

Again (March 24, 1860):

What a splutter and mess Whateley is making about Darwin's book in the *Spectator*; he is bent on widening the breach between science and religion. To me such exhibitions of fatuous prejudice are truly melancholy. What will be thought of them 50 years hence!

Against the attacks made at Cambridge, especially the impetuous assault of Sedgwick, full of *odium theologicum*, a firm stand was made by Henslow, as described in his letter which follows:

7 Downing Terrace, Cambridge: May 10, 1860.

MY DEAR JOSEPH,—I don't know whether you care to hear Phillips, who delivers the Rede Lecture in the Senate House next Tuesday at 2 P.M. It is understood that he means to attack the Darwinian hypothesis of Natural Selection.

Sedgwick's address last Monday was temperate enough for his usual mode of attack, but strong enough to cast a

[1] Cp. further letters of 1862: C. D. to J. D. H. (November 20, 1862), M.L. i. 212; and December 12, 1862, M.L. i. 222.

[2] The Rev. Samuel Haughton (1821–97) was a Fellow of Trinity College, Dublin, and from 1851–81 Professor of Geology in Dublin University; specially distinguished for his work in mathematical physics, and later in *Animal Mechanics* (publ. 1873), the outcome of his bold step in entering the medical school as a student when he was thirty-eight, in order to equip himself with anatomical knowledge for dealing with fossils. His vehement opposition to evolutionary doctrine no doubt sprang from his religious views.

slur upon all who substitute hypotheses for strict inductions, and as he expressed himself in regard to some of C. D.'s suggestions as *revolting* to his own sense of right and wrong, and as Dr. Clark,[1] who followed him, spoke so unnecessarily severely against Darwin's views, I got up, as Sedgwick had alluded to me, and stuck up for Darwin as well as I could, refusing to allow that he was guided by any but truthful motives, and declaring that he himself believed he was exalting and not debasing our views of a Creator, in attributing to him a power of imposing laws on the Organic World by which to do his work, as effectually as his laws imposed on the inorganic had done it in the Mineral Kingdom.

I believe I succeeded in diminishing, if not entirely removing, the chances of Darwin's being prejudged by many who take their cue in such cases according to the views of those they suppose may know something of the matter. Yesterday at my lectures I alluded to the subject, and showed how frequently Naturalists were at fault in regarding as *species*, forms which had (in some cases) been shown to be varieties, and how legitimately Darwin had deduced his *inferences* from positive experiment. Indeed I had on Monday replied to a sneer (I don't mean from Sedgwick) at his pigeon results, by declaring that the case *necessitated* an appeal to such *domestic* experiments, and that this was the legitimate and best way of proceeding for the detection of those laws which we are endeavouring to discover.

I do not disguise my own opinion that Darwin has pressed his hypothesis too far, but at the same time I assert my belief that his Book is (as Owen described it to me) the 'Book of the Day.' I suspect the passages I marked in the *Edinburgh Review* for the illumination of Sedgwick have produced an impression upon him to a certain extent. When I had had my say, Sedgwick got up to explain, in a very few words, his good opinion of Darwin, but that he wished it to be understood

[1] William Clark, Professor of Anatomy. In the *Life of Charles Darwin*, ii. 308, C. D., writing to Lyell, quotes Henslow as informing him that Sedgwick and then Clark attacked his book at the Cambridge Philosophical Society. To this Sir F. Darwin adds a note: 'My father seems to have misunderstood his informant. I am assured by [the late] Mr. J. W. Clark that his father (Prof. Clark) did not support Sedgwick in the attack.' The inference seems to be that he did not support Sedgwick's denunciations of the *Origin* on moral as apart from scientific grounds.

that his chief attacks were directed against Powell's [1] late Essay, from which he quoted passages as 'from an Oxford Divine' that would astound Cambridge men, as no doubt they do. He showed how greedily (if I may so speak) Powell has adopted all Darwin has suggested, and applied these suggestions (as if the whole were already proved) to his own views.

I think I have given you a fair, though very hasty, view of what happened, and as I have just had a letter from Darwin, and really have not a minute to spare for a reply this morning, perhaps you will send this to him, as he may like to know, to some extent, what happened.

To Henslow he replies:

I expect there will be before long a great revulsion in favour of Darwin to match the senseless howl that is now raised, and that as many converts on no principle will fall in, as there are now antagonists on no principle. Owen has done himself great damage in the eyes of independent literary men (who do not care a rush for the Scientific aspect of the question) whether for the gratuitous attempt to insult me, or the utter baseness of his conduct to his pretended friend Darwin.

And in June 1860:

I never see the *Literary Gazette* now, and am getting very tired of Darwinian Reviews; there is wonderfully little to the purpose in any but Gray's [2] and Owen's,[3] Huxley's [4] and Carpenter's.[5] All the rest seem ignorant prejudice. I like a *good* hostile review even if the tone and spirit are as bad as Owen's; but from all I hear, Phillips [6] at Oxford and Clark at Cambridge are mere twaddle, and the latter invective. All

[1] Dr. Baden Powell.

[2] *Amer. Journ. of Science and Arts*, April; reprinted in the *Athenæum*, August 4, 1860.

[3] To Owen was ascribed the review in the *Edinburgh Review*, April 1860, which also attacked Huxley and Hooker. Cp. M.L. i. 145, 149.

[4] *Westminster Review*, April.

[5] *National Review*, January, and *Med. Chirurg. Review*, April 1860.

[6] John Phillips (1800–74) imbibed his love of geology from his uncle William Smith, with whom he worked. Later he was Professor of Geology successively at King's College, London (1834), and Dublin 1844, migrating to Oxford 1853, where he was also Curator of the Museum (1857). President of the Geological Society 1859–60; Wollaston medal 1845; F.R.S. 1834.

show how powerful the book must be felt to be. You and Asa Gray are models of prudent dissentients. Clark, Phillips, Haughton, Sedgwick, Whately seem to me all to be *beside the mark*, they cannot appreciate the subject, are not naturalists, and have no real understanding of the fundamentals of Nat. Hist.

Edinburgh opinion, led by Balfour, the Professor of Botany, was also in opposition. The following extracts are from letters to Anderson, Hooker's Calcutta friend, who was then in Edinburgh.

Only think of five Reviews taking up Darwin in one month, viz., *Quarterly, British Do., Edinburgh, Frazer's, N. British.* Nothing but the super-excellence of the book and of its theory could command such attention; tell this to the Edinenses!

I hope you have read Owen's review in the *Edinburgh.* I should think it must add gall to the Balfourians' bitterness of spirit, for not content with snubbing me and spitefully entreating Darwin and Huxley, the cool fish hedges for a transmutation view of his own!

The following letters to his old friend Harvey illustrate his attitude towards a fellow botanist—perhaps a systematist rather than a generaliser—who could appreciate the scientific arguments involved, but who was strongly moved by questions of religious metaphysics and the suspicion that Darwin had ascribed too great efficacy to secondary causes and, as it were, deified Natural Selection. He had refrained from reading the 'Origin' until his lectures should be over and himself at leisure. He had, however, written in the *Gardeners' Chronicle*, February 18, 1860, on a monstrous sport of *Begonia frigida* so different from the normal type that it might have typified a distinct natural order. This he adduced as an objection to the theory of natural selection, which supposed changes not to take place *per saltum.* Hooker replied in the next number of the *Gardeners' Chronicle*, showing that a fallacy underlay this example.

Harvey had also written and privately printed a serio-comic squib on Darwin for the Dublin University Zoological and Botanical Association, which his friends thought rather unworthy of the occasion, and which in the following October he sent to Darwin 'with the writer's repentance.'

Kew: Tuesday, 1860.

My dear Harvey,—I send you an answer from Darwin, to whom I wrote for information as to Primroses, etc. I never went into the case myself; regarding it as one that wanted working out by Herbarium as well as garden. You will see that he offers you his MS.! He is a noble fellow; he little knows the coals of fire he is heaping on your head! Again let me caution you how you play with these questions. You have not the faintest conception of their difficulty, magnitude, and importance, I do assure you; study the question, experiment a little, or earnestly seek for light by taking up some great orders or groups etc. and endeavouring to understand the relations between all the tribes, genera, and varieties, leaving species *as species* out of view for a time. Do not snatch at superficial observations and commit yourself to superficial observations on them. Keep your opinion of species and confirm it, if you can, but if you are going to write about it, study it first; and behave like a Naturalist of 30 years' standing before the world, not like a superficial geologist or ignorant priest. I say ignorant advisedly, for I hold Whately and Sedgwick to be as really *ignorant* of the fundaments of Natural History as I am of Church History or you of fluxions. The eyes of the intelligent unscientific enquirers are now upon us, and I am most anxious that, for the credit of the age we live in, some naturalists at any rate should appear as earnest enquirers and honest workers, and should show that we have something more and better to show for our creed in the matter of species, than what satisfied us a quarter of a century ago, when the higher departments of Biology were nowhere. There is plenty to be said on both sides of the question, but nothing worth saying that is not the product of thought and study. Above all things remember that this reception of Darwin's book is the exact parallel of the reception that every great progressive move in science has met with in all ages; it is widely different from the reception of the 'Vestiges.' No good naturalist praised it, whilst seven of the ablest men of this day (and a host of smaller fry) pronounced Darwin's book to be the most remarkable of its generation, and, though not conclusive as to its own ultimate views, to have thrown the doctrine of original creation of *species* to the winds—this is my view of the question.

I really should like to have your opinion of what I have said on the subject; as you have only such opinions of my Essay as Haughton's to judge by, and I do not feel complimented by my friends' indifference to what I do, say, and think, though I am profoundly indifferent to the sneers and contempt I have received from the opposite side of the Channel and opposite side of your passage [the Irish Sea]. Asa Gray alone has treated me with candour and fairness; all other Botanists are either indifferent, hostile, or contemptuous. I venture to think that if you will read my Essay, and specially what I have said at p. xxiv (par. 34 and onwards to end of discussion) you will have a better opinion of my judgment and grounds for advocating Darwin than you now have. I do not suppose for a moment that anything I have said will alter your opinion of the main question, but I do think it *may* give you a higher opinion of the minds and consciences of your opponents, and at any rate prove to you that we may be earnest, truth-seeking, searching enquirers; candid in the exposition of our difficulties and cautious advocates too. I do not ask your praise or approval, and shall be quite content if you will say whether you think what Asa Gray says is fair or not.

One other point and I have done. I cannot bear your flinging away at Darwin and ignoring me; not because my dignity is hurt; not because you regard me as a mere disciple and copyist, but because we are both Botanists. I am sure fair generous friendship can stand any test; we shall not quarrel 'for an idea,' however hotly we may argue it. I threw down the gauntlet in G. C. when you attacked him, Darwin, from a Botanical redoubt.

Ever yours affectionately,
Jos. D. Hooker.

Kew: May 26, 1860.

Dear Harvey,—I thank you much for your last letter, which gives me great hopes of our coming to a mutual agreement as to the legitimacy and propriety of the line of study Darwin has opened up.

I believe we are all of us entirely at one about *miracle,* we all think variation miracle in the sense you accept (or propose), and we none of us think N.S. miracle in that or any other sense. I think I told Darwin over and over again that I thought his title a mistake and would mislead; his book by no means carries out his title. I think still, however, that you mistake his expressions and give an unfair interpretation of his expression '*efficient cause.*' Most people would say that moisture was the *efficient cause* of luxuriant foliage, without atheism being suspected, and in the present condition of English thought and language I see no objection whatever to the statement; at the same time, in another higher and the only true sense, moisture is not the efficient cause, nor is even the property imparted to the plant of being affected that way by moisture, but the will, or law, or call it what you will, of the supreme Governor of the universe of mind and matter.

I see now that your objections are widely different from what I supposed. I think they are peculiar to yourself *amongst naturalists*; and if you will kindly tell me how far you think I am right in my interpretation of your objection, I will re-read Darwin with the sole view of seeing how it may be remedied.

I doubt if any book that has discussed such questions is free from this real or supposed objection, and of what may be made out to be far worse. Throughout A. De Candolle's Geog. Bot., Physical causes are treated as efficient causes in the same sense; and I have always been taught to regard them as such, *but limited in their action to varieties!* a view which, if logically carried out, always seemed to me irreligious and nonsensical *in the abstract.*

I did not, I assure you, interpret the Gooseberry season to mean contempt. I wish I could join you, but have examinations all July and August.

Geol. Record meo sensu $=\pm 0$. I have turned it, heavily enough, against Darwin, as you will see. Pray do not accept Siluria as the beginning of creation yet.

Truly no, we are not obliged to accept either view *to the exclusion of any other*, nor do I do so; I only avow a preference for, not a belief in, Darwin's, and expressly state I am *ready* to lay it down for a better. There is a middle way, loosely much written about, often broached and attempted, of transmutation by saltus; Owen is hedging for it in his review of Darwin and snub of me in *Edinburgh Review*, and there is a deal to be said for it; I have often carefully examined it for plants, this 15 years; but have failed to find any reasonably cumulative support in facts, and none in Geog. distrib. or classification. Other views will turn up, but in the present state of science, I look to an advance on Darwin's *general* views as [the] most hopeful future.

Ever yours affectionately,
Jos. D. Hooker.

Kew: Tuesday, 1860.

Dear Harvey,—I sent Darwin the note of your objection to Nat. Selection as *the* efficient cause, that he might clearly see that I was not singular in my view that his words state far more than he means *if taken in the sense you and others take them*. He was anxious to write to you, and I told him I was sure you would be glad to hear, but *not* till after your Lectures. Do not be dragged into a discussion of the subject till you are at leisure. Thwaites has written an unconditional surrender to Darwin's view *under present aspect* of question.

Ever yours affectionately,
Jos. D. Hooker.

Dear Harvey,—I see you are going in for a transmutation doctrine after all! and evidently the one that Owen is hedging for in his review of Darwin (and snub of self) in *Edinburgh Review*.

I have enquired about Cowslips, etc., and will let you know. The battle will now be between transmutation by saltus and by slow measures. How you can deny N. S. in *either* case is to me incomprehensible! Every real naturalist owns N. S. to be a *vera causa*, though few admit the *plenary* power that Darwin gives it. In our Herb. there is every intermediate between Primrose and Oxlip.

You seem to confound variation with Nat. Selection. N. S. is not itself divarication; it no more accounts for divarication than 'gravity' *accounts* for the motion of planets. Give time, abate prejudice, and let your ideas clarify, which they will assuredly do in time Remember that I was aware of Darwin's views *fourteen years* before I adopted them, and I have done so *solely* and *entirely* from an independent study of plants themselves.

Bentham, Thwaites, and Thomson are all shaken to the bottom. Asa Gray writes as differently as *possible* now, from what he did on first reading Darwin and Wallace. Henslow is fast changing and defending ¾ at least of Darwin's book! at Cambridge against Sedgwick and Phillips, and is urgently recommending his students to buy the book and read it carefully. I have no wish to convert you, but I am extremely anxious that you should not commit yourself in your present state of very partial knowledge and strong feeling on a subject that requires years of thought and the calmest study, and above all a singleness of mind in seeking for truth at all hazards.

It is one thing to say that Darwin has gone *far too far* (though I do not think so), and another to defend the present weak illogical prejudices and ignorant attacks of geologists and theologians, or that worst of all class of scientifical-geological-theologians like Haughton, Miller, Sedgwick, etc., who are like asses between bundles of hay, distorting their consciences to meet the double call on their public profession. The difficulties (scientific) of Darwin's views are appalling, but of the old doctrine *insuperable.*

Ever yours,

Jos. D. Hooker.

As to the article in the July *Quarterly Review*, the secret of its authorship soon leaked out. It was written by the Bishop of Oxford, a frequent contributor to the *Quarterly*.[1] Internal evidence pointed to the prompter of his scientific ignorance. 'He and Owen,' writes Hooker to Anderson in July, 'have published a most ridiculous article in the *Quarterly* against Darwin, absurd for its egregious ignorance and blunders in Nat. Science.' To scientific readers the most significant point about it was that one of the printed pages had been cut out and another substituted. 'What gigantic blunder had been detected at the last moment?'

This ill-omened conjunction led up to the first decisive

[1] This was acknowledged in 1874, when the Bishop republished the article among his Contributions to the *Quarterly*.

encounter at Oxford, where the British Association met in 1860. Here the Bishop, a facile and persuasive speaker, primed he knew not how uncandidly on a subject outside his range, was put up to bring the meeting to a brilliant conclusion by 'smashing Darwin' before a popular assembly, mainly recruited from those who would have held themselves, in later phrase, to be on the side of the angels. The result was decisive, because it proved that men of high standing were ready to speak out, to prevent reasoned conclusions from being overwhelmed by impassioned prejudice and tasteless ridicule, to carry the war into the enemy's country, if need be, and demand that argument should be met by argument based on equal knowledge.

The scene has already been described at some length both in the 'Life of Darwin,' ii. 320, and in the 'Life of T. H. Huxley,' i. 179. The 'eye-witness' quoted in the former, will easily be identified from one of the letters which follow, as Hooker himself, who has minimised, after his manner, his own share in the contest. But I may be permitted to re-tell it briefly, in order to lead up to Hooker's own letter which tells the story of the day to Darwin.[1]

Feeling was already in a state of tension. A sharp passage of arms had taken place on the Thursday (June 28) as a sequel to a paper by Dr. Daubeny [2] of Oxford 'On the Final Causes of the Sexuality of Plants, with particular reference to Mr. Darwin's Work on the Origin of Species.' Huxley was called upon to speak by the President of the section, but tried to avoid a discussion: 'a general audience, in which sentiment would unduly interfere with intellect, was not the public before which such a discussion should be carried on.'

[1] My thanks have been already given elsewhere to Sir Francis Darwin for his friendly help in the telling of this episode; and they are warmly repeated here. But this is one small point only; the whole Life of his father and the 'More Letters' (with Prof. Seward's collaboration) which he has given to the world, to me are a continual pleasure to read and an endless storehouse of information.

[2] Charles Giles Bridle Daubeny, M.D. (1795–1867), was successively Professor of Chemistry, 1822–55, of Botany from 1834, and Rural Economics, 1840, at Oxford; especially dealing with the chemical side of his botanical and earlier geological work (on volcanoes); his paper 'On the Sexuality of Plants,' read at the Oxford meeting of the British Association in 1860, gave strong support to Darwin.

But this consideration did not weigh with Owen, who proceeded with the discussion, saying that he 'wished to approach the subject in the spirit of the philosopher,' and declared his 'conviction that there were facts by which the public could come to some conclusion with regard to the probabilities of the truth of Mr. Darwin's theory.' As one of these facts, he asserted that the brain of the gorilla 'presented more differences, as compared with the brain of man, than it did when compared with the brains of the very lowest and most problematical of the Quadrumana.'

Now this proposition, enunciated by him at the Linnean Society in 1857, had led Huxley to investigate the whole question afresh. Previous research, new dissections, even the specimens at the Hunterian Museum under Owen's charge, told the opposite tale.

Accordingly he rejoined with a 'direct and unqualified contradiction' to these assertions, and pledged himself to 'justify that unusual procedure elsewhere'—a pledge crushingly fulfilled by his article 'On the Zoological Relations of Man with the Lower Animals,' which appeared in the first number of the *Natural History Review*, January 1861. (See Huxley, 'Scientific Memoirs,' ii. 36.)

Battle was in the air. The encounter was renewed on the Saturday, June 30, when Dr. Draper of New York read a paper on 'The Intellectual Development of Europe considered with reference to the Views of Mr. Darwin.' It was not to hear his hour-long discourse, however, but the coming eloquence of the Bishop, that the crowd gathered. The Lecture-room of the Museum could not hold them; they moved to the long west room, since partitioned across for the purposes of the library. Even this was crowded to suffocation long before the speakers appeared. Seven hundred or more managed to find place; the very windows by which the room was lighted down the length of its west side were packed with ladies, whose white handkerchiefs, waving and fluttering in the air at the end of the Bishop's speech, were an unforgettable factor in the acclamation of the crowd.

Neither of the destined champions of the day had intended

to be present, knowing what an unscientific atmosphere they might expect. Hooker, as his letter tells, came at the last moment, *faute de mieux*; Huxley, who had meant to leave Oxford that morning, was only rallied into coming by Robert Chambers' appeal that he would not desert them.

In the chair was Henslow, wise and judicious, a man as universally beloved as respected. On his right were the Bishop and Dr. Draper; near the extreme left Hooker, beside Sir J. Lubbock; and nearer the centre, Huxley, beside Sir Benjamin Brodie.[1]

For an hour or more 'Dr. Draper droned out his paper'; then discussion began. The first three speakers embarked on theological and other denunciations; but were shouted down as irrelevant, and Henslow then demanded that the discussion should rest on scientific grounds only.

When the Bishop's turn came, he rehearsed various arguments from hostile reviews; but all his science was science at second-hand, its source and bias self-betrayed to those who knew, but applauded by the mass of the audience who had not the knowledge nor perhaps even the temper to discriminate. It was an audience that at the moment, Huxley felt, would hardly listen as a whole to cold scientific arguments or weigh them. He was astonished to find that the Bishop was so ignorant that he did not know how to manage his own case, and his spirits rose proportionately; but he saw no chance at first of delivering a telling counterstroke; they were carried away by the eloquence, the personality of the speaker. 'It was all in such dulcet tones,' says the eye-witness, i.e. Hooker, 'so persuasive a manner, and in such well-turned periods, that I, who had been inclined to blame the President for allowing a discussion that could serve no scientific purpose, now forgave him from the bottom of my heart.'

He spoke thus 'for full half an hour with inimitable spirit, emptiness, and unfairness. . . . In a light, scoffing tone, florid and fluent, he assured us there was nothing in the idea of

[1] Sir Benjamin Collins Brodie (1783–1862), sergeant-surgeon to William IV. and Queen Victoria, and President of the Royal Society, 1858–61. For physiological researches he was elected F.R.S. in 1810, and received the Copley Medal in 1811, at the age of twenty-eight only.

evolution; rock-pigeons were what rock-pigeons had always been.'

Then, passing from the perpetuity of species in birds, and denying *a fortiori* the derivation of the species Man from Ape, he tried to stir feeling; shall woman also be set on a level with the ape? 'Turning to his antagonist with a smiling insolence, he begged to know whether it was through his grandfather or his grandmother that he claimed his descent from a monkey.'

This was equally bad taste and bad tactics. It gave his opponent an opportunity not only of restating the true position of science in the theory of common descent and of showing how incompetent the Bishop was to enter upon the discussion, but of clinching the latter argument in a way easily understood by his hearers. The gibing descent to personalities was met by a thrust that staggered the orator's personal ascendency. For concluding his scientific reply, Huxley went on to this effect:

> I asserted—and I repeat—that a man has no reason to be ashamed of having an ape for his grandfather. If there were an ancestor whom I should feel shame in recalling it would rather be a man—a man of restless and versatile intellect—who, not content with an equivocal success in his own sphere of activity, plunges into scientific questions with which he has no real acquaintance, only to obscure them by an aimless rhetoric, and distract the attention of his hearers from the real point at issue by eloquent digressions and skilled appeals to religious prejudice.[1]

[1] This is from a letter of the late John Richard Green, the historian, then an undergraduate, to his frie_d, afterwards Professor Boyd Dawkins. It is fairly certain, however, that the word 'equivocal' was not used, and the sentence, as it stands, gives the impression of being 'much too "Green."'

Simpler and in many ways more characteristic in turn and balance, is the impression recorded in a letter to me by Mr. A. G. Vernon Harcourt, F.R.S., late Reader in Chemistry at the University of Oxford.

'"But if this question is treated, not as a matter for the calm investigation of science, and if I am asked whether I would choose to be descended from the poor animal of low intelligence and stooping gait who grins and chatters as we pass, or from a man, endowed with great ability and a splendid position, who should use these gifts" (here, as the point became clear, there was a great

A great commotion followed. Excitement rose high on either side. A lady fainted and had to be carried out. The hostile part of the audience was staggered and confused, not subjected. With doubt still hot and opinion shaken, this was the moment to strike anew with scientific argument, and Hooker, though he hated public speaking, nerved himself to come forward, and took his share in giving the Bishop 'such a trouncing as he never got before.'

Botanic Gardens, Oxford: July 2, 1860.

DEAR DARWIN,—I have just come from my last moonlight saunter at Oxford and been soliloquizing over the Radcliffe and our old rooms at the corner, and cannot go to bed without inditing a few lines to you, my dear old Darwin. I came here on Thursday afternoon and immediately fell into a lengthened reverie:—without you and my wife I am as dull as ditchwater, and crept about the once familiar streets feeling like a fish out of water. I swore I would not go near a Section and did not for two days, but amused myself with the College buildings and attempted sleeps in the sleepy gardens and rejoiced in my indolence. Huxley and Owen had had a furious battle over Darwin's absent body, at Section D, before my arrival, of which more anon. H. was triumphant; you and your book forthwith became the topics of the day, and I d—d the days and double d—d the topics too, and like a craven felt bored out of my life by being woke out of my reveries to become referee on Natural Selection, &c., &c., &c. On Saturday I walked with my old friend of the *Erebus*, Capt. Dayman, to the Sections and swore as usual I would not go in; but getting equally bored of doing nothing I did. A paper of a Yankee donkey called Draper on 'Civilisation according to the Darwinian Hypothesis,' or some such title, was being read, and it did not mend my temper, for of all the flatulent stuff and all the self-sufficient stuffers, these

outburst of applause, which mostly drowned the end of the sentence) "to discredit and crush humble seekers after truth, I hesitate what answer to make."

'No doubt your Father's words were better than these, and they gained effect from his clear deliberate utterance, but in outline and in *scale* this represents truly what was said.'

were the greatest; it was all a pie of Herbert Spencer[1] and Buckle without the seasoning of either; however, hearing that Soapy Sam was to answer I waited to hear the end. The meeting was so large that they had adjourned to the Library, which was crammed with between 700 and 1000 people, for all the world was there to hear Sam Oxon.

Well, Sam Oxon got up and spouted for half an hour with inimitable spirit, ugliness and emptiness and unfairness. I saw he was coached up by Owen and knew nothing, and he said not a syllable but what was in the Reviews; he ridiculed you badly and Huxley savagely. Huxley answered admirably and turned the tables, but he could not throw his voice over so large an assembly, nor command the audience; and he did not allude to *Sam's* weak points nor put the matter in a form or way that carried the audience. The battle waxed hot. Lady Brewster fainted, the excitement increased as others spoke; my blood boiled, I felt myself a dastard; now I saw my advantage; I swore to myself that I would smite that Amalekite, Sam, hip and thigh if my heart jumped out of my mouth, and I handed my name up to the President (Henslow) as ready to throw down the gauntlet.

I must tell you that Henslow as President would have none speak but those who had arguments to use, and four persons had been burked by the audience and President for mere declamation: it moreover became necessary for each speaker to mount the platform, and so there I was cocked up with Sam at my right elbow, and there and then I smashed him amid rounds of applause. I hit him in the wind at the first shot in ten words taken from his own ugly mouth; and then proceeded to demonstrate in as few more: (1) that he could never have read your book, and (2) that he was absolutely ignorant of the rudiments of Bot. Science. I said a few more on the subject of my own experience and conversion, and wound up with a very few observations on the relative positions of the old and new hypotheses, and

[1] Herbert Spencer (1821–1903) the philosopher, had set forth his scheme of evolutionary philosophy based on scientific data, independently of Darwin, whose *Origin of Species* was, so to say, a crucial test of the doctrine of evolution. Spencer was a life-long friend of Hooker and his scientific friends, and though he avoided the regular scientific societies, was one of the nine members of the informal circle of the *x* Club.

with some words of caution to the audience. Sam was shut up—had not one word to say in reply, and the meeting *was dissolved forthwith,* leaving you master of the field after 4 hours' battle. Huxley, who had borne all the previous brunt of the battle, and who never before (thank God) praised me to my face, told me it was splendid, and that he did not know before what stuff I was made of. I have been congratulated and thanked by the blackest coats and whitest stocks in Oxford.

CHAPTER XXVII

THE JOURNEY TO PALESTINE AND THE WORK OF 1860

In the autumn of 1860, with Daniel Hanbury [1] for his travelling companion, he spent a couple of months in the near East, joining Captain Washington, Hydrographer of the Royal Navy, in a scientific visit to Syria and Palestine. One of his chief objects was to ascend Mt. Lebanon and examine the decadent condition of the famous Cedars. This led to his publication, two years later, of a paper discussing the whole genus, from the cedars of Algeria, of Lebanon and Taurus, to the deodars of India, a relationship which had long interested him (*Nat. Hist. Review*, 1862, pp. 11–18). A paper on 'Three Oaks of Palestine' also was read before the Linnean Society (*Trans. Lin. Soc.*, 1862, xxiii. 381–387). Another result of this journey was the 'masterly sketch' of the botany of Syria and Palestine, published in Smith's Bible Dictionary in 1863.

He left Trieste on September 15 for Smyrna and Beyrout, arriving on the 25th; returning from Beyrout on November 5 and reaching Marseilles, by way of Malta, on the 14th. With wars and rumours of wars on every side, the journey promised to be more than a little hazardous; Italy was still engaged in the struggle for liberation from Austria; in Syria Moslem and Christian were at daggers drawn; the French as Protectors

[1] Daniel Hanbury (1825–75), F.R.S., was a partner in the firm of Allen and Hanbury. His keen interest in botany and pharmacology laid the foundation of a close friendship with Hooker. He was a member of the Pharmaceutical, Linnean, Chemical, Microscopical and Royal Societies. Apart from science papers, his chief works were 'Inquiries relating to Pharmacology and Economic Botany' (in the *Admiralty Manual of Scientific Inquiry*) and 'Pharmacographia,' 1874, written in conjunction with Prof. Flückiger of Strasburg.

of the Faith—a phase dated by the popular tune 'Partant pour la Syrie'—were chief in organising the Powers' campaign against the Druses. Hooker and his party reached Damascus only a day after the sacking of the Christian quarter of the city. Happily the English were not the object of popular resentment, and no untoward incidents happened to them, save that all the decent horses had been commandeered.

A few quotations from the diary illustrate things noted. Thus the Ionian Islands appear to dread the exchange of British administration for Greek misrule: the meanness of the Europeans' houses in Smyrna and the lack of hot country comforts are such as no one in India of far inferior rank would put up with. Indeed, the relative standard of native habits is higher in India.

Even under the deplorable conditions of Turkish rule, Rhodes is superb: and its 'old fortifications are far too grand, tumultuous, extensive, and picturesque to give any account of.' Two days after reaching Beyrout they were in the mountains. On Lebanon, at a height of 3000–4000 feet, the 'general character of scenery Tibetan and wretched.' On the 29th they reached the 'great shallow amphitheatre of bare, red, rounded sloping hills, at bottom of which the Cedars stand. These form one small clump, like a black speck in the great amphitheatre, and there is no other tree or shrub near them.'

The youngest of the trees standing appeared to be about fifty years old. Some seedlings were found, but all dead. Good cones there were in plenty, so that 'with very little care this grove may be indefinitely increased and made to cover all the moraines.'

Two days were spent here; the cedars were sketched and planned by the surveyors while Hooker botanised to the summit of the mountain.

Baalbek (October 2) was most impressive. A glorious sunset on the mountains was followed by bright moonlight. He notes: 'Magnificence of ruins in spite of earthquakes and Turks. Hanging keystone of Arch in Temple of Jupiter. Crawl into temple on hands and knees. Columns 7 feet through. Wolf among ruins.'

After the sterile desolation of Lebanon, the beauty of Damascus (October 4), set in its velvet green, was doubly striking. Owing to the illness of Captain Washington, they had to stay four days in their hotel in 'the street called straight' —'which is crooked and not 15 feet broad in parts'—and to give up the ascent of Mt. Hermon. Indeed there was much sickness in the city, especially among the Turkish troops, no doubt aggravated by the appalling conditions after the massacre, which took place the day before our travellers arrived, with a destruction estimated at five millions sterling and a slaughter of some 5000 persons. 'Ruins piled 4 feet deep in every lane, heaps of mutilated corpses, bones—stench! burnt books, pictures'—such is the impression of a visit to the Christian quarter under official escort.

On the return to Beyrout through the Anti-Lebanon country comes a note for the benefit of Darwin, who had asked him to keep a look out for special markings to compare with those of the zebra and other of the horse tribe: 'Saw two asses with forked end to shoulder stripe,' matching an earlier note at Syra: 'Saw 4 asses with banded legs both fore and hind down nearly to hoof.'

After three days' rest at Beyrout they left on October 14 for Jaffa. At Sidon Hooker paid a flying visit to M. Gaillardot, chief medical officer of the Turkish Government, and collated his botanical knowledge, which, not having been rubbed up for many years, was not very serviceable. At Haifa also a short excursion was made to the famous convent of Mar Elias.

Leaving Jaffa on the 16th, they visited Jerusalem, the Dead Sea, Bethlehem, Samaria, and Nazareth, the Lake of Galilee, Mt. Carmel, and so again to Beyrout.

The rounded steppe-like hills of the great limestone plateau between Ramleh and Jerusalem appeared 'very bare, except of cultivated terraces scarcely distinguishable at this season' of entire drought. Considering the 'good light red soil, admirable for Vine and Mulberry' into which the rock decomposed, 'in Lebanon every inch of this ground (except rock) would have been cultivated and most productive.' The only superiority appeared in the building of the houses.

I do not think the climate of this part of Judæa can have at all changed since Jews—safety of Jerusalem lay in its position in rugged country without much cultivation —if rain has washed soil from hills, as is supposed, why is it not in valleys? Character of country accords well with the account in the Bible. These hills of Judah being the East slope of a broad range whose West alone is exposed to rainy winds, and further being immediately facing the desert, the great depression of the Dead Sea must always have been very dry. The artificial pools are further evidence. Total absence of public works and Jewish remains is most remarkable. The Jews never were or will be an agricultural people, nor could they have been manufacturers, artisans. They were pastoral and great fighters—probably greatly exaggerated their own numbers and never enjoyed a settled Govt. without fighting with one another. The Western world owes them nothing in Art, Manufactures, Agriculture, Commerce, or Antiquities, and yet they arrogated to themselves the character of the finest people in the world. They were one out of many fighters for Judæa, and held it in part by fraud and cunning and in part by power of combination and bravery in the field.

Generally speaking, the Jews appeared to be at the bottom of the scale among the population of Palestine. The diary records:

Wretched and disgusting appearance of Polish Jews, who are very numerous—sallow, with long tress on each side of face and Old Clo hats—all squalid in extreme, very fair complexioned. Spanish Jews better. Arab Jew best. Of the latter there are some families near Safid who boast they have never left the country through all dynasties—these are wealthy and have splendid cultivation.

The various agencies for bettering the condition of the Jews or converting them to Christianity tried much but effected little. 'Rabbis of Jerusalem prevent Jews working, but very doubtful if they wish to. Sir M. Montefiore was stoned out of the city on last visit.'

The operations of the Christian Societies had brought their representatives to loggerheads over the question of using the funds subscribed for the conversion of the Jews for improving agriculture and bettering their temporal condition. Thus:

> There is a feeling rising that for conversion of Monotheists, Hindus, &c., a broader theology, free of all doctrine and less of the authorised Bible would be very efficacious,—the personal Trinity is the great stumbling-block, and many of the miracles that will not bear investigation—also the anomalous conduct attributed to Jehovah in the Old Testament, who is *not* there an unchangeable God of infinite goodness and truth, but an anthropomorphous being, subject to like passions with ourselves, and carrying out Divine purposes by means that are wholly opposed to Christianity. The questions of truth of Prophecy, so easily answered in England, here assume a very different aspect. Similarity of Jew to Arab and Mussulman in assuming God's authority for everything he wanted to have and God's approval for everything he wanted to justify, however wicked, as recorded in O.T. a great difficulty—another is progress of science—another (discord ?) of all Christian sects—another, disputed authority of many parts of O.T. as Jewish record—doubtful if Moses really was a person. Answers to all these and a thousand other practical difficulties, all learnt by heart and rule in England, and explained away variously, rarely satisfactorily. Jews and Mussulmen will not trouble themselves to discuss these things with protestant clergymen and missionaries because these are all bound to certain sects and doctrines—but will with secular Christians.

An enthusiastic lady had started the Garden of Solomon anew for their agricultural salvation. It was irrigated from the aqueduct that once went to Jerusalem from Solomon's pools, and under the foundress' vigorous management paid splendidly from vegetables. The stimulus to all this was the supposed near return of the Jews to Palestine and implicit faith in the literal interpretation of prophecy concerning it, of which Hooker remarks:

> I must read this subject up, for as the Jews have never yet possessed but a portion of the promised land, I do not see how a speedy realisation of the prophecy is possible.

But with all her zeal the agricultural missionary had only one convert to work under her, with his son. Their tenure of the land was quite patriarchal; Hooker 'witnessed' the drawing of lots between owner and tenant for the upper and lower half of the property.

The conclusion drawn from all these activities is: 'more money spent on Jerusalem in charity than any other place of size—no proportionate good done, especially to Jews.'

The plain of Jericho and the Moab Hills left an impression of great beauty; the Dead Sea was 'very grand' with shores much bolder and promontories more rocky than he expected and no visible white incrustation at the end. In camp on the supposed site of Jericho, 'At night the village Arabs, a scoundrelly set, came and performed an Arab war dance. Three Sheiks attitudinised with swords, and a dozen or two men crouched and grunted like camels and sang before them—utter barbarity.' Reascending the heights on the way back, he notes the 'curious effect of rising to level of plants of level of Mediterranean.'

At Hebron,

> turned off road to visit Abraham's Oak, about one mile out of town; a very fine tree, acorns larger than of the usual surrounding stunted Oaks,—probably not 300 years old, no dead twigs—24 ft. girth.

As to the reverence with which this tree was regarded he notes later:

> Dragoman says that he bought fallen limb of Abraham's Oak at Hebron for £1 from Mr. Finn [the consul], but that superstition so strong that any one cutting it would lose his first-born son that no one would cut it for a long time: it was load for 7 camels and cost £10 in all to transport.

To Charles Darwin

December 2 [?], 1860.

I paid particular attention to your query about the sudden appearance of plants on ascending Lebanon and made a good many observations to the effect that the more remarkable *forms* especially generic do appear very suddenly in great quantities, and am inclined to believe that the lower limit of these is far better defined than the upper limit of those that disappeared. This applies to *Astragalus, Acantholimon, Vicia,* and several other plants which are characteristic of the dry soil and climate that prevail above 7000 ft., but not to other plants which are equally peculiar to the elevation but which depend on some little moisture, as *Potentilla,* &c. The vegetation above 8000 ft. was extremely scanty, and I found but one Alpine or Arctic plant (*Oxyria reniformis*), and that was close to the tip-top and very rare. This absence of Alpine plants on the mountains of Asia Minor is a very characteristic feature, and is shared, I am assured, by the mountains of Algeria. Under these circumstances the presence of so very marked an Arctic plant as *Oxyria* is very interesting—it seems to say that an expulsion of other Arctics must have taken place, and the drought would effect this well enough.

The Cedars are going owing to the same causes. Every seedling dies, there are no trees under 40–50 years old, from which ages up to 500 (perhaps the oldest) there are trees of all (or many) ages.

Though the last of the Southern Floras was now published, 1860 did not bring a hoped for lull in the press of work ; pressure, if anything, increased ; official work at Kew, both correspondence and practical administration, grew steadily ; the Linnean and other learned Societies made considerable demands upon him ; to his own work he was always ready to add investigation and experiment for Darwin, especially on the fertilisation of perplexing Orchids, their structure and homologies, and the rationale of the curvature of the style in oblique flowers. As a successor to the Antarctic Flora he was now deep in the Arctic Flora, examining, comparing, speculating, and, as he tells Asa Gray (June 26, 1860),

horribly stumped by so many inosculating groups in America and Europe. What a deal there is to do in redoing N. temperate Flora . . . I can only account for peculiarity and paucity of Greenland Floras by plants having been driven out by Glacial cold and never got back.

He was well embarked on the vast undertaking of the Genera Plantarum with Bentham. And in March he tells Henslow:

> Murray and others are very anxious, I understand, that I should bring out a Darwinian book on Botany—a sort of elementary book on Classification, Distribution, and origin of species. I am dubious and considering. I think I could make it a good instructional one with woodcuts illustrating all sorts of transitional forms, independent of all theory.

This was the work referred to by Darwin (March 12, 1860) apropos of a visit from the Lyells.

> We talked over your Essays and agreed about the Book which you ought to make. What fine materials in all combined, including, as Lyell remarked, the Galapagos papers! But I see in the *Gardeners' Chronicle* that you have started on a gigantic task with Bentham.

And again, July 12:

> I have been thinking about your Book, and the more I think of it the more awfully difficult it seems, and therefore the more worthy of your attempting. One of the first points which seems naturally to occur is difference between plant and animal! and then, as I suppose, you will allude to unicellular plants, what makes an individual! And thirdly the difference between propagation by germination and sexual generation! Nice little simple subjects to discuss!

The work in hand was sufficient, however, to keep him from this Darwinian Botany book. His friends, too, noticed that he looked overworked, and Darwin added one of his solicitous warnings. Accordingly he turned over a new leaf, and tells Darwin (December 2?):

> I have taken up the Genera Plantarum with Bentham in earnest, and am going to mend my ways now and for ever: giving up all Societies but the Linnean, and every unnecessary excitement, keep early hours, cut off all correspondents (except those I love) with short letters—eat well, walk a good deal in the garden, and avoid all occasions of sin. I have not had a headache for three months now!

To Huxley he made a similar profession of good resolutions, and received a reply in kind.

Kew: December 12, 1860.

We are not likely to meet except at the Linnean, for I have inaugurated a new era in my life, and am going to take the world and all that is therein as coolly as I can. When perfect myself I shall commence operating on you. What is the use of tearing your life to pieces before you are 50? which you are (and I was) doing as fast as possible.

From T. H. Huxley

Jermyn Street: December 19, 1860.

And finally as to your resolutions, my holy pilgrim, they will be kept about as long as the resolutions of the anchorites who are thrown into the busy world. Or, I won't say that, for assuredly you will take the world 'as coolly as you can'—and so shall I. But that coolness amounts to the red heat of properly constructed mortals.

It is no use having any false modesty about the matter. You and I, if we last ten years longer—and you by a long while first—will be the representatives of our respective lines in the country. In that capacity we shall have certain duties to perform, to ourselves, to the outside world, and to Science. We shall have to swallow praise, which is no great pleasure, and to stand multitudinous bastings and irritations, which will involve a good deal of unquestionable pain. Don't flatter yourself that there is any moral chloroform by which either you or I can render ourselves insensible or acquire the habit of doing things coolly.

It is assuredly of no great use to tear one's life to pieces before one is fifty. But the alternative, for men constituted on the high pressure tubular boiler principle like ourselves, is to lie still and let the devil have his own way. And I will be torn to pieces before I am forty sooner than see that.

Fortified by a few months of this regime, he can point the moral to his friend Anderson in Calcutta (April 22, 1861) apropos of T. Thomson's break-down in India.

That cursed Society of Calcutta and Sunday *labour* in entertaining is at the root of all the mischief, and I do earnestly hope that you will follow my example here and demand the Sunday for yourself and those only of your own friends you choose to ask personally.

And a month later:

I can only repeat, for God's sake do not overtask yourself, proceed methodically and kick out the *Society* or the Bot. Gardens; cultivate *moral courage* as the first of all qualities in a man of business.

June 2: I shall transgress my rule of writing only once a month to give you a stave. . . .

'Servate animam aequam,' my dear fellow, and do not allow your frightful accumulation of work in hand to overwhelm you as it well may. Just now I can well appreciate your position and labour, for here have I been [at Hitcham] for 10 days emptying poor Henslow's house—such an accumulation. Tons have gone to Cambridge and Kew, and there are 150 boxes to go to Stevens' auction room.

You are right to give a few hours a day to each job till each is cleared off; if you can carry this through all will go well with you—and if you do not *tant pis* for you—but do I beg of you *servate animam aequam*. Do not be bothered—go steadily to work. Do not fret about the plants arriving dead at Calcutta. You had a better experience of our luck during your stay at Kew than is usual with us.

To his staunch helper, Professor Oliver, he also writes an emphatic warning against his overwork, and for himself, on bidding Bentham to Kew on September 3, 1861, he adds: 'We can talk over Genera and gamble in the evening, for I have reformed my habits of working at night, now that I have not to write so much as heretofore.' Though he managed to keep off the R. S. Council with its heavy work in 1862, he was compelled, much against his will, to accept a botanical Jurorship at the Exhibition of 1862; but despite the loss of income and regret at surrendering an outpost of Science, he was glad to give up the examinership at London University in 1864. Though he tried to resign his other examinerships at the same time, he was compelled to continue the work until he succeeded to the Directorship of Kew in 1865.

It was this engrossing pressure, common to Hooker and his closest scientific friends, that led to the foundation of the famous *x* Club. This has already been described at some length in the 'Life and Letters of T. H. Huxley,' i. 368 *seq.*, and in 'Sketches from the Life of Edward Frankland,' p. 148 *seq.* A further account may be added here, for the club was not only a unique constellation of intellects, but a notable factor in the personal life of its members. All were keen workers in science and progressive thought; all were friends of long standing. The growing pressure of work made meeting difficult save casually, perhaps dining at the Athenæum before important lectures at the Royal Institution or the regular gatherings of the Royal and other societies. These unpremeditated encounters suggested something more definite. 'I wonder if we are ever to meet again in this world,' Huxley had written to Hooker in his 'remote province' of Kew. Now in January 1864 Huxley proposed to him that they should organise some sort of a regular meeting. All the friends, with the exception of Herbert Spencer, being Fellows of the Royal Society, the date chosen for dining together was the first Thursday in each month (except July, August, and September) before the Society's meeting. The usual hour was six o'clock, so that they should be in good time for the meeting at eight. On December 5, 1885, Huxley, who was treasurer, notes in the minutes, 'Got scolded for dining at 6.30. Had to prove we have dined at 6.30 for a long time by evidence of waiter.' However, at the February meeting, 'agreed to fix dinner hour six hereafter.'

Eight members met at the first meeting, November 3, 1864; at the second, a ninth member was added in William Spottiswoode, but a proposal to add a tenth was never carried out. On the principle of *lucus a non lucendo*, this gave point to the symbol *x* for the name of the club, the origin of which is described as follows by Huxley in his reminiscences of John Tyndall in the *Nineteenth Century* for January 1894:

'At starting, our minds were terribly exercised over the name and constitution of our society. As opinions on this grave matter were no less numerous than the members—indeed more so—we finally accepted the happy suggestion of our mathematicians to call it the x Club; and the proposal of some genius among us, that we should have no rules save the unwritten law not to have any, was carried by acclamation.'

The meetings were at first regularly held at the St. George's Hotel, Albemarle Street, with Almond's Hotel, Clifford Street, and the Athenæum to fall back upon in case St. George's were not available. In the latter eighties, however, the Athenæum became the regular meeting place, and it was here that the 'coming of age' of the club was celebrated in 1885.

For some years also there was a summer week-end meeting in the country, which was attended by members and their wives. For this the Treasurer whose turn of duty it was, did not send out the usual postcard of invitation $x=2$, or whatever the date might be. The correct formula for the occasion was x's + yv's. The place of these meetings was sometimes the foot of Leith Hill, or Oxford, or Oatlands Park, but most usually Maidenhead, with possibilities of a drive to Burnham Beeches and Dropmore, and boats on the river. But this grew increasingly difficult to arrange, and in course of time was dropped.

Hooker, Busk, Spencer, and Tyndall [1] had all been close friends of Huxley's soon after his return from the voyage of the *Rattlesnake*; Frankland and Hirst [2] were yet older friends

[1] John Tyndall (1820–93), natural philosopher and Alpinist, after beginning life on the ordnance survey and as a railway engineer, went to Queenwood College as teacher of mathematics and surveying. Resolving to devote himself to science, he, with his colleague Edward Frankland, the chemist, went to Marburg, and then Berlin, studying chemistry and magnetism. He returned to Queenwood in 1851, but in 1853, Dr. Bence-Jones, having heard of the impression made by him in Berlin, invited him to lecture at the Royal Institution, with the result that he was immediately chosen as Professor of Natural Philosophy there, becoming the colleague and from 1867 the successor of Faraday, the superintendent. In addition to his researches on heat (for which he received the Rumford Medal 1867), light and sound and the germ theory, he was celebrated as a lecturer and expositor of science for the public. He was scientific adviser to the Trinity House 1866–83. He was a warm friend of his fellow members of the x Club, particularly of Huxley.

[2] Thomas Archer Hirst (1830–92), mathematician, was articled as surveyor, &c. at Halifax, Yorkshire. Taking his Ph.D. in 1852, he became Lecturer in Mathematics at Queenwood College, Hants, 1853–6, and University College School, 1860; F.R.S. 1861; F.R.A.S. 1866; Professor of Physics, University College, 1865, and of Pure Mathematics, 1866–70; Director of Naval Studies at Greenwich, 1873–83; Fellow of London University, 1882. He published various mathematical writings.

and allies of Tyndall; Sir John Lubbock and Spottiswoode[1] were later friends of them all.

The one purpose of the club was to afford a definite meeting point for a few friends who were in danger of drifting apart in the flood of busy lives. But it was in itself a representative group of scientific men destined to play a large part in the history of science. Five of them received the Royal Medal; three the Copley, the highest scientific award; one the Rumford; six were Presidents of the British Association, three Associates of the Institute of France, and from amongst them the Royal Society chose a Secretary, a Foreign Secretary, a Treasurer, and three successive Presidents.

> I think, originally [writes Huxley, *l.c.*] there was some vague notion of associating representatives of each branch of science; at any rate, the nine who eventually came together could have managed, among us, to contribute most of the articles to a scientific Encyclopædia.

As I have written elsewhere, they included leading representatives of half a dozen branches of science—mathematics, physics, philosophy, chemistry, botany, and biology; and all were animated by similar ideas of the high function of science, and of the great Society which should be the chief representative of science in this country. However unnecessary, it was perhaps not unnatural that a certain jealousy of the club and its possible influence grew up in some quarters. But whatever influence fell to it as it were incidentally—and earnest men with such opportunities of mutual understanding and such ideals of action could not fail to have some influence on the progress of scientific organisation—it was assuredly not sectarian nor exerted for party purposes during the twenty-eight years of the club's existence.

> I believe that the *x* [continues Huxley] had the credit of being a sort of scientific caucus, or ring, with some people.

[1] William Spottiswoode (1825–83) was an accomplished mathematician and physicist as well as a man of business. He succeeded his father as Queen's Printer in 1846, and after being Treasurer became President of the Royal Society 1878–83, following Hooker and preceding Huxley. His great personal charm endeared him to his friends.

In fact two distinguished scientific colleagues of mine once carried on a conversation (which I gravely ignored) across me, in the smoking room of the Athenæum, to this effect: 'I say, A., do you know anything about the *x* Club?' 'Oh yes, B., I have heard of it. What do they do?' 'Well, they govern scientific affairs, and really, on the whole, they don't do it badly.' If my good friends could only have been present at a few of our meetings, they would have formed a much less exalted idea of us, and would, I fear, have been much shocked at the sadly frivolous tone of our ordinary conversation.

Thus, in the minutes of December 5, 1885, already quoted, when Huxley as treasurer revived the early custom of making some notes of the conversation, we read: 'Talked politics, scandal, and the three classes of witnesses—liars, d—d liars, and experts. Huxley gave account of civil list pension. Sat to the unexampled hour of 10 P.M., except Lubbock, who had to go to Linnean.'

In the minutes of the sixties and early seventies the notes of talk usually record the more serious subjects, especially the progress of science through education in schools, learned societies, and research. Thus at the first meeting there was discussed the reorganisation of the *Reader*, a journal in which the Young Guard of science were seeking a literary mouthpiece. Again 'the claims of several candidates now proposed for admission to the Athenæum and Royal Society formed one of the subjects of conversation.' Later 'the present unsatisfactory mode of election of the Council of the Royal Society was discussed. Frankland, Hirst, and Spottiswoode expressed their intention of bringing the subject before the Council as soon as possible.' The subject recurs more than once in the minutes, and indeed it was subsequently 'agreed that the R.S. Council should form a subject of consideration at the October meeting of the Club each year'; while, when Huxley was President Elect of the British Association, the choice of presidents of the sections was discussed and a provisional list made out.

So too 'Lubbock's proposition was discussed of the foundation of a Christie Lectureship at the Royal Institution' and 'the advisability of Tyndall's acceptance of the Professorship of Physics at Oxford.'

'Spottiswoode informed us that the Liberal party at Oxford were about to try to utilise the present movement for university extension, originated by the theological party. The former would be glad to receive support from the friends of Science outside the university. A conversation ensued relative to the changes which ought now to be introduced into school education generally.'

'Huxley's forthcoming lecture at St. Martin's Hall, and the Sunday League generally, were subjects of conversation. Spencer spoke of some of the results of his late botanical inquiries.'

'Frankland proposed that we should take into consideration some method of hastening the publication of papers in the "Phil. Trans." Hirst read a letter from the Secretary of the Sunday League in reference to the late suppression of the Sunday lectures by the Sabbatarians.'

'One of the principal subjects of conversation was the President of the British Association for 1868. We all requested Hooker to allow himself to be nominated, but he declined, on the ground that it would interfere too much with the scientific work he had in hand.'

'The constitution of Section D of the British Association was discussed.'

'Sir John Lubbock having been asked by a number of graduates of the University of London to come forward as their representative in Parliament, we decided to give him our support by expressing our unanimous opinion that scientific men would regard him as a most appropriate representative in Parliament.'

'The relation between Faraday, when young, and Davy. Tyndall observed that we must not judge Davy's treatment of Faraday by the light of subsequent experience of Faraday's powers, but must remember that Faraday came to the Royal Institution simply as a bookbinder's boy, who wanted to change his business for some other occupation. Hooker mentioned that the menial position in which Faraday travelled with Davy, in 1813, was owing to the fact that the French Government would allow only a maid or valet to accompany Sir H. and Lady Davy on their journey.'

'In the beginning was the atom, and the atom was without form and void, and darkness was on the face of the substance. And the spirit of Frankland moved on the face of the substance, and he said, Let there be an atom: and there was an atom; and he saw that it was good. And the atom and its shadow were the first edition; and Frankland said, Let there be a bond, &c., &c.'—*Hirst's minute.* (See Hooker's recollection of this incident, ii. 359, and also 112.)

'The conversation turned on Tyndall's discoveries in chemical composition, &c., &c., due to light. On Huxley's new observations on microscopic organic forms and on the possible bearing of these on one another. Also on some arrangement for the publication of English scientific works in America.'

'It was resolved to add Lubbock's name to the B.A. Committee on Scientific education, in order that he might consult that committee on points arising in the public school committee. The subject of State assistance to original experimental research was discussed; an extension of the Government grant through the Royal Society was thought by the majority to be the best means. Huxley reported that the question of Sunday evening lectures had been revived independently by the Sunday League; and will report further hereafter.'

'The conversation turned, during the larger part of the evening, on Tyndall's discoveries in the reflecting of blue rays from the molecules of attenuated vapours. We were more than once called to order by Spencer for allowing the conversation to become broken up instead of remaining general.'

'The conversation was very metaphysical; Spencer *v.* the field. Airy was spoken of as a possible future President of the Royal Society. It was suggested that five years would be a suitable period for the tenure of the presidentship, as well as for membership of council.'

Later the period of ten years of office of P.R.S. was discussed.

The prevailing opinion was in favour of 'no restriction.' On another occasion, when matters of scientific organisation had filled up the evening to the exclusion of general subjects or the bearing of special work undertaken by individual members, Spencer, the guardian of strict justice, 'protested against the transaction of so much business.' It was more satisfactory when debate turned 'on the merits of Bacon as the originator of the method of induction in science,' or on the opinions expressed at 'the meeting of clergy at Sion College, where Huxley delivered a discourse,' [1] or on the occasion when 'Professor Masson dined with us. Masson and Spencer fought the battle of the ladies.'

Finally, after Hooker's retirement from Kew, 'discussed Linnean presidency, which Hooker positively declined.'

These quotations are typical, but typical only of part of the *x* Club meetings. As Professor Frankland writes (*l.c.* p. 161): 'It must not be supposed that the talk at the meeting was by any means confined to such topics. There was always a judicious admixture of ordinary dinner-table talk, with a by no means sparse sprinkling of witticisms, good stories, and, perhaps occasionally, though very rarely, a little scandal.'

Guests were not excluded from the club dinners; men of science or letters of various nationalities came by special invitation from time to time. Among the twenty-nine whose names are recorded in the archives are Darwin, Colenso, Richard Strachey, Tollemache, Helps; Professors W. K. Clifford, Bain, Masson, Robertson Smith; Bentham the botanist, John (Lord) Morley, Francis Galton, Jodrell, the founder of several scientific lectureships; Dr. Klein; the Americans Marsh, Gilman, A. Agassiz, and Youmans, who met here several of the contributors to the International Scientific Series organised by him, and Continental representatives such as Helmholtz, Laugel, and Cornu.

[1] This meeting took place on December 12, 1867, under the auspices of Dean Farrar and the Rev. W. Rogers of Bishopsgate, 'Hang Theology' Rogers. The bearing of recent science upon orthodox dogma was discussed; some denounced any concessions as impossible; others declared that they had long ago accepted the teachings of geology, whereupon a candid friend inquired, 'Then why don't you say so from your pulpits?'

The club met 240 times, the average attendance being seven up to 1883, and the whole nine assembling on twenty-seven occasions. Hooker himself attended 169 times. The original circle remained unbroken for nearly nineteen years. Spottiswoode died in 1883; Busk in 1886. The meetings, wrote Huxley, 'were steadily continued for some twenty years, before our ranks began to thin; and one by one, *geistige Naturen* such as those for which the poet so willingly paid the ferryman,[1] silent but not unregarded, took the vacated places.' Proposals were often made to fill up the gaps, especially when ill-health drove other members to live out of town; but as the *x* really had 'no *raison d'être* beyond the personal attachment of its original members,' it seemed to some that new members, however personally welcome, could not be admitted without destroying the unique relationship of friendship joined to a common experience of struggle and success. This feeling was expressed by Huxley in a letter to Frankland of 1886:

> Nobody could have foreseen or expected twenty years ago when we first met, that we were destined to play the parts we have since played, and it is in the nature of things impossible that any of the new members proposed (much as we may like and respect them all), can carry on the work which has so strangely fallen to us.
>
> An axe with a new head and a new handle may be the same axe in one sense, but it is not the familiar friend with which one has cut one's way through wood and brier.

And to Hooker two years later:

> The club has never had any purpose except the purel personal object of bringing together a few friends who did not want to drift apart. It has happened that these cronies had developed into bigwigs of various kinds, and therefore the club has incidentally—I might say accidentally—had a good deal of influence in the scientific world. But if I had

[1] Nimm dann Führmann,
Nimm die Miethe
Die ich gerne dreifach biete;
Zwei, die eben überfuhren
Waren geistige Naturen.

to propose to a man to join, and he were to say, Well, what is your object? I should have to reply like the needy knife-grinder, 'Object, God bless you, sir, we've none to show.'

The matter at last was wittily disposed of. No proposition of the kind was to be entertained 'uuless the name of the new member contained all the consonants absent from the names of the old ones. In the lack of Slavonic friends this decision put an end to the possibilities of increase.'

After the death, in February 1892, of Hirst, a most devoted supporter of the club, who 'would, I believe, present it in his sole person rather than pass the day over,' only one more meeting took place, in the following month. With five of the six survivors domiciled far from town, meeting after meeting fell through, until the treasurer (Hooker) wrote, 'My idea is that it is best to let it die out unobserved, and say nothing about its decease to any one.'

Thus it came to pass that the March meeting of the club in 1892 remained its last. No ceremony ushered it out of existence. Its end exemplified a saying of Hooker's: 'At our ages clubs are an anachronism.'

END OF VOL. I.

AT THE BALLANTYNE PRESS
PRINTED BY SPOTTISWOODE, BALLANTYNE AND CO. LTD.
COLCHESTER, LONDON AND ETON, ENGLAND

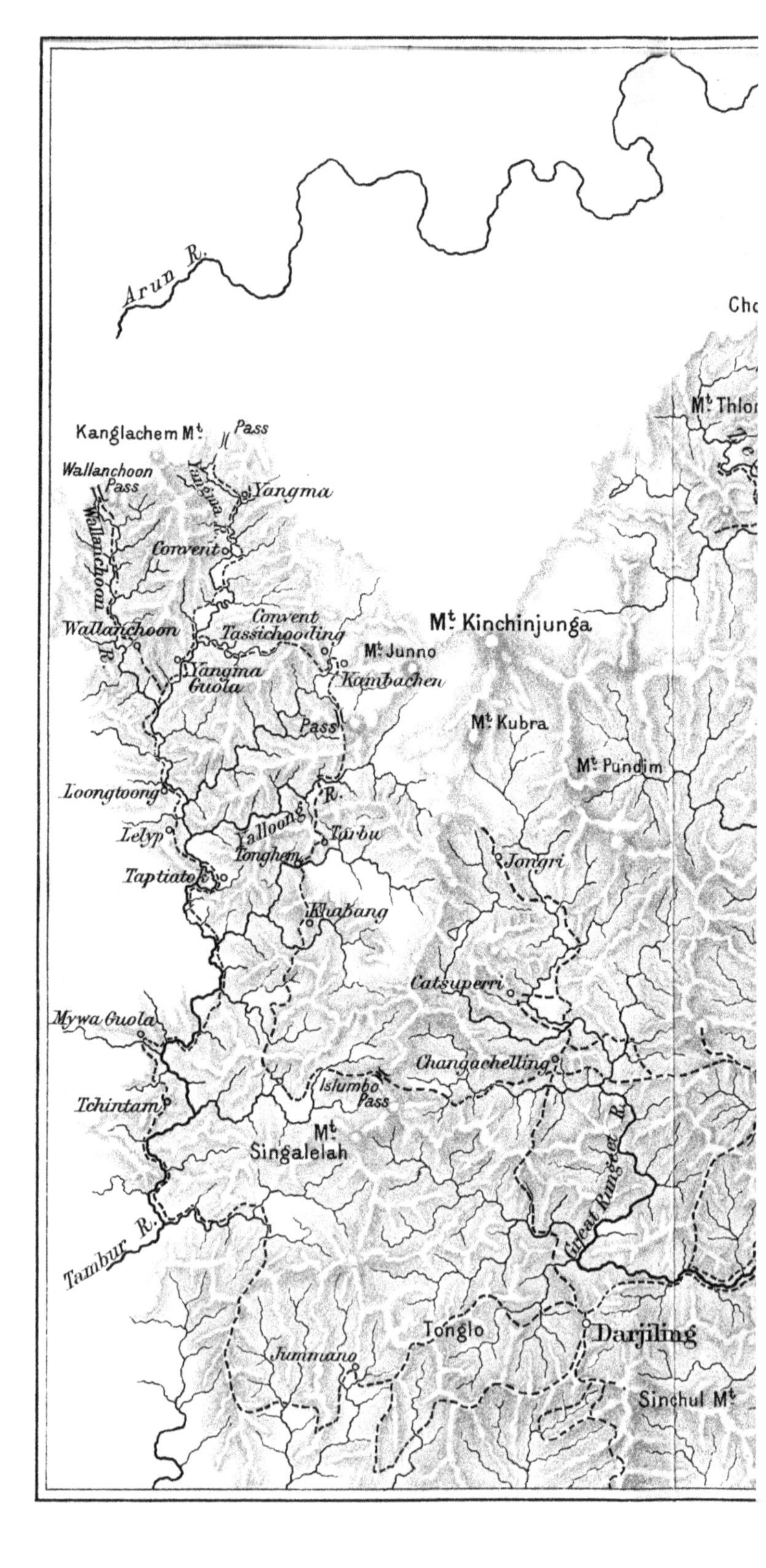

Arun R.
Cho
Mt. Thlo
Kanglachem Mt.
Pass
Wallanchoon Pass
Yangma R.
Yangma
Wallanchoon R.
Convent
Convent Tassichooding
Wallanchoon
Mt. Kinchinjunga
Mt. Junno
Yangma Guola
Kambachen
Mt. Kubra
Pass
Mt. Pundim
Loongtoong
R.
Yalloong
Lelyp
Tarbu
Tonghem
Jongri
Taptiatok
Khabang
Catsuperri
Mywa Guola
Changachelling
Islumbo Pass
Tchintam
Mt. Singalelah
Great Rungeet R.
Tambur R.
Darjiling
Tonglo
Jummano
Sinchul Mt.

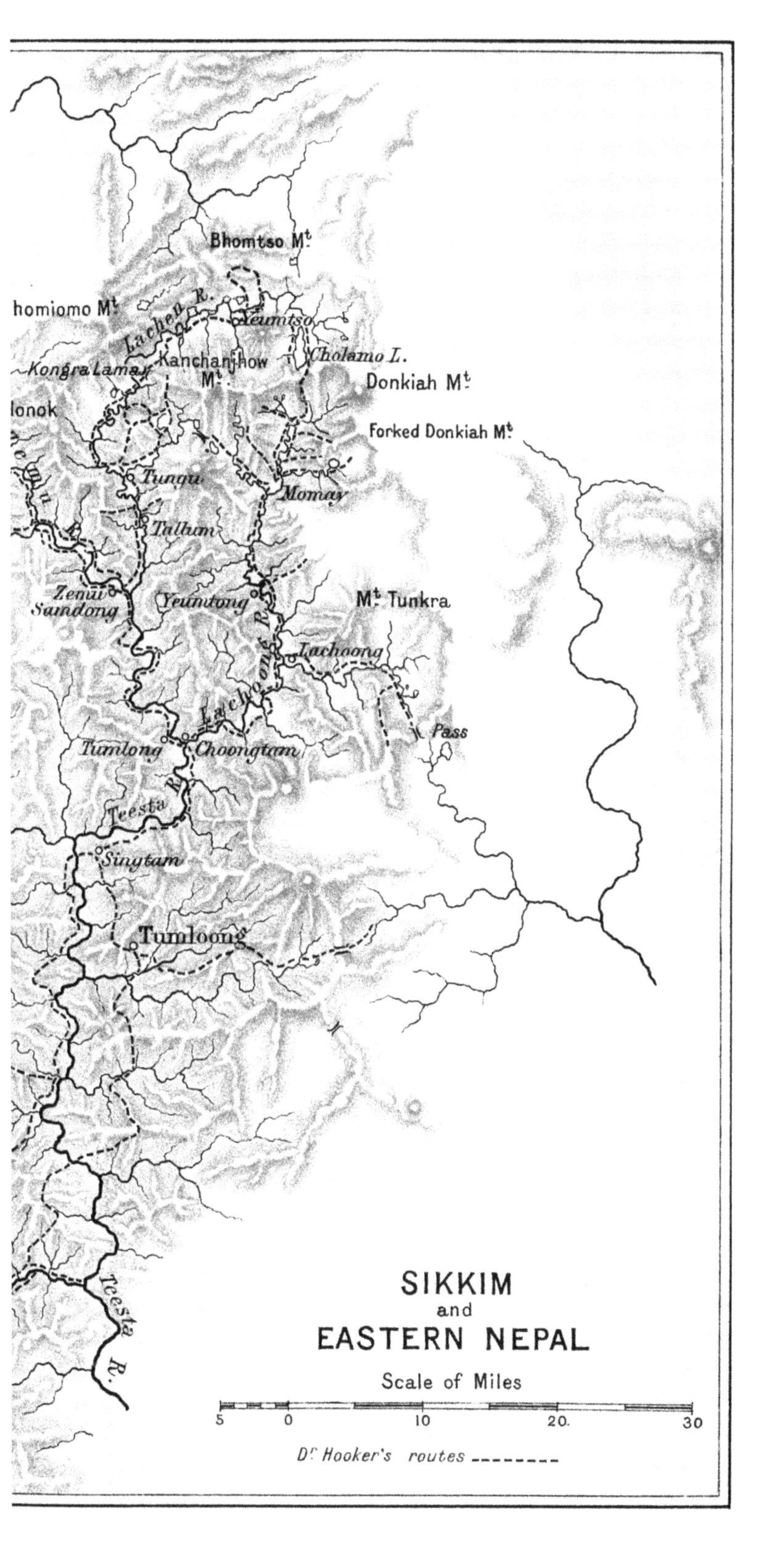
Bhomtso Mt.
homiomo Mt.
Lachen R.
Yeumtso
Cholamo L.
Kongra Lama
Kanchanjhow Mt.
Donkiah Mt.
lonok
Forked Donkiah Mt.
Tungu
Momay
Tallum
Zemu Samdong
Yeumtong
Mt. Tunkra
Lachoong R.
Lachoong
Pass
Tumlong
Choongtam
Teesta R.
Singtam
Tumloong
Teesta R.
SIKKIM
and
EASTERN NEPAL
Scale of Miles
5 0 10 20. 30
Dr. Hooker's routes